Nuclear Strategy in a Dynamic World

Nuclear Strategy in a Dynamic World

AMERICAN POLICY IN THE 1980s

Donald M. Snow

THE UNIVERSITY OF ALABAMA PRESS
UNIVERSITY, ALABAMA

This book was keyboarded for scanner typesetting by Chellia Holder.

Copyright © 1981 by
The University of Alabama Press
All rights reserved
Manufactured in the United States of America

FOURTH PRINTING 1983

Library of Congress Cataloging in Publication Data

Snow, Donald M. 1943—
  Nuclear strategy in a dynamic world.

  Bibliography: p.
  Includes index.
  1. United States—Military policy. 2. Atomic warfare. I. Title.
UA23.S526   355'.0335'73   80-13634
ISBN 0-8173-0044-9
ISBN 0-8173-0045-7 (pbk.)

To
My wife Donna, for her constant support
and
My son Eric (Rickey), in the small hope that his generation may escape the potential horrors herein described

Contents

Introductory Note	ix
1: Nuclear Strategy in a Dynamic World	1
2: Deterrence: The Elusive Art	23
3: The Evolution of American Strategic Doctrine	48
4: American Strategic Forces	86
5: Soviet Strategic Doctrine and Forces	130
6: The Arms Control Process	161
7: The Uncertain Future: American Policy in the 1980s	204
Glossary	243
Notes	249
Cited Bibliography	274
Index	282

Introductory Note

A fundamental and far-reaching debate is currently raging in the professional and academic communities concerned with the formation of strategic nuclear doctrine. The basic doctrine around which nuclear strategy is formed was articulated almost fifteen years ago and has endured, with modification, since, despite demonstrable changes in the conditions affecting the efficacy of that doctrine. Colonel Richard G. Head stated the need succinctly: "Pressed by the Soviet buildup and the altered military balance on the one hand and the opportunities presented by the new technologies on the other, there is a critical need to review U.S. strategic and doctrinal concepts."[1] This book is a response to that challenge.

Strategic nuclear doctrine is not a subject that the reader comes to easily or without trepidation. It is an extremely technical and complex area, requiring a grasp of physical and engineering concepts and applications, as well as abstract and often obtuse theoretical constructs that sometimes threaten to overwhelm even the most knowledgeable student. At the same time, the subject matter is shrouded in a jargonized language that is eerily sanitary in describing the truly awful. As a veteran in the field, Fred Ikle, passionately writes: "The jargon of American strategic analysis works like a narcotic. It dulls our sense of moral outrage about the tragic confrontation of nuclear arsenals. . . . It fosters the current smug complacence regarding the soundness and stability of mutual deterrence. It blinds us to the fact that our method of preventing nuclear war rests on a form of warfare universally condemned since the dark ages—the mass killing of hostages."[2]

In a very real sense, Ikle is correct. The lexicon of strategic studies is replete with euphemism: countervalue targeting really means targeting innocent civilians who are hostages to be summarily executed should conditions be "appropriate"; mutual assured destruction is really mass genocide with a less repugnant name; and so on. This material is heady and potentially macabre, and it is not entirely surprising that even the informed citizen would not be drawn readily to it.

To say that the stuff of strategic nuclear doctrine is technically and theoretically complex and that the language encasing it is arcane is not to suggest that the material is inherently mystical or beyond the grasp of the novitiate. The success of strategic doctrine lies at the heart of national survival, and that is reason enough for anyone interested in national defense to know about it. The implementation of strategic doctrine consumes billions of dollars annually that are consequently unavailable for other priorities, and that should be sufficient cause for the concerned student of politics to be able to make some reasoned judgments about the policies.

This concern is all the more important because strategic doctrine has historically been made and analyzed by a fairly tight and narrow group of people. The government experts in the Congress and the bureaucracy (including the professional military) and the civilian experts, both in universities and the so-called "think-tanks," comprise a small, interconnected group with a great deal of movement among the constituent parts. The scholarly community regularly lends its resources to government service, and the defense agencies provide the funding that allows scholars to conduct strategic nuclear research. That the combination is potentially incestuous was suggested by Philip Green in 1966: "Perhaps it is not the responsibility of mere consultants—which is the position of most of the deterrence theorists—to give advice about the propriety of American political values or to judge the state of their employers' intellects."[3] One does not have to suggest any sinister or insidious nature to the relationships between those who make and those who critique and analyze strategic policy to argue that the broadest possible public and scholarly understanding is beneficial to that policy process and product. The following pages represent a modest attempt to contribute to the possibility of that understanding.

An examination of strategic doctrine is timely. As the title of this volume suggests, the environment in which strategic doctrine is fashioned is dynamic, rapidly changing technological possibilities intermix with factors in the environment external to the United States to create the setting and circumstances to which strategic nuclear doctrine must respond. Many factors in that environment are coming together as America looks toward the 1980s, and the need for a comprehensive, integrating view of the conditions currently present and likely to emerge in the upcoming years is clearly appropriate.

Both the academic and decision-making communities are cleaved with regard to the current situation. That there is major disagreement on specific policy matters, the directions of the strategic system, and what should constitute the basis of American strategic doctrine clearly demonstrates a lack of consensus on strategic matters such as largely existed in the 1960s. In the absence of conceptual agreement, decisions often have been made in a piecemeal, discrete, and incremental fashion that cumulatively have contributed to the present doctrinal malaise. Swirling around focal points such as the ongoing Strategic Arms Limitation Talks (SALT), the nature of the Soviet strategic threat, and various potential new weapons systems, an examination of the situation that attempts to stand back, sort out, and arrange the various considerations in an ordered, systematic way can contribute to understanding and dealing with the strategic environment.

It is toward that goal that the present work attempts a modest first step. In Chapter 1, "Nuclear Strategy in a Dynamic World," I will discuss

Introductory Note

the dynamic nature of the nuclear system, overview some of the theoretical approaches that have been employed to organize understanding that system, and present a conceptual framework around which various concerns can be analyzed. Chapter 2, "Deterrence: The Elusive Art," looks at basic concepts about the deterrence system, its dynamics, the stability of the system, and limitations on theorizing in the area. Chapter 3, "The Evolution of American Strategic Doctrine," is an overview of the tenets of American doctrine from the Eisenhower doctrine of massive retaliation through the current debate, organized around conceptual categories developed in the first chapter.

Chapter 4, "American Strategic Forces," looks at American nuclear capabilities, beginning with an examination of the technological processes by which weapons systems enter strategic inventories, continuing with discussion of the TRIAD concept of organization and the various TRIAD "legs," comparative measures of nuclear capability (known as "bean counting"), areas connected to American strategic forces (active and passive defenses and the tactical nuclear arsenal), and ending with a consideration of future American strategic systems. Chapter 5, "Soviet Strategic Doctrine and Forces," represents a parallel exercise to Chapters 3 and 4 for the Soviet Union, presenting an overview of Russian deterrence thought as it is publicly available, examining the structure and composition of the Soviet nuclear arsenal, and concluding with a preliminary assessment of the Soviet threat. Because a major forum in which judgments about the Soviet threat occurs is SALT, Chapter 6, "The Arms Control Process," attempts to view efforts to stem the strategic armaments spiral. The discussion begins with the conceptual bases and dynamics of arms control processes, looks historically and conceptually at the dual arms control thrusts of SALT and testing ban–nonproliferation efforts, and concludes with some general observations regarding lessons from past arms control experience that are instructive for future endeavors. The final chapter, "The Uncertain Future: American Policy in the 1980s," has two basic purposes: first, identification and discussion of the three most important issues confronting the system (controlling the qualitative arms race, the continuing Soviet challenge to the United States, and the dangers of nuclear weapons proliferation); and second, an attempt to demonstrate and apply a systemwide approach to strategic issues employing the framework developed in Chapter 1 and utilizing the major problems in the system identified.

This work would not have been possible without the generous assistance of others, and I would like to express my thanks to those who were of help. Particular thanks for reading earlier drafts and making helpful comments and criticisms are given to my colleague in the Political Science Department, Victor H. Gibean, to Joseph I. Coffey of the

University of Pittsburgh, and to Lt. Colonel William Wuest of the Air Command and Staff College, Air University. Their technical and conceptual criticisms helped add clarity and correctness to the final effort and are gratefully acknowledged. The writing of the manuscript was made possible by the grant of a sabbatical leave from The University of Alabama, which is due thanks for making the opportunity available. My secretary, Chellia Holder, showed enormous patience in deciphering my illegibilities and in typing numerous drafts of materials that were at best obtuse and at worst arcane to her, and her cheerfulness and determination are gratefully acknowledged. Last but by no means least, my family had the task of putting up with me during the manuscript's gestation and development, and their ability to do so was no small achievement.

Nuclear Strategy in a Dynamic World

Chapter 1:
Nuclear Strategy in a Dynamic World

Setting the Context Although well over half the world's population have lived only in the era of nuclear weapons, this age of potential thermonuclear obliteration is little more than a tick on the clock that marks the history of man. The first fission reaction, the so-called Manhattan Project that produced the first nuclear event under the University of Chicago football stadium, occurred in 1942, and the first nuclear bomb lit the New Mexico sky in the spring of 1945. On August 6 and 9 of that same year, the United States initiated the use of nuclear weapons in anger when it attacked Hiroshima and Nagasaki in a successful attempt to convince the Imperial Japanese leadership that continued prosecution of World War II was futile and potentially cataclysmic.

To say that the discovery of nuclear energy was serendipitous would be an overstatement. In the heat of World War II, the United States had commissioned nuclear research because it knew Nazi Germany was engaged in the same effort and did not want to be outdone in an important new technology. At the same time, it would be equally misleading to suggest that anyone had a clear idea of the "genie" that was unleashed when the nuclear "bottle" was opened. As Edward Teller, one of the scientists in the team that first successfully split the atom, testified before the Senate Foreign Relations Committee on August 20, 1963: "You may not know it, but on the day when the first nuclear explosion was fired, no serious prediction had succeeded in guessing at the real size of the explosion. All of us underestimated it. After four years of strenuous effort, of theoretical calculations, of careful design, we did not succeed in predicting what was going to happen."[1]

By the standards of today, the explosions that could be created were primitive and inefficient. "Fat Man," the first atomic bomb used against

Japan, weighed about five tons and was "just under five feet in diameter."[2] Its yield was fifteen to twenty kilotons (KT), the equivalent of fifteen to twenty thousand tons of TNT. When "Big Boy," its mate that leveled Nagasaki, was exploded, the world's arsenal of nuclear weapons had been depleted. By contrast, the Soviet Union currently has nearly three hundred weapons now deployed with yields of twenty-five megatons (the equivalent of twenty-five million tons of TNT), and the United States nuclear arsenal has grown from two to nearly ten thousand warheads in the strategic arsenal (weapons targeted at the Soviet Union), with an additional estimated seventy-five hundred so-called theater nuclear weapons (TNWs) for use in Western Europe in the event of a Soviet invasion. Originally the only means of delivering a nuclear warhead to its target was a propeller-driven aircraft strained to its limits by the weight of the device; today literally thousands of these instruments of mass destruction can be sent several thousand miles on ballistic missiles fired from missile silos or from under the ocean's waters with confidence they will land within several hundred feet of their intended destinations. The "deadly dance" of nuclear armament has, in the words of a popular tobacco commercial, "come a long way, baby," and there is little on the horizon to suggest that progress (if that is an appropriate descriptor) will not continue.

These introductory remarks are intended to be dramatic and seek to underscore the basic thesis that will be developed in subsequent pages: the environment in which strategic nuclear doctrine is framed is extremely dynamic, subject to quantum leaps in capability that tax human capacity to comprehend effectively. "Thinking about the unthinkable," to borrow Herman Kahn's famous phrase, has been a matter of human concern for less than a third of a century and a matter of critical and concerted effort for only about twenty years (since the advent of ballistic missiles). In this short period of time, basic notions about the military utility of weapons and the meaning of military victory have had to undergo fundamental adjustments, and a basic body of thought about preventing nuclear holocaust has emerged, all conditioned by a tremendous uncertainty about what would happen should another nuclear weapon be employed in anger.

Redefining Military Concepts Though new military techniques from the bow and arrow to the catapult to gunpowder were, at the time each was discovered, heralded as the ultimate weapon, and though yet more awful and efficient means of killing doubtless wait and beg to be found, the thermonuclear revolution has forced a profound rethinking of concepts of war. The Carthaginian solution has always been available to some extent, although its instances were infrequent, limited, and extremely strenuous and time-consuming. Today, however, the ability effectively to end world civilization is at the command of a handful of humans and can be accomplished within hours. Deterring the destruction caused by war has always been a concern shared by men, but, as Bernard Brodie puts it, the advent of massive thermonuclear arsenals "is one respect in which the world is utterly different now from what it was in 1939 or 1914, when deterrence, however effective temporarily, had the final intrinsic weakness that one side or both did not truly fear what we would now call general war."[3] The modern world is one in which "general war" is truly to be feared and in which its prevention is arguably not only the first but the only true military objective. Some cherished military doctrines have had to be adjusted, sometimes wrenchingly, to this reality. Among these are the relationship between offense and defense in warfare, the meaning of military victory, and the utility of thermonuclear weapons themselves.

Military weapons systems have traditionally been thought of in terms of their offensive and defensive capabilities, and calculating outcomes of contemplated or actual military engagements has been the result of analyzing the success of a given offensive action against a particular defense, or vice versa. Nuclear weapons in combination with delivery systems against which there are no effective defenses radically change that calculus. Brodie stated the point succinctly in 1959: "Among the major changes we have to cope with today, perhaps the most significant militarily is the loss of the defensive function as an inherent capability of our major offensive forces."[4] Nuclear weapons cannot be used to defend or protect the population. Rather, they can be used only to attack and destroy. Thus, their only defensive function is at best convoluted: "Deterrence theory proposes that we 'defend' ourselves . . . by the threat or reality of nuclear retaliation against the enemy, destroying indiscriminately those engaged in attacking us, and those who are not so engaged in any meaningful sense of the word. To equate such an event to the traditional morality of 'heroic' self-defense is to commit a solecism."[5]

Nuclear weapons are weapons of mass destruction with little other meaningful use (a possible exception being their tactical or theater employment, which will be discussed in Chapter 4), which points to another way their existence alters military concepts: the notion of victory and the relationship between victory and the ability to inflict hurt and destruction. As Thomas C. Schelling has forcefully and convincingly pointed out, in previous times attainment of military victory over an adversary's forces was prerequisite to working one's will on a defeated populace.[6] Before Carthage could be sacked and leveled, the military might of Rome had to defeat Carthaginian forces in the field.

Nuclear weapons change all that. The ballistic missile arsenals of the United States and the Soviet Union are aimed at the populations of the other and can be delivered quite apart from military outcomes on the battlefield. The ability to destroy no longer requires defeating the enemy. The relationship between victory and destruction has become independent. As Green states: "The threat of destruction in all previous times that one can think of was the threat of a stronger party against a weaker, usually a helpless one. In none of the historical examples that are usually given was the deterrent threat of national destruction a mutual or dual threat."[7] Populations are thus the "mutual hostages" of one another (discussed at length in Chapter 2) because each can be destroyed as surely as if they were being held at gunpoint, and there is nothing in the world that military forces can do to prevent this from happening. The consequences of this fact are that "nuclear weapons threaten to make war less military, and are responsible for the lowered status of 'military victory,' " because, as pointed out, "deterrence rests today on the threat of pain and extinction, not just on the threat of military defeat."[8]

Thermonuclear weapons have truly genocidal characteristics that are terribly efficient in dispatching human life, and they are capable of carrying out that grotesque mission independently of traditional calculations of military requisites. Despite sophistications that offer the "promise" of more selective usage (discussed more fully in Chapters 4 and 7), the obvious utility of nuclear weapons is specific: mass destruction on a scale unprecedented in human history. This knowledge has led many to question whether nuclear weapons have a military role in a civilized world. As we shall see in Chapter 3, the debate on this subject has been and continues to be lively. Most observers, however, agree that the major utility of having nuclear weapons is to keep someone else from using theirs on you. Secretary of Defense Harold Brown summarizes the point: "We no longer seriously believe (if we ever did) that we can credibly deter most hostile action by the threat of nuclear retaliation. *Nuclear forces are useful primarily as a deterrent to nuclear actions* and to overwhelming non-nuclear attacks" (emphasis added).[9]

Deterrence as the Primary Value Secretary Brown's statement points to the central anomaly of the strategic nuclear balance: the awesome destructive ability possessed by the United States and the Soviet Union serves the primary (many would argue sole) purpose of deterrence. The heart of deterrence is keeping someone from doing something you do not wish him to do, "through fear, anxiety, doubt."[10] In the case of nuclear armament, that which discourages the use of nuclear weapons against one is the promise of a sure and dreadful retaliation that would be far worse than any gain one could hope to enjoy.

To some extent, deterrence is and always has been a part of military preparation and military force. Traditional military forces have always

In a Dynamic World

had dual purposes: deterrence by convincing an adversary that attack is senseless because the goal of the attack cannot be accomplished (denial) or that the attacker will be punished if he chooses aggression (punishment); and war fighting (either offensive or defensive) should deterrence fail. The uniqueness of nuclear weapons is that there are real questions about whether they have any war-fighting utility, or whether their role is exclusively deterrent, as the following suggests: "In the nuclear era it is necessary to be guided not by considerations of military force and the various forms of its utilization, but by the unconditional striving to avoid any kinds of occasions for military confrontations, and first of all by the rejection of searches for 'acceptable' scenarios of nuclear conflict which would give the appearance of 'legality' to the utilization of nuclear weapons as an instrument of policy or politics."[11]

These words also set the parameters of much of the current debate about nuclear weapons, phrased in terms of how nuclear strategy and planning can best provide for nuclear war avoidance and, should deterrence fail, how nuclear devastation can be minimized. To presage later discussions (especially in Chapter 3), the debate has developed two schools of thought: the "deterrence-only" position, associated with the strategy of mutual assured destruction (euphemistically nicknamed by detractors MAD from its acronym); and the "deterrence-plus" position, which adds war-fighting planning to nuclear concerns (and is thereby condemned in the citation above).

Because they provide the basis for so much of the debate that follows, these positions need to be described briefly. The deterrence-only position emphasizes the enormous qualitative change that nuclear weapons have introduced and implies that any nuclear usage would be extremely difficult to control short of a cataclysmic exchange in which the Soviet and American homelands would be largely decimated. Given that possibility, the purpose of deterrence (and thus nuclear weapons) is strictly deterrent—to keep the potential cost of initial weapons usage as high as possible. The method for doing this is mutual assured destruction: the threat that a nuclear attack will be met by a massive counterattack guaranteeing the effective destruction of the attacking state. Implicit in this analysis, according to Klaus Knorr, is the assumption that the usefulness of nuclear weapons "is narrow and specific, for it rests primarily on the ability to deter nuclear attack."[12]

Deterrence-plus theorists disagree. They feel that, in a world of massive nuclear arsenals, the threat to destroy another society is an inadequate definition of nuclear purpose. Their critique rests on two basic points. First, the MAD threat is too inflexible. As Richard Rosecrance says, "If the choice was solely between inaction and Armageddon, there had to be another alternative."[13] Since a nuclear attack could come in a wide variety of ways, carefully planned and proportional means are needed to meet that attack (the "plus" in deterrence-plus). This logic leads to the second criticism of MAD: that it is not believable. Deterrence-plus theorists argue that nuclear war is not necessarily a general exchange between homelands destroying both, because both sides know the awful consequences of such attacks. As a result, general nuclear attack is the least likely form of nuclear aggression (because it is obviously suicidal), and thus the assured destruction threat is primarily a deterrent against the least likely form of nuclear war.

These are complex, controversial, and important points that will be given thorough consideration in subsequent pages. They are presented in summary form here to illustrate two different points. On the one hand, the fact that there is this level of disagreement about nuclear weapons and their utility illustrates their uniqueness. On the other hand, and possibly more important in the context of the present endeavor, the fact that thoughtful analysts could reach quite different conclusions about the usefulness and uses of nuclear weaponry is some indication of the state of theorizing about nuclear deterrence. Before presenting a framework within which to analyze nuclear strategy and doctrine, it is useful briefly to overview some of the approaches that have dominated the analytical literature.

Uncertainty in Theoretical Formulation As Rosecrance noted in a paper for the International Institute of Strategic Studies in 1975, "Fundamental notions of deterrence have continued to be held with little change since the 1950's."[14] Although there were seminal works in the latter 1950s and the 1960s by such authors as Herman Kahn,[15] Bernard Brodie,[16] Thomas Schelling,[17] Kenneth Boulding,[18] Henry Kissinger,[19] and others that sought to explore this new and burgeoning area and to provide a bridge in defense thinking from a non-nuclear to a nuclear world, none has generated an enduring theoretical framework within which to view the strategic nuclear system or to allow orderly assessment of changes as they occur and the impact of changes on the overall system. This problem is particularly important because of the extreme dynamism of the system and the effects of incremental changes on the overall balance. The urgency of this problem, particularly given the current doctrinal debate and the potential effects of some developments currently within sight, requires conjecture.

In important ways, the theoretical uncertainty that typifies the literature is a direct consequence of the subject matter it seeks to study. Two aspects stand out particularly: the qualitative changes that nuclear weapons have brought to military thinking and the apparent discontinuity between traditional and nuclear military thought; and the dynamism of the subject matter itself.

In a Dynamic World

The qualitative uniqueness that nuclear weapons have brought to theorizing has two major consequences for theoretical development. On the one hand, the introduction of nuclear weapons raised fundamental questions about the relevance of traditional military concepts for organizing an understanding of a nuclear-armed world. Although Colin S. Gray has argued that there may in fact be more considerable continuities than initially meet the eye,[20] there has been relatively little exploration of those commonalities and thus extrapolations of traditional conceptions to the nuclear situation (which Gray admits). Rather, it was assumed early on that nuclear thinking would require its own distinctive conceptual set, a task made more difficult by the second consequence of uniqueness. That consequence is that, since nuclear weapons represent an unprecedented phenomenon in military experience, there is essentially no experiential, empirical base on which to build theory about nuclear consequences. For example, it is either true or false that nuclear war would inevitably escalate to general exchange, but since there has never been a nuclear war in which both combatants possessed these weapons, what would happen remains speculative (this concern is treated in depth in Chapter 2).

The dynamism of the subject matter has also presented difficulties for developing adequate theoretical perspectives. The pace of change in the third of a century that has marked the nuclear weapons age is intellectually breathtaking: the world's arsenal has expanded from two nuclear gravity bombs delivered by propeller-driven bombers to many thousands of warheads on the tips of ballistic missiles capable of traveling thousands of miles and delivering their payloads within fractions of a mile of the targets at which they were aimed. Exacerbating the problem of comprehension is an extremely dynamic technological process that is constantly producing new and highly complex and sophisticated innovations that must be understood and encompassed in theoretical formulations.

Despite these impediments to development of adequate theoretical perspectives, the field has been very active. Making no pretense at inclusiveness, two of the most prominent conceptual organizing devices that have been employed in the scholarly community to deal with nuclear matters, the game-theoretical and action-reaction phenomenon (ARP) approaches, will be discussed briefly. Because it has been so important in defense planning and thus will recur in many future discussions, the method of "worst-case analysis" will be introduced as part of the theoretical milieu.

Game theory is an analytical tool devised by mathematicians to analyze situations possessing three basic common properties: two or more participants (usually referred to as "players"); two or more decision options available to each of the players; and an outcome dependent on the mutual (though normally independently arrived at) decisions of all of the players.[21] Derived originally from the play of games like poker (hence the name game theory) where the three basic conditions are met (there are two or more players; in each hand they have multiple choices; and who wins the pot depends on what they all do), the concept has been extended to the study of strategic situations, notably the phenomenon of the arms race.

The arms race in the buildup of strategic nuclear arsenals has, of course, been the most vexing problem facing deterrence thinkers, along with the problem of avoiding the use of the arsenals developed as a result of the arms race. The search for explanations of this phenomenon has thus been a central concern of strategic thinkers, and a permutation of game theory known as the "prisoner's dilemma" has been used as an analytic device to organize understanding the arms race phenomenon. Often combining analysis with ARP (through its basis in so-called Richardson processes, named after the late mathematician Lewis S. Richardson[22]), the prisoner's dilemma has some explanatory value.[23] Although the prisoner's dilemma, and particularly how to break out of that dilemma, is treated more systematically in the context of arms control (Chapter 6), its basic elements and criticisms of those elements can be described here.

Briefly put, the prisoner's dilemma is a situation wherein if two players engage in cooperative behavior each will gain slightly, but, if they both act in an uncooperative manner, both will lose slightly. The game is "rigged" toward uncooperative behavior, in that if one cooperates and the other does not, the noncooperator can make substantial gain at the expense of the cooperative player.

The prisoner's dilemma analogy has a straightforward application to the strategic arms race, with the United States and the Soviet Union as the players. Cooperative behavior is the cessation of arms buildup (or, more drastically, unilateral arms reduction or disarmament), and uncooperative behavior is the continuation of the arms race through additional strategic systems procurement. Mutual gain accrues if both engage in cooperative behavior, because resources devoted to the arms race can be diverted to other (and presumably more productive) ends. Noncooperation means continuation of the arms race and incremental loss, since both continue to expend resources without gaining advantage (in the current context, as we shall see, this latter point is debatable). The rub is that if one acts cooperatively (stops the arms race or rolls back weapons stockpiles) while the other continues to build arms, the noncooperator gains military advantage, conceivably ultimately in the form of possessing a first-strike capability (a concept defined and discussed in the next chapter).

In a Dynamic World

Basic to the game-theoretical approach to understanding the nuclear arms race is the initial assumption of much of game theory: the so-called minimax principle. The minimax principle defines rational behavior in game-theoretical situations like the prisoner's dilemma and says that a rational player will act to minimize his maximum possible loss in a given situation (technically, the principle is limited to a single play of a game). Minimax has, for some planners, become a basic operating principle. It is an essentially conservative strategy, as can be seen by viewing the arms race prisoner's dilemma from the effect it has on player choices.

Minimax applied to the arms race dictates the continuation of building arms and can be shown by looking at the potential outcomes for either player. Cooperative behavior has the possible outcomes of modest gain (saving resources) or high loss (having the other player gain strategic advantage). Continuation of the arms race has the possible results of high gain (getting a strategic advantage) or a small loss (expending resources that are matched by the other). Minimizing the maximum possible loss dictates following the strategy wherein one's losses are at a minimum and in this case makes continuation of the arms race rational behavior.

Minimax is thus a conservative strategy, and using it as the definition of rationality has aroused considerable criticism. Among the most virulent critics has been Philip Green, who states: "Those who talk about rational behavior have merely found a high-flown way of justifying the policy they favor, while ignoring any discussion of important political problems, and of the possibility that their policy may ultimately be self-defeating."[24] That conformance with minimax arising from the prisoner's dilemma analogy places a conservative bias militating toward the continuation of strategic arms development is undeniable. That this bias can have an impact on the thinking of military planning also follows, as Green points out: "Deterrence theorists might claim to borrow from game theory the conservative definition of rationality and define 'rational' as expecting and planning for the worst."[25]

A related notion is the action-reaction phenomenon. Derived from the work of Richardson,[26] the heart of this explanation of the arms race is that for every arms action by one side, there will be a reaction by the other. In turn, this reaction will trigger a further action (or reaction to the initial reaction) by the first party, and so on. From this perspective, the arms race gains a certain automaticity and reinforcement of the internal dynamic of the strategic spiral. An exemplary study using this mode of analysis is Ronald L. Tammen's *MIRV and the Arms Race: An Interpretation of Defense Strategy*.[27]

Closely related to and in some cases implicitly or explicitly contained in analyses based on ARP is the notion of mirror imaging: the psychological projection of our self-images, ways of doing things, perceptual sets, and the like to the adversary. The two notions are interrelated because if one analyzes one's own and an adversary's reactions to events as more or less identical (when we act, he reacts; when he acts, we react), then it is only sensible to assume that both parties act out of some common basic perceptual set about the arms race and possibly with symmetrical (if not necessarily identical) political processes.

The ARP–mirror-imaging mode of analysis enjoyed considerable vogue in the 1960s and into the early 1970s, but it has met with increasing criticism as an adequate analytical form. One criticism is that the analogy is too simplistic in the face of an overwhelmingly complex reality. Tammen points out some of the limitations of employing ARP (the major reason his analysis is framed in terms of ARP is his contention that it was the framework within which the MIRV decision was originally made): "The process of action-reaction is marked by a number of uncertainties such as (1) the characteristics of the adversary's weaponry, (2) the number of deployed weapons, (3) technological innovations, (4) research and development lead times, and (5) attempts to be deceptive about the numbers and characteristics of weaponry."[28] Tammen means that, given the complexity of measuring weapons balances, the length of time and uncertain turn of weapons systems' histories from drawing board to actual deployment, and the clandestine and purposely deceptive nature of major portions of the developmental process, it is difficult to determine in any specific case what action sparked any given reaction.

A second, and related, criticism is that the framework fails to explain too many phenomena. For instance, it is difficult to explain the massive buildup (particularly in throw-weight) of the Soviet ICBM force in ARP terms unless one does one of two things. On the one hand, one can define ARP as so universal and overarching as to be virtually meaningless: "Since every reaction is the result of some action, the Soviet buildup was a reaction to an American action." Obviously, such an analysis reduces ARP to little more than tautology. On the other hand, one can painfully sift back through all possible actions that might have caused the reaction and select one or a combination of sufficient causes (as reasons). The danger, of course, is picking the incorrect sufficient cause (picking a reason other than that which motivated the Russians). The only way to demonstrate the correctness of the selection, in the absence of normally unavailable data on Soviet decision processes, is commission of the logical fallacy of affirming the consequent.

Referring specifically to the failure of ARP to explain the size of the Soviet force, William T. Lee offers a third objection: the mirror-imaging concept is fallacious. Lee argues: "The most pervasive, pernicious, and subtle—but least obvious—error to which all US analysts of the USSR are prone is 'mirror imaging,' that is, the implicit or explicit assumption that Soviet objectives are the same as ours, that they react the same way we do to common problems and experiences, even if their modus operandi is somewhat different."[29] As will be shown in Chapter 5, there is a body of literature showing that the Soviets in fact have different conceptions of the strategic balance and the purposes of strategic nuclear arms than do Americans and raising questions about the mirror-image concept. Raymond L. Garthoff, however, warns against "throwing the baby out with the bathwater" in totally rejecting mirror imaging with respect to its opposite extreme, which he terms double mirror imaging: "Less readily recognized than the mirror image fallacy is the 'double mirror image.' In avoiding projection of American thinking to Soviet leaders, there is the risk of the equally fallacious assumption that their ways of thinking and intentions always necessarily differ from our own and always in worse-case ways."[30]

Arising partly from a definition of duty and partly from sympathetic (at least compatible) predilections in the analytical literature is a tendency within military force planning toward what is known as worst-case analysis. A pervasive phenomenon at all levels of military planning, worst-case analysis involves looking at all possible military (in our case strategic nuclear) situations, determining which of those exigencies is most dangerous to American security, and planning forces and strategies effectively to combat that worst case. In the process of planning for the worst possibility, there is the implicit assumption that less dangerous possibilities will be dealt with as well. John Newhouse describes the process and some of its consequences: "Planners on both sides tend toward gloomy calculation; they reckon with all contingencies, including the most remote, and they become victims of an internal rhetoric, now exaggerating the qualities of the opponent's forces, now doubting the performance of their own."[31]

One of the most common manifestations in which worst-case analysis arises is the relationship between an adversary's strategic intentions and his capabilities. Can one impute the reasons (intentions) of an adversary on the basis of force configurations (capabilities)? Or can one understand capabilities only on the basis of knowing why a nation has them? Brodie summarizes the most common assumption: "It is often alleged . . . that the opponent's military capabilities are the best clue we have to his intentions, but the core of the truth in that assertion depends on the discrimination and sensitivity with which we scrutinize his military expenditures and capabilities."[32]

The tendency to impute intentions from capabilities arises from the fact that, in more or less precise ways, we know what the capabilities are, but not the intentions. This lack of knowledge occurs either because an adversary finds advantage in not divulging intentions, and, even if he does, there is always the possibility that he is lying or that his explanations do not comport with our conceptions of the utilities of his capabilities. When combined with the mind-set of worst-case analysis, the result of implying intentions from capabilities is obvious: there will always be a tendency to place the assessment of intentions in the worst possible light and to assume the direst and most hostile intentions.

The effects of worst-case analysis are not necessarily bad, and it is well that someone views strategic developments this way. Military planners would be irresponsible not to prepare for the worst possible case, in the event it does represent the intentions of a potential adversary. The conservative cast of some theoretical formulations, most notably those associated with derivations from game-theoretical formulations, moreover, is supportive of this mode of analysis.

To say that worst-case analysis is an appropriate mode is not to say it is the only method that should determine policy. During the McNamara years (see Chapter 3), one of the results of analysis starting from this base was projection of the Greater Than Expected Threat, a purposeful overestimation of Soviet capabilities that drove some procurement programs and resulted in force levels that, for the time, were far in excess of any Soviet threat. Rather, worst-case analysis is one of a number of perspectives that should go into policy making and policy analysis.

The purpose of this analysis has been to show that the area of strategic nuclear studies does not have an adequate conceptual framework for theory building and conceptual analysis. The field has been dominated by a conservative bias reflecting the legitimate concerns of military planners and more or less permanent participants in the policy process. It has developed piecemeal and reactively, focusing on narrow concerns, technological innovations, and environmental changes, and as such has never developed the analytical detachment necessary for dispassionate assessment. Reviewing the literature emerging from the first two decades of study, Tammen concurs: "In retrospect, the literature of the 1950s and 1960s was provocative and misleading; provocative because many new subfields were opened and explored; misleading because almost no one stopped to examine the underlying assumptions on which these subjects were built."[33]

A Framework for Analyzing the Strategic Policy Process Having stated the need for an analytical framework within which to organize study of the strategic nuclear system, in this section I will present the first step in constructing such a framework. I am not attempting to build a rigorous theoretical construct from which hypotheses can be derived deductively and tested, because I feel such an exercise would be premature given theoretical knowledge of the subject and because so much of the deterrence field is nonempirical and thus not readily amenable to these techniques. Rather, my purpose is more modest: to specify a series of analytical categories of factors involved in the making of strategic nuclear doctrine and to suggest some of the ways these factors interact to produce strategic doctrine. Hopefully, the exercise will produce a workable framework within which doctrine and specific issues affecting doctrine can be placed and their effects assessed.

The basic framework is a modified process model and is depicted in Figure 1.1. It consists of three sets of factors (or variables) that influence the making of strategic doctrine. The analytic paradigm is a rough kind of systems analysis, with the categories similar and implicitly based on man-milieu analysis. The series of causal arrows suggest ways the variable categories interact to produce strategic doctrine and ways doctrine may influence environmental factors. The remainder of the chapter will be devoted to describing each analytical category and linkages between them in order to establish the basis for applying the framework to substantive discussions in subsequent chapters.

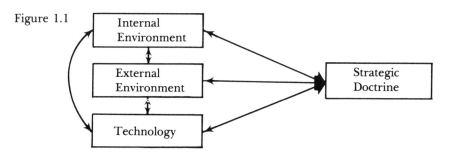

Figure 1.1

The Internal Environment The internal environment refers to the domestic factors that are influential in formulating strategic doctrine. Internal environmental influences will vary among societies, reflecting the degree of openness or political constraint or opportunity for strategic debate and thus broad input into decisions and the relative ascendancy of certain groups interested in strategic outcomes. Other factors include governmental structure (by whom and how controls of defense are organized), leadership attitudes, historical experience, and political ideology. Similarly, nations develop general outlooks toward national priorities, including national defense and security.

In the Soviet Union, evidence (discussed in Chapter 5) suggests that the professional military is predominant in strategic policy formulation, and the hierarchical command structure of the Soviet state allows substantial suppression of strategic disagreement with military preferences and the economic effects these have on realization of other priorities. In the United States, however, the nature of an open society facilitates and encourages public discussion, as witnessed by the extensive debate over the SALT II treaty. Although the technical nature of much of the subject area discourages general public input on specific issues, bureaucratic (including professional military), congressional, and private sector defense intellectuals debate vigorously the content of strategic policy and its effects on other priorities. Newhouse describes the polar extremities of this debate in the context of bureaucratic interplay within SALT as illustration: "At one end are the ritual disarmers, eager to promote or acquiesce in any agreement to reduce nuclear weapons, the very existence of which they see as a threat to all life. At the other extreme are the 'worst-case' force planners, armed with their fantasies and their circular pocket-sized bomb—damage-effect computers—the working tool."[34]

National societies also vary in their general orientations toward defense and security matters, a phenomenon that Jack L. Snyder refers to as "strategic culture."[35] The Soviet Union, conditioned by a long history of foreign invasions and an ideology proclaiming a hostile external environment implacably committed to the demise of the communist system, tends to place a high emphasis on strategic security that facilitates diversion of considerable economic and human resources into the defense effort. In the United States, on the other hand, the idea of a large standing military in peacetime and the continuous commitment of considerable resources to defense is a post–World War II phenomenon, making criticism and skepticism about security issues a more prominent aspect of the domestic setting for American defense officials than for their Russian counterparts.

These factors are, of course, exemplary rather than exhaustive of the ways the internal environment can have an impact on strategic policy making. The influences are often issue-specific, resulting from particular external environmental or technological stimuli that arise in a particular context over a particular concern or that episodically enter the strategic debate over a given concern (for example, the anti-nuclear power lobby and questions about breeder reactors). In assessing domestic impacts on particular doctrinal matters, one must be sensitive to and search for the appropriate factors from a multifaceted set of potential influences.

The External Environment Domestic considerations on strategic doctrine are in large measure conditioned by circumstances, or, more precisely, perceptions of circumstances, in the relevant international milieu. That factors in the international milieu affect the formation of strategic doctrine is stated clearly by former Defense Secretary Robert McNamara: "What is essential to understand here is that the Soviet Union and the United States mutually influence one another's strategic plans. Whatever their intentions or our intentions, actions—or even realistically potential actions—on either side relating to the buildup of nuclear forces necessarily trigger reactions on the other side."[36] Klaus Knorr adds that this influence is dynamic and that any assessment must be made in a specific context: "In other words, there is no such thing—as is too often assumed—as an absolute ability to deter, fixed in power and constant at all times regardless of changeable circumstances. Rather, the power to deter is the power to deter a particular adversary in a particular situation."[37]

In the postwar world, this external environment has largely been the dyadic nuclear relationship between the United States and the Soviet Union, with an assessment of the Soviet situation the main external variable. At the same time, the impact of nuclear decisions is conditioned by effects on allies (notably NATO). The emergence of additional nuclear weapons states increasingly promises to alter the strategic environment. Of more recent vintage, the entire momentum of institutionalized arms limitation discussions has added a conditioning factor to both the internal and external environments.

The Soviet Union Although the nuclear weapons "club" now numbers six (seven if Israel is counted), the only state capable of attacking the United States with nuclear weapons is the Soviet Union, and only the superpowers possess massive nuclear arsenals. The heart of the strategic nuclear system remains the Soviet-American dyad, and the largest international environmental impact on American strategic development is the nature and evolution of the Soviet threat. A good deal of the internal analysis and debate about the nature of American doctrine and appropriate force configuration is based on estimates of the Soviet posture, often in the forms of the capabilities-intentions debate and worst-case analysis planning. The enormous growth of Soviet strategic forces in the last decade has been the primary force in the ongoing debate about American doctrine. At the same time, it is necessary to realize that the debate and strategic decisions have a reciprocal effect, conditioning at least to some extent the nature of the Soviet challenge (the action-reaction phenomenon).

Allies and Other Nuclear States The United States considers Western Europe to be vital to its national interest and has consistently extended the nuclear umbrella over its NATO allies. Since a conflict in Western Europe almost certainly could not be successfully prosecuted without early resort to at least theater nuclear weapons (thereby raising the possibility of escalation to strategic exchange involving the American and Russian homelands), strategic doctrine must be formed and decisions reached with a sensitivity to the impact on NATO. The early SALT I discussions about so-called "forward-based systems" and attempts to include French and British nuclear capabilities in American numerical force levels were clearly influenced by this factor, and the furor over deployment of theater enhanced-radiation warheads with NATO forces and the Pershing II missile offer evidence that it continues.

The prospect of additional nuclear states raises another potentially important factor. As discussed more completely in Chapter 7, a number of states will be able to exercise the "nuclear option" by the end of the century. Some of these states pose potential threats to one or both of the superpowers and may force both to reassess strategic doctrine. Moreover, developed notions about nuclear deterrence, as translated into doctrine, have been formulated almost exclusively on the basis of the American-Soviet dyad and may be forced to undergo change.

Institutionalization of Arms Limitation Within the context of SALT, formal discussions on limiting nuclear arms have been going on, with periodic interruptions, for roughly a decade. In the process, they have attained a certain momentum and international level of expectation that probably make them a fairly permanent part of the strategic equation. Most obviously, the continuation of some level of commitment and activity (apart from achievement) is necessary in continuing efforts to dissuade non-nuclear states from obtaining nuclear weapons (an external environmental factor). At the same time, there have developed bureaucratic, pressure group, and political interests in both countries (notably the United States) that militate toward continuation of the dialogue.

Technology The state of American military technological development is actually a part of the internal environment affecting policy making (just as Soviet technology levels are an external environmental factor). Technological development has, however, played such an important, and in many cases independent, role in the formulation of strategic nuclear doctrine that it has been singled out as a separate category influencing that process. To understand the independent role of technology in strategic doctrine, it is necessary to look briefly at the so-called "R and D" process to examine the relationship between that process and policy making and to compare the role of technological innovation in the United States and the U.S.S.R.

The Technological Process The process by which new weapons systems enter the strategic arsenal contains four steps: research, development, testing, and engineering (R,D,T,&E) (discussed in Chapter 4). The first two stages (often referred to as R&D) encompass the process from initial problem identification and ideation through basic theoretical research and application to the actual design of a system and construction of a prototype. The last two stages refer to the process of repeated systems testing in more or less realistic circumstances, evaluation of testing results, and finding and applying engineering solutions to correct deficiencies discovered during testing. For a major strategic weapons system, this process typically takes eight to ten years from project initiation to weapons delivery (initial operating capability or IOC).

R,D,T,&E, particularly in its early stages, is a process of ideation and discovery, and, like any creative intellectual endeavor, is difficult to control and monitor. A major innovation with enormous strategic consequences can result from individual original thought or borrowing a development from another seemingly unconnected technology (the principle of the MIRV bus, for instance, came from space technology designed to allow placing multiple satellites in orbit from a single rocket launch). Much scientific advance, however, is incremental, aimed at gaining more precise knowledge and applying that knowledge. As Harry G. Gelber puts it: "In an important sense, the thrust of developmental work is less toward the creation of 'better systems' per se than toward the identification and definition of deficiencies in existing policies, structure, and weapons inventories."[38] This incremental nature gives technology a self-sustaining dynamic that is also self-justifying. "The solution of one R&D problem therefore often means pressures for further, and usually more, R&D."[39]

Technology and Strategy In an ideal and totally rationalized system, the fruits of technology would flow from well-conceived doctrinal decisions: we would decide we needed a particular weapons system and would tell the scientists to build it. In the real world, this is not always the case. According to Gelber, "Discoveries and particular innovations can be 'made to order' in the sense of being a response to some external demand. Or they can come without external or organizational stimuli."[40]

The relationship between technology and strategy is thus reciprocal: sometimes policy dictates invention, and at other times technological innovations drive policy. Robert J. Pranger and Roger P. Labrie describe this process: "The relationship of technology to strategy may be symbiotic, with each lending legitimacy to the other. The degree to which either dominates the other may change with circumstances and the character of political leaders."[41] Technology has something of an advantage in this regard, in that political leaders change more frequently than do the scientists in weapons laboratories. There is also within the R&D process a phenomenon known as technological determinism: "A hypothetical situation wherein a given weapon system is initiated, developed, and possibly deployed as a result of engineering curiosity, serendipity, or energetic engineering (science for science's sake) rather than as a response to defined military requirements."[42] The result is often weapons systems for which doctrinal justifications must be created after the fact. Jack Snyder concludes: "Weapons systems frequently develop and then search for missions (sometimes passing through several in the course of their evolution). Bargaining chips get deployed, accuracy improves in part because of the momentum of technology, and so on."[43]

Comparing American and Soviet Processes The reciprocal, often serendipitous relationship between technology and doctrine is more characteristic of the United States than the U.S.S.R. and in many ways reflects the differences between an open and a closed society. In the United States, science is decentralized, with work going on independently in government weapons and academic laboratories and coordinated mainly by the requirements of research contracts. In the Soviet Union, the system is more hierarchical and authoritarian, thereby facilitating control of scientific investigation. Robert Perry summarizes the results of these differences on technological flow: "The USSR ordinarily invests in an R&D process that relies on incremental, sequential improvement of previously developed and deployed weapons—with certain important exceptions—and the United States normally prefers technological thresholding—also with some significant exceptions."[44] This Soviet preference reflects in part their insistence on subsuming other considerations below political purposes. Richard Pipes adds, "Soviet military theorists reject the notion that technology (i.e. weapons) decides strategy. They perceive the relationship to be the reverse: strategic objectives determine the procurement and application of weapons."[45] Colonel Richard G. Head well summarizes various ways in which technology influences strategic concerns: "The impact of technology on the military balance may be revolutionary or incremental—depending on how it is exploited—and the military balance may be changed by improvement of older technology as well as by development of the new. . . . The effective exploitation of technology for military purposes is dependent upon a range of essentially nontechnological factors such as military doctrine, tactics, training, resource allocation preferences, organizational processes, R&D style, budgets, and arms control."[46]

Strategic Doctrine The result of these sets of considerations is the formulation of strategic nuclear doctrine. That doctrine in turn is one of the two major elements of overall defense policy (the other being conventional or general purpose doctrine, referring to non-nuclear force uses). The general strategic doctrine is translated into a series of operational strategies and imperatives to implement the general posture. The strategic missions defined by operationalizing general doctrine are then translated into force requirements, composition, and configuration. This process is depicted in Figure 1.2.

The factors in conventional doctrine do not concern us here (the processes are analytically similar), but those affecting strategic nuclear doctrine do.

Strategic Doctrine The general strategic doctrine is the broadest and most overarching statement of the way strategic nuclear forces guarantee deterrence against a nuclear attack, including the ways those forces

Figure 1.2

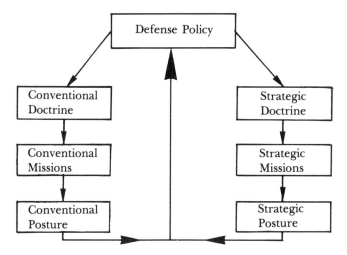

would be used in the event of such an attack. This formulation reflects the security philosophy of the nation and its view of the world environment in which it exists.

The doctrine that became known as mutual assured destruction is exemplary of the kind of statement that forms general doctrine. Based on an assessment of Soviet capability that recognized the inability of American forces to defend against a strategic nuclear attack by the Russians, the doctrine maintained that deterrence was guaranteed by the mutual ability of each of the superpowers to absorb any conceivable attack by the other and retain sufficient nuclear forces to be able to launch a counterattack resulting in unacceptable levels of damage to the attacking state. Under this set of circumstances, any attack was considered irrational because it would be suicidal.

Strategic Missions General doctrine states the basic position on the strategic system and the basic view of deterrence, but requires operationalization to allow application to situations. This is done by developing a series of statements about the strategic missions and force uses that are required to implement doctrine. These statements, in turn, allow formulation of force levels and characteristics.

The key to MAD doctrine as a stabilizing element was creation of the situation in which launching a preemptive (or first) nuclear strike would never be profitable. Among the strategic missions arising from that posture was the adoption of a second-strike strategy (a position that the United States would fire its forces only in retaliation to an initial attack by the Soviet Union). The key element in being able to sustain a second-strike strategy was to have forces that could survive an attack in sufficient numbers to carry out the assured destruction mission (discussed more fully in Chapters 2 and 4).

Strategic Posture Use of strategic nuclear forces is the ultimate method in which doctrine is implemented. The actual nuclear forces—numbers, configurations, and projected missions—derive from the strategic needs of the nation as described by general doctrine and translated through strategic missions. Ideally, force requirements are arrived at deductively and designed to carry out prescribed and predetermined missions. As the discussion of technology indicated, these processes are sometimes reversed in practice.

In implementing mutual assured destruction, the United States relied on an incremental force configuration known as the TRIAD. Consisting of land-based intercontinental ballistic missiles (ICBMs), sea-based submarine-launched ballistic missiles (SLBMs), and strategic bombers, force development had as its goals to ensure that each separate component, and thus the entire force, was as invulnerable as possible to attack and that there was a maximum probability that the weapons, if launched, would reach their destinations. The actual size of the force (which still governs deployed launcher levels) was determined by calculating the number of warheads necessary to guarantee unacceptable damage.

Interaction among the Factors As the causal arrows in Figure 1.1 indicated, factors within each of the environments interact in the formulation of strategic doctrine, and that doctrine in turn affects the other variables. The diagram identifies six sets of reciprocal relationships among the four variable groups. Although actual situations are often more complex (two or three variable sets influencing a doctrinal decision, one independent variable influencing another, which in turn causes a change in doctrine or one of the other independent variables) than the simple relationship between each pair, and recognizing that factors within each category affect one another, there is utility in examining how the various factors affect one another.

Internal Environment–External Environment These factors can affect one another in many ways. A change in political leadership in the United States accompanied by new strategic conceptions or economic conditions constraining perceptions of the ability to sustain levels of strategic spending can have a positive or negative impact on the Soviet Union. At the same time, attitudes expressed by the Soviet Union, such as their at least rhetorical rejection of MAD as the basis of deterrence or their refusal to accept so-called "linkages" between their behavior in arms control negotiations and their support of "brush-fire" wars in Africa, have fueled large doctrinal debates within the various active sectors of the American domestic environment.

Internal Environment–Technology These interactions are, at heart, all internal. Those factors identified as composing the internal environment influence technological development primarily through the resource allocation process. By establishing resource priorities for the R,D,T,&E process generally and for various programs and thrusts, the internal environment can enhance or impede the flow of weapons technology. For example, the enhanced-radiation warhead emerged as a battlefield weapon to some extent because resource priorities had placed developmental responsibility with the army (the warhead was originally developed as part of another army program, the anti-ballistic missile or ABM). At the same time, political control of technological output is highly imperfect, and weapons occasionally emerge independent of planning processes. These systems can, as in the case of the enhanced-radiation warhead, result in considerable bureaucratic interplay and legislative-executive discussion.

Internal Environment–Strategic Doctrine The impacts of the internal environment on doctrine are obvious: doctrine is formulated within that milieu, is shaped after often fierce internal bureaucratic bargaining and extensive legislative-executive debate partially conditioned by interest group activity, and must rely for implementation on resource allocations from the political process. The doctrine itself, however, helps to condition and form parameters around the internal debate. As example, the current doctrinal debate about the adequacy of MAD uses as its point of departure the doctrine itself, and detractors must begin from the defensive position of demonstrating its inadequacies before their own counterproposals become relevant.

External Environment–Technology The case of the multiple independently targetable reentry vehicle (MIRV) neatly describes the reciprocal relationship between these variables. MIRV was originally developed as a means of countering projected Soviet ABM programs (its multiple warheads were designed to confuse and overwhelm such defenses), a technological advance spurred by an external environmental source. When the Soviet ABM threat failed to materialize, the United States found itself with an enormously sophisticated and potent offensive weapons system that threatened the symmetry of the nuclear balance and apparently spurred the buildup of Soviet arms (technology causing external environmental change). This situation was thus a kind of two-stage action-reaction phenomenon,[47] with external environmental considerations leading the first stage and technology the second.

External Environment–Strategic Doctrine Along with the great advances in targeting accuracy, the physical expansion of Soviet armaments to a point of parity (and many argue potential superiority) has been the major cause of the ongoing strategic debate. The relationship is, however, reciprocal: although the Soviets maintain that MAD doctrine is at best an unpleasant and ephemeral basis for deterrence (see Chapter 5), nonetheless its tenets and the forces arrayed against them that result from the doctrine are central to their planning. As example, the Soviet civil defense system, aimed at evacuating cities, dispersing industry, and providing shelters for key personnel, makes most obvious sense in reaction to a doctrine in which urban areas are a major targeting priority (which is the case under MAD).

Technology–Strategic Doctrine As discussed above, ideally, technological innovation should flow from directions prescribed by strategic doctrine. The developmental efforts leading to increasing sophistication and capability of the atomic submarine fleet culminating in the current Trident program is exemplary of this ideal. Mutual assured destruction dictates highly survivable forces to guarantee retaliatory capacity, and the submarines are our most invulnerable weapon. At the same time, incremental improvements in targeting accuracy, carried on without any specific doctrinal mission, may have enormous and potentially destabilizing effects on doctrine (discussed more fully in Chapters 4 and 7).

Chapter 2:
Deterrence: The Elusive Art

What Is Deterrence? As suggested in Chapter 1, the deterrence purpose of military forces is not a new concept, and if nuclear weapons and forces occupy a "unique" place in the continuum of weaponry, it is because their utility has been conceptually limited to its deterrent effect. The deterrent–war-fighting dichotomy of weapons potential is more clearly focused and more sharply drawn regarding nuclear forces than so-called conventional forces. Although we may fervently hope that all weapons will deter and thus not be used, the deterrent role of nuclear forces gains an ascendancy not associated with other weapons systems. For this basic reason, understanding the nature and dynamics of deterrence is prefatory to other insights into the nuclear balance.

At the outset, it is necessary to define two basic terms: deterrence and what constitutes "strategic" in the nuclear context. Within the literature, there is widespread agreement on the definition of nuclear deterrence, but considerably less agreement about how it works and what are its dynamics. These latter concerns will occupy most of the rest of this chapter and will be central to assessing doctrine, nuclear forces, and associated concerns throughout the discussion. What constitutes "strategic" has a hazier consensus.

To reiterate the distinction made in Chapter 1, the key aspect of deterrence is keeping an adversary from doing something you do not want him to do, and the basic mechanism is the threat that creates in an adversary "fear, anxiety, doubt."[1] The deterrent effect can be in the context of a specific situation or can be extended to cover broad classes of phenomena (such as NATO defense). In nuclear terms, this means possession of enormous stockpiles of deadly weapons that can be hurled without danger of interdiction by enemy defenses and that are capable of widespread societal devastation. These weapons have the major purpose of "persuading an enemy not to do what he would otherwise do."²

The basis of deterrence is thus dissuasion. Dissuasion is accomplished through the basic mechanism of the threat of pain and frustration. Threat, in turn, can either be that of frustration of a proposed activity such that its accomplishment is so unlikely as to make the effort obviously futile, or the promise of such painful retribution that the hurt one would receive would far outdistance any gain from the original act. The first threat method has been associated with historical military activities (for example, deterring an aggression with the promise that it would fail). Since the ability to inflict great damage and suffering is no longer predicated on attaining traditional military objectives (such as winning the war), effectively deterring nuclear aggression relies on the promise that such action would be painfully retributed. Nuclear deterrence thus rests on saying something like, "Although you can hurt us terribly, if you do, we will pay you back by hurting you worse."

In practice, the two threatening methods may be intertwined, as Bernard Brodie points out in discussing the escalatory process: "The control of escalation is an exercise in deterrence. We try first and foremost to deter the opponent from doing that which will oblige us to threaten or resort to any use of arms; if he nevertheless persists and a conflict starts, we try to deter him from enlarging it, or even from continuing it. Deterrence, at any level, thus naturally means inducing the enemy to confine his military actions far below those delimited by his capabilities."³ In terms of nuclear deterrence and the grim consequences should it fail, this entreaty has meant avoiding direct military confrontation between those capable of massive mutual nuclear destruction (the United States and the Soviet Union) by keeping massive conventional and nuclear arsenals to persuade one another of the impossibility of achieving conventional military objectives (such as conquering Western Europe) and elaborate, albeit conjectural, planning to avoid escalation to general nuclear exchange should initial deterrence efforts fail.

There is less agreement about what constitutes "strategic" as opposed to "tactical" (or theater) doctrine. Henry S. Rowen offers a representative list of how the term is used: "What does 'strategic' mean anyway? The word may refer to 1) attack by United States or Soviet forces on opposing homelands; 2) attack on population (and/or industry) as distinct from military targets; 3) attacks on missiles in silos and other long-range forces versus attack on general purpose forces; 4) attack on 'deep' targets; 5) nuclear as opposed to non-nuclear attack; 6) attacks using long-range vehicles against any target; or 7) any attack launched from outside the theater."⁴ Coming to grips with these distinctions is important in establishing parameters and context around the meaning of the term "nuclear deterrence."

Deterrence

Clarifying these distinctions will assist in reaching sensible definitions of what constitutes strategic doctrine by analyzing the two-by-two matrix represented by Figure 2.1:

Figure 2.1

	Nuclear	Non-Nuclear
Strategic	Direct attack with nuclear forces on opposing homelands	Invasion of opposing homelands by general purpose forces
Tactical (Theater)	Limited nuclear use in defense of European invasion	Invasion of homelands of allies in Europe with general purpose forces

The distinctions are obvious: strategic attacks are defined as attacks on nuclear powers' homelands, whereas tactical attacks refer to military actions against objectives lesser than (although possibly prefatory to) superpower homeland attacks; nuclear as opposed to non-nuclear refers to the kinds of weapons used (as pointed out, in tactical weapons systems, this distinction is becoming blurred because some "conventional" weapons now have yields higher than the smallest nuclear weapons). The distinctions are also perspectival in terms of the U.S.–U.S.S.R. dyad: the West Germans would certainly consider any attack on their homeland strategic.

Examining the cells helps to resolve the tactical-strategic definitional issue. There is little doubt that an attack on the homeland of either superpower would be considered strategic. A non-nuclear (conventional) attack is so unlikely, however, as to be discountable: before it could be even conceivable, it would have to be prefaced by massive conventional combat (the failure of the first method of deterrence) that would probably include the use of nuclear weapons, minimally as a means to secure invasion routes.

On the other hand, the distinctions are not so clear in the tactical categories, particularly in the European theater which is the subject of most discussion. The cloudiness relates to the escalatory process, and debate arises over two basic issues connected to the vitality of interest in Europe to both sides. First, if maintaining something like the current European balance is vital to superpower interests (both sides say it is), then there is a very real question about whether either side would (or could) accept a conventional European military defeat or would feel compelled to employ nuclear weapons to prevent such a defeat. Moreover, both sides (particularly NATO) have contingency plans that include tactical nuclear weapons use as an early method of stopping a conventional attack. Second, there is great uncertainty about whether a tactical nuclear exchange could be limited at that level or would degenerate into general exchange involving homelands.

This latter concern is hotly debated, although on dubious empirical grounds (see discussion later in this chapter). Arising from these distinctions and recognizing that it is not universally acceptable, "strategic" in the context of nuclear doctrine (and thus the purposes of nuclear deterrence) will be defined as any potential direct attack with nuclear weapons on the homeland of one nuclear power by the other or any employment of nuclear weapons in support of military objectives that significantly threatens the possibility of homeland exchange. What "significantly threatens" triggering escalation is complex and controversial. Before attacking that debate, it is useful to examine some more basic elements of the deterrence system.

Unconditional and Conditional Viability In *Conflict and Defense,* Kenneth E. Boulding coined a distinction that is a useful starting point toward understanding the unique contribution of nuclear weaponry to national security concerns and thus the absolute primacy of deterrence as a military value.[5] These terms are unconditional and conditional viability.

Boulding defines a state as unconditionally viable if it cannot be destroyed as an independent source of decisions. In other words, an unconditionally viable state is one that cannot be conquered or subjugated, or one that cannot be obliterated short of successful invasion and subjugation. More recently, and in view of the emergence of enormous self-deterring arsenals, Boulding has modified that definition, implicitly on the basis of how secure the threat system regulating relationships is: "A two-country system is a system of unconditional viability if . . . both countries may threaten each other but neither is able to threaten the other without fear of great loss or defeat in carrying out the threat."[6] This latter distinction gains meaning depending on how "successful" a threat system is, because "successful threats are those that do not have to be carried out."[7]

Boulding's conceptual modification muddies the analytical waters, because it introduces intangible perceptual questions about what deters and how. These questions are empirical, not tautological, and are questionably intersubjective (what constitutes "fear of great loss" to one may not to another). Without imputing motivations to Boulding, however, the modification is necessary if the category of unconditionally viable states is not to be an empty category: the United States and the Soviet Union can destroy one another, and individually or collectively they can destroy any other state or states. The original conceptualization is neat and determinate, however, and the fact that no state currently falls into it does not dilute its analytical (to say nothing of dramatic) utility.

The companion concept is conditional viability: a state is said to be conditionally viable if it can be destroyed, but if the state with that capability refrains from doing so. As we shall see later in this chapter, there are gradations of the claim to conditionally viable status. William H. Kincade describes the operation of conditional viability between the United States and the Soviet Union: "The viability of each of the superpowers as superpowers continues to be conditional. It is dependent on the other's ability to do it deep and lasting economic injury, in other words to remove it for a more or less long period from the list of major powers."[8] That smaller nuclear powers may have the ability to threaten the viability of the major powers is suggested by the French strategist Pierre Gallois: "With a limited stock of highly destructive nuclear weapons, it became possible for one state to inflict extensive, even unconscionable, damage upon an enemy, even if the enemy was incomparably stronger. All countries, no matter what their size, became vulnerable to losses that they cannot absorb with impunity."[9]

Historically (before the full impact of the thermonuclear revolution), a number of states probably were unconditionally viable. A minimal list would include the United States (by virtue of physical location and size) and the Soviet Union (by virtue of size and climate—lessons learned all too vividly by Napoleon and Hitler, among others), with the possible addition of a united China. Today all states are conditionally viable: all can be destroyed with nuclear weapons, but none have been. To understand the critical dynamics that created a world exclusively of conditionally viable states, one must look at the qualitative as opposed to quantitative aspects in the development of the nuclear revolution.

Quantitative and Qualitative Aspects of Change The discovery and application of nuclear fission had a profound impact on destructive capabilities, but the impact was more quantitative than qualitative. In some important aspects, the result was to bring munitions capabilities in line with the purposes of strategic saturation bombing proposed before World War II by the Italian General Giulio Douhet and operationalized by the Allies against Nazi Germany: massive destruction could now be carried out much more efficiently than even the advent of incendiary explosives allowed.[10] The early weapons were deliverable only by conventional means (aircraft). Although the quantum leap in destructive power each aircraft could carry made defensive measures more urgent, the possibility of defense and the calculation of limiting damage below societal devastation were conceivable. The change, then, was quantitative rather than qualitative.

The advent of the ballistic missile equipped with thermonuclear warheads profoundly changed conventional notions. At the time these missiles were introduced, there were no defenses against them, and the signing of the ABM Treaty as part of the SALT I Agreement effectively signaled the end of the attempt to devise such defenses, at least temporarily. This change required fundamental alterations of notions about military defense that were qualitative in nature: "The coming of the ballistic missile is enormously discouraging to the defense against strategic bombing, not only because it is so difficult to cope with itself but also because it calls further into question the usefulness of costly defenses."[11]

The major impact of the ballistic missile was to bring to an end the situation of unconditional viability because, as Harland Moulton points out, "The achievement of the ICBM brought under effective nuclear threat virtually any priority target on earth."[12] Massive societal destruction had become both possible and undeniable, with the necessity of prior military victory removed from the calculus: "The innovation consists of the fact that nuclear weapons, coupled with intercontinental missiles, can by themselves carry out strategic missions which previously were accomplished only by means of prolonged tactical operations."[13] In other words, strategic targets (those associated with the capability to sustain military activities) can now be attacked directly and early on during hostilities, rather than after "prolonged tactical operations" have rendered them vulnerable.

Recognition of the inability to do anything but watch the bombs fly made much more real the prospects of nuclear holocaust and led to grim descriptions. Brodie summarizes these calculations:

> There are at least four reasons why casualty rates with nuclear weapons are likely to be far greater in relation to property destroyed than was true of nonatomic bombing: (1) warning time is likely to be less, or nonexistent, unless the attacker deliberately offers it before attacking; (2) the duration of an attack at any one place will be literally a single instant, in contrast to the several hours' duration of a World War II attack; (3) shelters capable of furnishing good protection against high-explosive bombs might be of no use at all within the fireball radius of a large ground-burst nuclear weapon, or within the oxygen-consuming fire-storm that such a detonation would cause; and (4) nuclear weapons have the distinctive effect of producing radioactivity, which can be lingering as well as instantaneous, and which causes casualties but not property injury.[14]

This kind of calculation combined with the realization that unconditional viability was lost to create an urgency to the study of nuclear deterrence that had not existed prior to the ballistic missile. The change in military calculation was qualitative, with a devolution of the war-fighting function and ascendancy of the deterrence notion. A more careful assessment of what it meant to be conditionally viable and how to stay that way was part of the new concern.

Secure and Insecure Conditional Viability Although all nations are conditionally viable, they are not coequal in their claims to that status. In order to lend some discrimination to the basic concept, the categories secure and insecure conditional viability were added to describe differences in status.

Secure conditional viability exists when a state can be destroyed as an independent decision unit, but it does not pay a potential attacker to do so. In the contemporary nuclear balance, the obvious situation in which it would be counterproductive to destroy another state is when doing so would invite one's own destruction. At present, the states with the strongest claims to this status are the United States and the Soviet Union. As we shall see, medium nuclear powers and either nuclear or non-nuclear allies of the superpowers have some claim to secure status, but their situation is somewhat ambiguous.

Insecure conditional viability exists when a state can be destroyed, but is not because of the goodwill or lack of interest of those states that can destroy it. This category describes states that are defenseless against nuclear attack because they have no ability to deter a nuclear strike. These states, however, are not necessarily in a desperate condition. Quite the contrary, many insecurely viable states have found that their vulnerability is, in many circumstances, at worst irrelevant and at best advantageous.

This apparent anomaly arises from the basis of insecure conditionally viable status. Goodwill or lack of interest on the part of a potential nuclear attacker translates realistically into an assessment that it is not worth the costs to destroy a nuclearly helpless state. First, to do so would be such a patently atrocious and barbaric act as to bring the proper disapprobation of world opinion down upon the attacking state in a manner unprecedented since the Nazi Holocaust. Second, there is enough uncertainty about what happens after the first nuclear weapon is used to make the risk unacceptable given likely gains. In other words, no matter how annoyed the United States may have become at the Amin regime in Uganda, it would scarcely have been worth the risk (however slight) of initiating World War III to remove that regime with a nuclear strike.

This fear of the possibility of nuclear escalation has been recognized and exploited by many states (Third World and otherwise) that fall into the insecure category. As nuclear arsenals have grown, the superpowers have been forced to evaluate their potential involvement in situations not only in terms of the effects of various outcomes on their interests, but also in terms of the likelihood that their involvement could lead to direct superpower confrontation that could, in turn, devolve into nuclear war. The result is what can be called nuclear muscleboundedness: a reluctance to become involved in situations arising from calculations about the probability of nuclear escalation. This caution in turn gives less powerful states additional discretion in conflictual situations: they know that nuclear threats are hollow and the possession of nuclear weapons constricts rather than expands the ability of the superpowers to influence some events.

The situation of allies and small nuclear powers is also interesting. The claim of non-nuclear superpower allies to secure status (for example, West Germany, Poland) is based on the guarantee that the dominant nuclear power would use its nuclear weapons to defend the ally in the event of a nuclear or overwhelming non-nuclear attack (the so-called nuclear umbrella). Unfortunately, in a world where the defense of allies by using nuclear weapons is potentially suicidal, some have questioned whether such a commitment would be honored, and thus whether secure viability can be inferred from superpower guarantees. Gallois, who has been a leading critic of the credibility of the nuclear umbrella, maintains that because of the mutual ability of the United States and the U.S.S.R. to destroy one another, "the American guarantee which only yesterday was unconditional is now conditional, and, hence, under some circumstances, uncertain." He reaches this conclusion because "if resort to force no longer merely implies risking the loss of an expeditionary army but hazards the very substance of national life, it is clear that such a risk can only be taken for oneself—and not for others, including even close allies. This conclusion signifies the complete collapse of the collective-defense systems upon which the security of the Western world has hitherto been based."[15]

Whether the United States or the Soviet Union, declaratory statements notwithstanding, would in fact defend its allies with nuclear weapons is an open question, as are many important issues in the deterrence field. Since the only way to resolve the controversy is a situation where such a resolve would be tested and the response could be viewed, the situation, and thus the basis of ally claims to securely viable status, remains unresolved.

One response (for example, the French) to this ambiguity has been to develop independent nuclear forces. Geoffrey Kemp describes the characteristics such forces must possess to be effective: "Medium nuclear powers wishing to develop a nuclear capability to 'deter' either of the superpowers from overtly hostile acts will need to develop forces capable of inflicting some level of 'unacceptable damage' in a second-strike mode against selected strategic targets, which might include military targets as well as industry and population centres."[16] These requirements define the claim to secure viability: "some level of 'unacceptable damage' " makes it not pay to attack such a state. As pointed out in a subsequent section of this chapter, however, attainment of a force capability that can be employed in a "second-strike mode" is not an easy task, and the failure to do so has destabilizing effects on the nuclear system. To

Deterrence

understand why this is the case requires defining and discussing some of the basic concepts of the nuclear balance and how they interact.

Basic Concepts and Dynamics The dynamics of the nuclear balance are largely defined by the nuclear capabilities a nation possesses and the employment doctrine it adopts for firing those weapons. In this section, these concepts will be defined and discussed in terms of first-strike and second-strike options and basic targets in a nuclear war. The interrelationship between capability and employment doctrine will be analyzed, as will their relationally specific context. The section will conclude with a discussion of the ways in which various combinations of capability and employment doctrine contribute to overall stability in the nuclear system.

Nuclear Capability and Employment Doctrine Nuclear capability refers to the amount and kind of nuclear power a nation possesses and is defined in terms of the physical size of the force measured in a number of ways (see the discussion of "strategic bean counting" in Chapter 4), its means of delivering to potential targets (known as penetrability, or the ability assuredly to reach its destination), and its invulnerability to preemptive attack (or survivability). Since deterrence rests on the guarantee that a nation retains sufficient capacity to inflict great damage on a hostile adversary, states will seek to optimize each of these characteristics. The characteristics are, at the same time, interactive: the more vulnerable a force is to preemption (survivability) or interception (penetrability), the larger it must be to carry out its assigned mission, and vice versa. Richard Rosecrance puts this relationship slightly differently: "Deterrence is the product of capability and credibility. If it is to be maintained, one of three situations must exist: (1) both elements must attain some minimum level; (2) if credibility falls, the ability to punish an adversary must be enhanced; (3) if useable capability declines, the credibility of its employment must increase."[17] In capability terms as described, what Rosecrance refers to as capability equates to size, and vulnerability and penetrability are analogous to credibility. His notion of credibility also relates to the psychological notion of whether it is believable that a nation would use its nuclear forces in particular circumstances. This consideration is key to the current doctrinal debate and will be discussed in Chapter 3.

Nuclear employment doctrine refers to the plan for use of nuclear weapons in a war-fighting situation and is normally described as "first-strike" (the intention to fire one's nuclear forces before an adversary has taken a similar action) or "second-strike" (the determination to use one's nuclear weapons only in response—or retaliation—to a nuclear attack by another state) at least in regard to initial usage in a nuclear environment.

The first-strike–second-strike dichotomy is also used to describe a nation's nuclear capability. The meaning in the capability context is somewhat different, and thus care must be exercised to avoid confusing the terms when attached to the capability and employment doctrine concepts. Before relating capability to the first- and second-strike notion, it is useful to describe the conventional distinction between the types of targets in a nuclear war because target selection is closely related to strike and capability postures.

The deceptively antiseptic distinction normally made is between counterforce and countervalue targeting. Counterforce targeting refers to aiming and firing one's nuclear weapons at the military forces of the adversary. For the offensive purpose of disarming a potential adversary, counterforce targeting implies attacking strategic nuclear retaliatory forces, and such strategies "are closely connected to and encourage notions about actually fighting and winning a nuclear war."[18] In a retaliatory (second-strike) mode, counterforce targets include military bases, submarine facilities, airfields, and empty missile silos (to avoid their reloading and to destroy residual missiles held in reserve). Obviously, forces capable of carrying out counterforce missions can be used either offensively of defensively (as in damage limitation by reducing reserved enemy forces). In Chapter 7, we will confront the curious argument that the attainment of counterforce capability is stabilizing if the United States has it (because we would use it only for retaliation) but destabilizing if the Soviets achieve it (because they would use it offensively).

Countervalue targeting refers to aiming one's nuclear weapons at the things people most value, notably their lives and their productive capabilities (Philip Green refers to this targeting doctrine in more descriptive terms as "counterpeople warfare"[19]). Obvious countervalue targets include population centers (cities), industrial facilities (factories), and energy-producing and transportation systems. The openly stated goal of a countervalue strategy is to deprive a nation of the illusion of being able to "win" a nuclear war by guaranteeing its inability to survive and recover from such a war's effects. This purpose is important to understand in terms of Soviet declaratory policy about nuclear war as discussed in Chapter 5.

In many, if not most, cases the distinction between countervalue and counterforce targeting is more apparent than real. Many counterforce targets are in or adjacent to population centers and could not be attacked without large civilian casualties (countervalue targets). For instance, Wright-Patterson Air Force Base cannot be attacked without causing extensive damage and loss of life from immediate effects in Dayton (to say nothing of down-wind effects from residual radioactivity). That such interconnections exist is admitted by writers on counter-

Deterrence

force targeting within the sanitary-sounding concept "collateral damage."

With this background in mind, ideal definitions of first-strike and second-strike capability can be defined and discussed. First-strike capability is the ability to attack another state with nuclear forces and to destroy that state and its ability to retaliate, with special emphasis on the capacity to eliminate the ability to strike back. Such a capability naturally emphasizes the offensive element of counterforce targeting. To attain such a capability requires a sizable force that is extremely accurate. The need for accuracy follows logically: any retaliatory forces that one misses can be used to counterattack; thus, if one cannot destroy all forces, one by definition does not have the capability. Given the invulnerability of nuclear submarines, neither the United States nor U.S.S.R. possesses this capability toward the other.

A state possessing first-strike capability would be unconditionally viable, because by launching a strike it would disarm an opponent, who consequently could not retaliate and destroy it. As the previous discussion indicated, no state is unconditionally viable, indicating that no state possesses absolute first-strike capability (although increasing missile accuracy, as discussed in Chapter 4, will endow both sides with limited first-strike capability against some forces in the near future). Since capability is meaningful only in relation to a specific adversary, however, it is possible for a state to possess first-strike capability against some but not other adversaries. It also follows that a state against which one possesses a first-strike capability cannot lay claim to secure conditional viability. By virtue of the fact that it can be disarmed, it loses the ability to make the superior state "pay" for its actions. Thus, its status must be insecure conditional viability.

Second-strike capability is the capacity to absorb any conceivable nuclear attack by an adversary and retain sufficient retaliatory forces to inflict unacceptable levels of damage on the original attacker. Such capability implies reliance on countervalue targeting (the guarantee of meting out unacceptable damage) and, to a lesser extent compatible with but not necessary to the capability, limited defensive counterforce capacity (to guarantee that the attacker cannot calculate advantage through subsequent military actions). This type of force relies primarily on the characteristics of survivability (to ensure that forces will be around with which to retaliate) and penetrability (to assure destruction). John Newhouse describes the ideal second-strike force: it "should be capable of delayed response; it should be invulnerable; and it should be unambiguously deprived of what is called a first-strike, or damage-limiting, capability."[20] This latter characteristic relates to the contribution of weapons to strategic stability and will be discussed subsequently. A state with a second-strike capability is said to be securely conditionally viable, because under no circumstance can a potential attacker calculate that an attack does not entail the prospect of a devastating counterattack.

The Relational Nature of Capability and Employment Doctrine Although notions of capability and employment doctrine are defined somewhat differently in the first-strike–second-strike context, the concepts are essentially relational in two important ways. First, there is a reciprocal relationship between the kind of nuclear employment doctrine a nation follows and its nuclear capabilities. Second, the calculation of first- or second-strike capability or employment doctrine is meaningful only in comparing the dyadic relationship between two parties.

Relationship between Capability and Employment Doctrine The relationship between capability and employment doctrine is analogous to that between general posture and force characteristics discussed in Chapter 1, where employment doctrine and general posture are equated and force characteristics and capabilities occupy parallel positions. The nuclear doctrine a nation adopts defines its preference and intentions for nuclear weapons employment (and serves as a threat to deter aggression). This preference is implemented by developing capabilities (force characteristics) compatible with that intention. At the same time, the kind of force characteristics a nation has present employment opportunities and limitations: adoption of a firing doctrine is meaningful only if one has the capabilities to carry it out. The possible combination of capabilities and strategies is depicted in Figure 2.2:

Figure 2.2

Employment Doctrine	Capability		
	1st Strike	2nd Strike	Less than 1st or 2nd Strike
1st Strike			
2nd Strike			

The only category that has not previously been defined is "less than first- or second-strike capability." This category refers to possession of nuclear forces that are neither capable of carrying out a disarming first strike nor capable of absorbing a first strike and retaliating in a devastating manner. Examining each cell in the matrix will add meaning to the concept.

First-Strike Capability and Employment Doctrine: A nation that has determined that its interests are best served by adopting the posture that it will fire first must develop the capability successfully to preempt for this strategy to be meaningful. It makes relatively little sense to launch a massive first strike if one cannot effectively disarm the opponent in the process (the "scenario" of a limited "surgical" strike will be discussed in Chapter 3), because the failure to knock out an enemy's retaliatory forces makes such a strategy potentially suicidal. Thus, a nation desiring to adopt a first-strike doctrine must develop forces numerically larger than the enemy's (at least in terms of deliverable warheads, including enough in reserve after an attack to be able to threaten additional destruction as an ongoing threat) capable of reaching the enemy's weapons and accurate or large enough to destroy defending forces (given the destructive capacity of even a few surviving weapons, there is virtually no margin for error). Force invulnerability, on the other hand, is a lesser concern, since the determination to fire before being fired upon means by definition that forces would not have a primarily retaliatory role. At the same time, capability can either impede or enhance adoption of this doctrine: if forces are not pinpoint accurate and thus truly first-strike capable, adopting the employment doctrine makes comparatively little sense.

Second-Strike Employment Doctrine–First-Strike Capability: Under some circumstances, a force capable of carrying out a first strike may also be employed in support of a retaliatory (second-strike) doctrine. In order to accomplish this mission, the force would have to possess the characteristics of a first-strike capability (size and accuracy) and the additional characteristics of a second-strike force (invulnerability and penetrability). Such flexibility is more expensive and complex than a capability compatible with a single doctrine because it adds additional (and costly) requirements to the capability. In addition, it is extremely difficult (probably impossible) to convince an adversary that a force with first-strike capability would be used only in a second-strike mode: the advantage of first-strike capability is, after all, in being able to avoid the devastation inflicted before one retaliates.

First-Strike Employment Doctrine–Second-Strike Capability: In a sense, this combination is contradictory, because the requirements for a second-strike capability do not include the necessary elements for first-striking. If one has enough confidence in the survivability of second-strike forces, they do not have to be numerically larger than (or as large as) an opponent's, and the countervalue targeting congenial with the capability does not require enormously precise targeting accuracy. In fact, to make this combination meaningful (allowing it to carry out the disarming function associated with the first-strike option) requires "upgrading" the forces to first-strike capability. At the point that is accomplished, the capability is no longer second-strike, and the same problem of convincing an adversary of intent as described above arises.

Second-Strike Employment Doctrine and Capability: This combination is compatible and is the basis of the American nuclear deterrent. The viability of following the strategy relies upon possessing forces in which one has confidence in terms of surviving a first strike and being able to retaliate effectively. Albert Wohlstetter described, in some detail, the requirements of this combination in a 1958 article: "Such deterrent systems must have (a) a stable, 'steady-state' peacetime operation within feasible budgets.... They must also have the ability (b) to survive enemy attacks, (c) to make and communicate the decision to retaliate, (d) to reach enemy territory with fuel enough to complete their mission, (e) to penetrate enemy active defense systems ... and (f) to destroy ... in spite of any 'passive' civil defense."[21]

The basis of this capability, and thus the meaningful option to follow the doctrine, is survivability, as underscored by Joseph Kruzel in terms of MAD doctrine and the relationship to Soviet forces, with emphasis on the SLBM force: "This invulnerability is a cornerstone of the assured destruction doctrine. Not only is it in the interest of each side to maintain an invulnerable sea-based deterrent; it is also in each side's interest to ensure that the other side's deterrent force is invulnerable."[22] Norman Polmar goes a step further, maintaining that a second-strike force defines superpower status: "We can define a 'superpower' as a nation that could be the victim of a surprise, first-strike attack by any other nation and still inflict massive retaliatory destruction on the aggressor."[23] Only the United States and the Soviet Union have the technological and economic capacity to adopt this combination of doctrine and capability and thus, in effect, to be able to threaten all other nations with annihilation should they attack.

First-Strike Employment Doctrine–Lesser Capability: This combination represents the situation facing a nuclear state whose forces are capable of being preempted and demonstrates the situation in which capability dictates strategy. Wohlstetter describes this situation and its limitations: "There remains a third capability—to strike first and do damage without precluding retaliation. Against a major power, only this third sort is likely to be at all general. But this is a good deal less useful to its possessor than either a deterrent or a preclusive first-strike capability."[24] The lesser capability is dictated by technological and economic limitations (for example, the French cannot afford more than a limited second-strike or first-strike capability). In turn, these limitations require launching first because the failure to do so could result in preemption and thus the inability ever to strike. Technological (accuracy) and size characteristics require countervalue targeting, because the tenuous deterrent value such a capability possesses is based on the threat of causing enough destruction and suffering in a first launch as to be greater than any gain from annihilating the possessor (for example, would the Soviets gain enough by destroying France to justify the potential devastation of major Russian cities?).

Second-Strike Employment Doctrine–Lesser Capability: If whatever size forces a lesser nuclear power has are deployed in an invulnerable mode (for example, hardened or mobile land-based missiles, sea-based), then a second-strike doctrine may be possible to adopt. Because, by definition, such forces are not large enough to destroy the attacking state, they must employ countervalue targeting to maximize potential damage. Deterrent value is based on the identical rationale as the case above predicated on a first launch. This combination of capability and doctrine describes the French *force frappe* (strike force), and its doctrinal rationalization normally goes under the designation "minimum deterrence."

Comparison of Forces Whether a given capability level attains first- or second-strike capability can be determined meaningfully only by comparing one nation's forces with another's. Force levels and characteristics do not inherently have first- or second-strike capabilities (although they can be and are designed to maximize one or another doctrinal-capability preference). Rather, the attainment of one or another capability depends on comparing one's own forces to the forces of another state. As a result, a given level and set of characteristics of a particular capability may endow it with first-strike capabilities against some nations, second-strike capabilities against others, and lesser capabilities against yet others. The purported Israeli nuclear and the American forces can be used to illustrate this phenomenon.

Israel has been reported to have had nuclear weapons for some time, although the Israeli government has never admitted such possession.[25] Israel also possesses mobile and fixed missile launchers and fighter-bombers capable of being equipped with nuclear warheads that have effective ranges encompassing the territory of adversaries in the area but not outside it (thus making them strategic in the regional sense but tactical in the global sense). Assuming that the Israelis have ten to twenty nuclear warheads (the most commonly cited figures) for the present illustrative case, we can examine this capability relationally (no position on whether the Israelis in fact do have such weapons is implied).

Against her Arab neighbors, such an Israeli arsenal would currently represent a first-strike capability: no other states in the region have nuclear weapons or offensive missiles, and Israeli weapons can be delivered accurately enough to destroy opposing air forces on the ground. Even when neighboring states gain nuclear capabilities (Egypt and Syria are often mentioned as potential weapons states), the mobility of parts of the force (surface-to-surface missiles) makes it invulnerable enough to constitute a credible second-strike force against the nuclear forces that can reasonably be projected for Israel's adversaries. At the same time, Israel has neither a first- nor second-strike capability against the Soviet Union, because her forces cannot reach the U.S.S.R. and, in a massive attack, the Russians could probably destroy the Israeli forces anyway. In the Soviet context, the Israelis do not have even a minimum deterrent.

U.S. capabilities represent a more limited example. All analysts would agree that the United States currently has a second-strike capability vis-à-vis the Soviet Union and vice versa. At the same time, the size and increasing accuracy of the American arsenal endow it with first-strike capabilities against virtually all other states. For instance, the Chinese nuclear force could not retaliate after an American first strike, partly because the Chinese lack delivery systems capable of reaching American territory (their first-generation ICBMs are projected for deployment in the mid-1980s) and partly because, despite recent efforts to the contrary, their forces remain relatively vulnerable to preemption. Since the United States has second-strike capability against the strongest challenger (the U.S.S.R.), there are no states against which it has a lesser capability.

Contribution of Capability and Employment Doctrine to System Stability The overwhelming goal of strategic nuclear doctrine is deterrence: the prevention of nuclear war. Some would add controlling the level of such a war should it occur as a subsidiary purpose (so-called "extended deterrence"), but the core value, deterrence, remains. As a result, capability, employment doctrine, or a combination thereof that contributes to reducing the likelihood of nuclear war is stabilizing to the nuclear system, and any condition that increases that likelihood is destabilizing.

A basic corollary proposition relates to whether nations have incentives or disincentives to contemplating initiation of nuclear war. Brodie states this proposition succinctly: "Stability is achieved when each nation believes that the strategic advantage of striking first is overshadowed by the tremendous cost of doing so."[26] This additional consideration in calculating deterrent stability is thus a matter of attempting to make nuclear usage as costly and thus as unattractive as possible. The two most meaningful combinations of capability and doctrine, with their dyadic implications, can be analyzed in these terms.

In the dyadic relationship, the possible combinations of strategy and capability identified in the previous section can be reduced to two major and two minor combinations. Two combinations can be removed because they are not practically meaningful (they were included earlier to exhaust all possibilities). The first-strike capability–second-strike doctrine combination is simply not believable to an adversary against whom one has that capability. Faced with capacity to be disarmed by a first strike, one would be a fool to assume that the superior state would

Deterrence

unnecessarily absorb a first strike it could avoid by preempting before it fired its missiles. As a result, the state not possessing the combination would have to assume and plan on the basis that the capability implied a first-strike doctrine. Against other than a weak state, a first-strike doctrine combined with a second-strike capability is similarly meaningless. By definition of force characteristics and targeting implications, a second-strike force is not suited to counterforce targeting. Giving it that additional capability means, at heart, that it is no longer a second-strike capability: it becomes a first-strike capability. A third combination, one side with a first-strike capability and doctrine and the other with a lesser capability and a second-strike doctrine, is self-contradictory. If the lesser power indeed could sustain the doctrine, whatever forces it has are invulnerable. If that is the case, the state might be destroyable, but it could retaliate, meaning by definition that the other state does not have first-strike capability. Since in a dyadic relationship, one side cannot, by definition, have a first-strike capability and the other have a second-strike capability, any combinations where that dichotomy is present are similarly eliminable.

This process of elimination leaves four possible combinations. Two of these are most likely to occur: where one side has a first-strike capability and doctrine while the other has a lesser capability and a first-strike doctrine; and where both sides have second-strike capabilities and strategies. Two other dyads are technically possible but unlikely: where both sides have first-strike capabilities and doctrines; and where one has a second-strike capability and doctrine and the other a lesser capability and a second-strike doctrine.

First-Strike Capability and Doctrine versus Lesser Capability and First-Strike Doctrine At first blush, possession of a first-strike capability wedded with a first-strike doctrine would seem an ideal situation: one could "win" a nuclear war by using the capability in a preemptive launch and avoid devastation in return. This situation connotes "superiority" and would thus seem desirable. Since the capability is held in relation to a specific adversary, however, reflection will reveal that the situation may not be as ideal as it seems.

Thomas C. Schelling captures the anomaly of effective counterforce capability (the accuracy factor necessary for first-strike capability): "The question is often raised whether a counterforce strategy is not self-contradicting; it depends on a decisive military superiority over the enemy and yet to succeed must appeal equally to the enemy, to whom it cannot appeal because he must then have a decisive inferiority."[27] An opponent against whom one has a first-strike capability by definition has a lesser capability that can be disarmed. Such capability dictates a first-strike firing doctrine (you fire first or not at all), and thus instability enters the picture. The heart of that instability is that, in the apparent imminence of a nuclear confrontation (need it be pointed out that misperception is highly possible in such circumstances?), the incentive for both sides is to initiate a nuclear holocaust: the dominant side to disarm the weaker and the weaker to ensure it gets to fire its weapons. Since disincentives to using weapons are stabilizing, this situation is maximally destabilizing.

Both Have Second-Strike Capabilities and Doctrines This combination is both meaningful and consensually agreed to be maximally stabilizing. It is meaningful because adoption of the doctrine, since it implies countervalue targeting, does not pose a threat to retaliatory forces. At the same time, the major imperatives of second strike are the survivability of forces and their penetrability to eventual targets, which simply reinforces the retaliatory capacity.

This situation is maximally stabilizing because when it holds, neither side has any incentive to begin a nuclear war. Instead, there are major disincentives; any potential aggressor knows the attacked state maintains the capacity to retaliate and inflict unacceptable levels of destruction, thereby making an initial strike suicidal. This situation and set of mutual disincentives describes the so-called "balance of terror." Tammen describes the perceptual basis of this phenomenon: "The degree of perceived terror is a critical figure in the calculus of deterrence. A policy-maker must make three measurements: the degree of self-terror, the degree of self-terror of the adversary, and the degree to which his own terror is thought credible by his adversary."[28] The addition of perception adds the possibility of misperception, but the objective fact of mutually held second-strike arsenals is leavening: it still never makes sense to strike first. If one is confident in one's capability, an element of reflective ability is added: "In fact, there is no basic technical reason why any retaliation would have to be swift; a great deal of technical, political and diplomatic effort during the past two decades has gone into measures to prevent just that compulsion."[29] This basis of stability and deliberation arises from the confidence each side has in its deterrent. Much of the subsequent discussion will be devoted to this relationship and challenges to it.

Both Have First-Strike Capabilities and Doctrines This situation is technically possible under either of two circumstances. If both sides developed large but vulnerable forces, either might be able preemptively to attack the other. Since the United States and the U.S.S.R. have gone to considerable lengths to protect their deterrents, this situation is unlikely. Second, and less implausible, is an implicit or explicit move toward counterforce capability through radical improvements in targeting accuracy to the point where previously invulnerable targets become vulnerable. This possibility (discussed as part of the doctrinal debate in Chapter 3) is disturbing, because the incentives to fire first would be the same as in the case where one was superior and the other inferior. The implications are dealt with extensively in Chapter 7.

Second-Strike Capability and Doctrine versus Lesser Capability and Second-Strike Doctrine This situation, which could describe the Soviet Union and France, is, in terms of stability and incentives, a variation of the circumstance facing two countries with second-strike capabilities and doctrines. The only source of potential destabilization is that damage would be asymmetrical, and thus disincentives for the more powerful state could be lowered, depending on the level of asymmetry and what each side was willing to absorb. No potential adversary of the United States, however, is nearly in this position with the United States (China could conceivably become one sometime in the future), so that this possibility is of minor interest in assessing American doctrine.

The Fragility of Deterrence Thought The thought and writing about nuclear strategy and deterrence have an elegance and eloquence that reflect the considerable efforts and obvious talents of those who have labored over the subject. The tight and reasoned analyses that have emerged create an almost narcotic sense that all is well because it seems so well thought out. Because it is easy to draw such a conclusion, a note of caution must be sounded.

As the subsection heading suggests, the basis of deterrence thought is fragile, because, unlike most knowledge areas in which we have some confidence, a great deal of the analytical "territory" is nonempirical and will remain so until the reason for its study fails: unless there is a nuclear war, key concepts and relationships between concepts remain hypothetical. Viewing the body of deterrence literature, Green concludes, "The 'method of science' has not been shown to be relevant to the study of our nuclear future, nor is there anything 'empirical' about available studies of nuclear war."[30] Because propositions about nuclear deterrence cannot be subjected to the crucible of testing and verification, widely variant conclusions on issues are possible. John C. Culver admits these deficiencies: "Despite all the think-tank analysis and war plans based on notions of deterrence, however, the present debate seems to show that we have really not defined our objectives; we are not sure how to measure deterrence."[31]

The theoretical uncertainty surrounding deterrence notions is not limited to the broad, "soft" theoretical notions of systems dynamics and human motivations. As then Secretary of State Henry A. Kissinger pointed out in a speech in Dallas, Texas, on March 22, 1976, the uncertainty is broader in scope: "No nuclear weapon has ever been used in modern wartime conditions or against an opponent possessing means of retaliation. Indeed neither side has ever tested the launching of more than a few missiles at a time; neither side has ever fired them in a North-South direction as would be required in wartime."[32] John D. Steinbruner and Thomas M. Garwin amplify the uncertainty that surrounds what one would assume to be the relatively "hard" area of weapons performance raised by Kissinger: "The United States has never fired an intercontinental range missile at a target in the Soviet Union, has never exploded a nuclear warhead at the end of an intercontinental missile flight, has never fired a strategic missile on 15 minutes warning from an operational silo randomly chosen and has never fired more than a very few missiles simultaneously or in close coordination."[33]

The limitations on deterrence thought are conditioning on the confidence one should place in scholarship and debate on the subject and also help explain the wide variation in conclusions arising over general and specific issues. The nonempirical nature of the field affects the most basic concepts. For exemplary and conditioning purposes, it is therefore worthwhile to look at four key concerns in this light: the nature of human behavior in a nuclear environment; the firebreak or nuclear threshold; the nature of the escalatory process; and the danger of accidental nuclear war.

Human Behavior in a Nuclear Environment How people would act in the imminence or actuality of nuclear war is obviously important in predicting and planning for that situation. The event, quite obviously, has never been observed, and it is difficult to find analogous situations to the extreme stress that would undoubtedly surround the occasion and thus provide the basis of theoretical transferrence.

Given these severe limitations, theorizing has had to progress indirectly. One method has been the formulation of elaborate simulation and gaming exercises seeking to replicate the conditions presumably surrounding nuclear war. The basic difficulty is in reproducing comparable emotional and psychological stress since much of that stress is produced by awareness of the consequences of decisions that are absent in the simulated environment (the participants know they are not really going to blow up the world). Brodie states this shortcoming: "Experienced persons agree that one simply cannot reproduce among the players in a gaming environment the kind and degree of emotional tension and feeling of high responsibility bound to be present in the nuclear era among decision-makers in real-life crises, where decisions have to be made whether and by what means to fight a war."[34]

The key element in projecting likely behavior is whether decision makers would act in a calm and rational manner faced with the awesome import of decisions they must make and the terrible surrounding emotional strains. There is considerable disagreement in the literature about what constitutes rationality (is it the conservative minimax of game theorists or some riskier form of behavior?) and thus whether the concept is useful. The common thread of agreement is that the nuclear environment is extremely stress-producing and unpredictable, leading one observer, Kenneth E. Boulding, to a pessimistic conclusion about systems viability: "Both historical and logico-mathematical analysis lead one to believe that deterrence is associated with malevolence and is likely to be unstable in the long run."[35] That one can eventually adapt to the uncertainties and stresses (develop a high tolerance for ambiguity) is at least a possible and hopeful conclusion.

The Nuclear Firebreak The point at which a conventional war becomes a nuclear conflagration or at which a hitherto nonviolent confrontation degenerates into nuclear exchange is obviously a critical deterrent concern. It is generally agreed, for instance, that the Cuban missile crisis is the closest mankind has come to reaching and crossing the firebreak, but that fact is not analytically very helpful. The simple fact is that we have no real idea where the threshold is because we have never reached it. Thus, there is no standard against which to measure our approach: the world may have come its closest to the firebreak over the Cuban missiles, but no one knows how close we were.

No one, of course, truly wants to gain empirical knowledge of the location of the firebreak, and the desire to avoid the discovery is a positive and restraining influence on superpower relations. As Alain Enthoven writes: "There is and will remain an important distinction, a 'firebreak' if you like, between nuclear and non-nuclear war, a recognizable qualitative distinction that both combatants can recognize and agree upon. . . . And, in the nuclear age, they will have a very powerful incentive to agree upon this distinction and limitation because if they do not, there does not appear to be another easily recognizable limitation on weapons . . . all the way up the destructive spectrum to large-scale thermonuclear war."[36] There is not universal agreement on the existence of a single firebreak or the automaticity of the escalatory process, and weapons in the theater inventory have muddied somewhat the easy recognizability and distinction between nuclear and non-nuclear weapons. That the perception of and uncertainty surrounding the firebreak have been useful in restraining the superpowers is widely supported. Colin S. Gray's assessment is typical: "The threshold of the casus belli has risen markedly. Indeed, with respect to the relations between nuclear weapons states, the very notion of the casus belli has lost nearly all meaning. Nuclear weapon states do not go to war with each other: they might slide fearfully into combat via limited strikes for diplomatic purposes, but they could not afford the traditional unlimited implication of going to war."[37] Before taking too much comfort from all this, it should be pointed out that it is much easier to avoid a danger spot if one knows where it is than if one does not.

There is also some debate about whether, as Enthoven implies, there is a single threshold. This discussion surrounds both the use of theater nuclear weapons (such as the "neutron bomb") in support of conventional military purposes and defense of the doctrine of limited nuclear options (see Chapter 3). Herman Kahn argues implicitly the possibility of more than one firebreak, speculating, "The first use of nuclear weapons . . . is likely to be less for the purpose of destroying the other side's military forces or to handicap its operations than for redress, bargaining, or deterrence purposes."[38] Recognizing the possibility that the firebreak might be crossed in less than a total manner, David Carlton suggests: "The real firebreak, so far as the Americans are concerned, may therefore be not between nuclear and non-nuclear weapons but between the use of nuclear weapons on the heartlands of the superpowers and all other military actions, nuclear or non-nuclear."[39]

The Nature of the Escalatory Process Should deterrence fail and the firebreak be crossed, what happens next is crucial to attempting termination of the conflict short of mutual societal destruction. The early literature in the field proceeded from the assumption that a nuclear exchange would almost certainly be uncontrollable and would move inexorably toward general destruction. More recently, there has emerged a body of thought suggesting the controlled mutual use of these weapons and the ability to terminate that usage at various points. As one might imagine, speculation tends to center around the "deterrence-only" and "deterrence-plus" poles identified in Chapter 1.

The common view that a nuclear war, once begun, would likely become general is summarized by Barry R. Schneider: "Once the nuclear threshold has been broken, there is no easy stopping point on the way to the top of the escalation ladder. . . . Once nuclear weapons are fired in battle, the clearest perceived 'firebreak' on the path toward nuclear holocaust will have been crossed."[40] Panofsky agrees, adding that because of the great uncertainty attendant to nuclear weapons, "escalation toward full-scale nuclear war is exceedingly difficult to prevent."[41] Just as uncertainty about the nuclear threshold is purported to breed caution, however, so does the danger of the escalatory process. Klaus Knorr states: "The existence of strategic nuclear weapons casts a shadow on the value of hypothetical limited conflicts in which strategic weapons are not employed. Any limited conflict . . . between nuclear powers carries the risk of escalating to the strategic level at which, as long as the mutual balance to terror prevails, the payoffs are hugely negative for both participants."[42]

The view that the escalatory process is not necessarily either automatic or uncontrollable was first raised by then Secretary of Defense Robert S. McNamara in his Ann Arbor speech of 1962 and was given added emphasis by former Secretary James R. Schlesinger in 1974 (both discussed in Chapter 3). The essence of this position is that nuclear war might be conducted on a limited basis, including the possibility of breaks in the hostilities during which negotiations leading to war termination could occur. Schlesinger added the notion of limited nuclear options. The heart of this conception is that the destructiveness of general exchange is so high and well recognized that it is the least likely form of exchange. More limited exchanges, he reasoned, were more likely, and thus the United States should plan proportional responses.

The suggestion that nuclear escalation can be controlled has met with vociferous opposition. Brodie seemingly foresaw the shortcomings of the advanced, intricate nuclear war planning associated with limited options in 1959 with a stern rejoinder: "The most important strategic decisions concerning that nuclear war must be made in the preceding period of peace. This is a sufficiently frightening thought when we consider the poor record for predictions of impending wars, including those embodied in war plans, and even more frightening when we consider how novel and strange are the technological conditions we are dealing with today."[43] Soviet writers Michael A. Milstein and Leo S. Semeiko, responding directly to the limited options proposal, conclude tersely that "the possibility of unleashing a 'small' and 'painless' rocket-nuclear exchange, and of containing it within safe limits, is a myth that in no way corresponds to the realities of nuclear war."[44]

Just as there is a point at which the firebreak is crossed, it is also either true or false that a nuclear conflict could be controlled. Uncertainty about the ability to be able to do so has probably had a restraining influence and sobered any thoughts of purposeful nuclear exchange. There remains, however, the possibility of nuclear war beginning without prior design or intention.

The Danger of Accidental War It was early recognized in the literature that nuclear conflict could break out accidentally, for instance, as the result of an accidental missile launch or a bomber pilot's misinterpretation of an order. The United States and the Soviet Union have both taken elaborate precautions in the form of "fail-safe" devices to prevent such an occurrence, and one purpose of the "hot line" between Washington and Moscow is to try to communicate and thus control events should an accident happen. No matter how unlikely, the stakes involved make the possibility important to consider: "The risk of the inadvertent outbreak of strategic nuclear war, while the balance of terror prevails, is hard to estimate; and when the consequences involved are so fatal, even a small risk is a serious matter."[45]

Unfortunately, accidents are sometimes difficult to prevent, particularly if their causes cannot be predicted. One can specify all the pitfalls one can think of and try to deal with them, but sources one does not think of are the most nettlesome. Recognizing that one rarely thinks of everything, crisis management should an accident occur becomes vital because, as Schelling notes, "With today's weapons it is hard to see that there could be an issue about which both sides would genuinely prefer to fight a major war rather than accommodate. But it is not so hard to imagine a war that results from a crisis getting out of hand."[46] Predictable crises, it goes without saying, are a great deal easier to manage than those that are accidental and thus unpredictable.

The Elusive and Changing Problem As the preceding sections have attempted to demonstrate, there are genuine issues surrounding nuclear deterrence about which reasoned and reasonable people can and do come into genuine, and sometimes animated, disagreement. There is universal agreement that preventing nuclear war is the first goal of strategic doctrine and nearly universal accord that deterrence is the only legitimate end of nuclear policy.

Within that framework of general agreement, however, there is disagreement about what deterrence is and, as a consequence, how it is best maintained. The world has lived in a balance of terror defined by the mutual American-Soviet possession of second-strike nuclear capabilities for at least a decade and a half (symptomatically, there is some disagreement about when the Soviet Union achieved that status), but issues regarding the balance continue to be debated, as Tammen points out:

> Within the BOT there is disagreement over the following points relating to balance: (1) Is a multiple balance the same as a bilateral nuclear balance? (2) If it is possible for there to be disequilibrium in the Balance of Terror, what numbers constitute a balance? (3) Can a balance be redressed by counting allied (British or French) nuclear weapons? (4) Can a small nuclear power balance out a major nuclear power (minimum deterrence)? (5) Does the balance apply only to the question of major nuclear war? (6) Does the balance shift back and forth between nuclear powers?[47]

These questions, reflecting many of the issues and problems addressed in this chapter, are illustrative of the concerns and contentions in the field.

There are at least three major causes of this disarray, two of which have been discussed here. First and least important, there is disagreement on some basic concepts. What, for instance, is "strategic" in the nuclear context? What is the firebreak (or possibly, rather, where is it)? Lack of conceptual clarity on key issues confuses the debate.

Second, and more fundamental, is the problem of the nonempirical nature of much of the material with which one has to deal. It would be far simpler, and would end many disputes, if one could observe, and thus answer such fundamental questions as how people would behave in a nuclear environment and whether the escalatory process is controllable. In the absence of evidence about critical concerns, and with the

Deterrence

understandable but confounding purpose of engaging in an enterprise designed specifically to avoid finding empirical evidence about its subject, the debate is bound to remain conjectural and subjective.[48] This is an inherent limitation about which little can be done, but it also precludes orderly scientific study and theory construction.

Third, and finally, is the problem of the dynamism of the balance itself. This problem is the recurrent theme of this work, but its impact can hardly be overstated. Few aspects of the postwar world have changed as often and as rapidly as the strategic balance, and it has been difficult for even the best of minds to deal effectively with the changing problems and requirements that confront them. The ways in which American doctrine has attempted to accommodate with its changing environment are the subject of the next chapter.

Chapter 3:
The Evolution of American Strategic Doctrine

In the three and a half decades that mark the nuclear age, American strategic nuclear doctrine has developed and evolved. Reacting to environmental changes and often developing a momentum of its own, the strategic nuclear system has grown from its modest beginnings when Franklin D. Roosevelt offered a crash nuclear development program in 1941 to the behemoth it is today. Doctrine, in turn, has been shaped and reshaped by the winds of nuclear change, while at the same time seeking to order and influence the pace and direction of that change.

It is important to understand the nature of change within the nuclear system. As argued in Chapter 1 and as applied in this chapter, strategy results from the reciprocal interaction of doctrinal preferences with the various environments affecting it. These environments (the internal and external environments and technology) operate in different ways at different times, producing opportunities and constraints for those charged with the development of and implementation of strategic doctrine.

In terms of the doctrine itself, what appear as major shifts in emphasis tend to be associated with presidential politics. Paul Doty and his associates point out, "As the nuclear age lengthens and the opportunity for viewing it in perspective grows, its essential features seem increasingly related to eight-year American presidential administrations."[1] At a basic level, this assertion is true and will be used as the organizing concept for analyzing the evolution of doctrine: major changes in designation and emphasis have been identified with presidential administrations. The Eisenhower administration developed the new look and nuclear strategy of massive retaliation, the Kennedy-Johnson years produced the policy of flexible response and its nuclear arm, the doctrine of mutual assured destruction, and the Nixon-Ford period produced the doctrine originally called strategic sufficiency, which evolved into essential equivalence. The early years of the Carter administration, as best as can be determined, represent a continuation of the basic tenets of its predecessor, with some modifications, that has come to be known as the countervailing strategy.

The dramatic names attached to different strategic doctrines can appear to represent more fundamental shifts than indeed they do. Change in the nuclear environment does not conform neatly to eight-year presidential cycles but is evolutionary and incremental. Decisions about nuclear weapons systems, for instance, take eight to ten years to implement (as pointed out in Chapter 1), meaning that important influences helping to shape doctrinal options for one administration are the legacy of its predecessor, and decisions being reached today will have their most profound influence in the 1980s and beyond. Moreover, the forces shaping doctrine are often reactive and self-sustaining, dragging doctrine along in their wake.

These introductory remarks serve as preface to the central purpose of this chapter, which is a description and examination of the basic strategic doctrines the United States has adopted and the influence of environmental factors on changes in that doctrine.

The Early Years and the Doctrine of Massive Retaliation The Truman administration was the first to have to deal in a concerted way with the problem of nuclear weapons, inheriting the results of the program begun at the onset of American participation in World War II. That there was a nuclear policy at all demonstrates the often incremental evolution of such policy. Roosevelt started the program because he was "convinced that Nazi scientists were engaged in a similar effort. He continued the program *after it became clear that Germany had abandoned its research*" (emphasis added).[2]

The results of the initial program that continued despite losing its original purpose are all too familiar: the only use of nuclear weapons in anger against the Japanese cities of Hiroshima and Nagasaki on August 6 and 9, 1945. Driven by internal environmental pressures, technological availability, and the failure of the Baruch Plan of 1946 to create the basis of a nuclear disarmed world (see Chapter 6), the United States remained a nuclear-arms state throughout the rest of the 1940s. As Samuel Wells, Jr., describes it: "The combination of budget pressures and the availability of new technology caused the United States to adopt a nuclear strategy. The success of the 1948 atomic tests promised a large stockpile of new weapons, which were cheaper, smaller, had a wide range of destructive force, and used much less fissionable material than the original A-bombs."[3] The most salient fact about the nuclear situation during this period, of course, was that the United States was the sole possessor of nuclear weapons. Deterrence was a simple matter of self-restraint, and elaborate policies and doctrines were not needed.

The impetus to develop such policies emerged from the external environment, when the Soviets successfully tested an atomic device in 1949. The United States was caught somewhat off guard; we had assumed that the Russians were several years away from accomplishing this feat. The result, at the initiative of the State Department, was a planning process that produced the first formal strategic statement in 1950. This document, which became known as NSC-68, did not reach the level of articulation of a formal doctrine. Norman Polmar captures the prevailing nuclear sentiment of the period: "The prevalent strategic view in the Truman phase of the 'atomic age' was that long-range bombers carrying nuclear weapons against enemy cities or military forces could defeat any nation or force hostile to the United States and its interests."[4] It is less than coincidental that the United States, thanks to its European-based bomber forces, could during this period reach targets in the Soviet Union, which had no reciprocal capability. This fact took some of the urgency out of the need for strategic nuclear doctrinal development, and attention to the Korean conflict acted as a further damper.

Environmental Factors in the 1950s The period between 1945 and the inauguration of Dwight Eisenhower in 1953 was largely formative in terms of the nuclear arms balance. It was marked by an initial American nuclear monopoly that "degenerated" into a massive superiority of nuclear arms combined with possession of the only strategic delivery systems (as we have defined that term). Although there were some abortive discussions about employing nuclear arms in the Korean War, strategic doctrine stultified during that conflict as attention naturally shifted to its prosecution. When the Eisenhower administration came to power, each of the environments affecting defense policy and thus its strategic nuclear component had changed sufficiently to warrant reassessment.

The Internal Environment The most obvious and overwhelming domestic factor influencing defense policy was public reaction to American participation in the Korean War. That conflict became very unpopular and expensive, and it was dragging along to an inconclusive ending. Domestic anticommunism was still at a zenith, and the new administration thought it could see at least two clear implications of public opinion that translated into constraints on future defense posture and would help shape doctrinal choices. The first of these factors was economic: "The U.S. defense budget was swollen by expenditures for the Korean war. The President and his chief advisors were firmly convinced of the need to drastically lower the level of defense spending."[5] Belief that return to a balanced budget was the key to a healthy American economy (an issue on which he had compaigned vigorously) meant that an effective, but economical, defense policy was needed.

The second major influence was somewhat more ambivalent. On the one hand, virulent anticommunism in the United States dictated a strong defense posture that would effectively "contain" presumed attempts at communist expansion. On the other hand, the unpopularity of the Korean conflict, both because of the human sacrifices it entailed and the fact that the United States did not "win" it in a conventional sense, strongly indicated American unwillingness to become involved in similar conflicts. The American population wanted a defense posture that would vigorously frustrate communist aggression, but not with American troops. The result was the need for a defense policy that was not only economical but minimized the number of Americans in uniform who were thus potential combatants. Since manpower costs (salaries and retirement benefits) represent the largest part of defense spending, these latter concerns were not necessarily inconsistent. The problem was to devise a defense posture that would adequately deter the Soviet Union within these constraints.

The External Environment The major external factor facing the United States throughout the 1950s was, of course, the Soviet Union. Because the international communist movement was viewed as a monolith controlled and directed by Moscow, the key to stopping communist aggression everywhere was to frustrate and deter the Soviet Union.

The Eisenhower period began with the Soviet Union possessing nuclear weapons but with marginal capability to deliver them outside the European theater. The sheer fact that the Soviets had these weapons, however, changed the rules of the strategic game, as Harland Moulton notes: "In a very few years following 1950, the Soviet Union achieved sufficient nuclear capability of its own to usher in the age of mutual deterrence."[6]

The major Soviet activity impacting on American strategic policy in the 1950s was technological and, as argued earlier, profound: the development of ICBMs. Both the Soviets and the Americans worked on developmental missile programs in the 1950s. Kahan points out the Soviet influence on American programs: "U.S. missile developments from 1954 through the middle of 1957 were largely motivated by the desire to prevent the USSR from acquiring a lead in the strategic missile field."[7] The pace of this action-reaction phenomenon quickened visibly when U-2 overflights revealed Soviet test sites for ICBMs early in 1957, followed shortly by the launching of Sputnik and successful ICBM tests by the Russians. A major impact was to call into question the Eisenhower strategic doctrine enunciated in 1954.

Technology The Eisenhower years were a period of great technological advancement in the strategic nuclear field. When President Eisenhower entered office in 1953, the state of the nuclear "art" was still comparatively primitive: the Soviets had yet to test successfully a thermonuclear (fusion) bomb (the United States had done so in 1952), and movement away from conventional delivery by aircraft was still confined to research laboratories. The changes that had entered the system were, to borrow the arguments developed in the last chapter, more quantitative than qualitative.

The technological situation changed radically during the 1950s. In retrospect, it was probably the most influential technological decade in the history of the arms race, when the earliest versions of the weapons systems currently in strategic arsenals were designed and in many cases deployed. Warhead design improved radically, largely in terms of making bombs smaller and lighter while increasing yields simultaneously. This improvement resulted from the discovery and application of fusion techniques to weaponry (the Soviets successfully tested a thermonuclear or hydrogen bomb in 1953), which in turn allowed development of high-yield fission-fusion-fission weapons (see Chapter 4) and spin-offs like the enhanced-radiation warhead (the "neutron" bomb), developmental work on which was begun by the U.S. Army in 1958.

Delivery system improvement was even more dramatic, although in important respects tied to greater warhead sophistication (warheads had to be made considerably smaller and lighter than the earliest designs before ballistic delivery became possible). During the 1950s, intercontinental ballistic missiles were designed, tested, and began to enter strategic inventories. R&D efforts produced the first atomic missile submarines, which were on order when the decade ended, intercontinental jet aircraft entered strategic arsenals, and the first cruise missile research programs were begun. The ballistic missile, of course, had the most profound impact: it made all states conditionally rather than unconditionally viable and thus qualitatively altered nuclear calculation.

Not all of these technological developments or their strategic implications were apparent when the Eisenhower administration entered office and formulated its strategic doctrine. By the time John Kennedy entered office (and even before), many of them had become apparent and were partial causes of a strategic debate about doctrinal adequacy not unlike that going on in the United States today. The Eisenhower doctrine is important both for its inherent interest and because it was a parallel phenomenon to the current U.S. debate.

Content and Criticisms of the Doctrine of Massive Retaliation The nuclear doctrine of massive retaliation was the cornerpiece of the new look defense policy of the Eisenhower administration. Shaped by the popular

Evolution

new president and his evangelical secretary of state, John Foster Dulles, the new look's basic thrust was to reduce reliance on American ground forces in the fight against communist expansion, replacing American conventional presence with indigenous forces in countries victim of such aggression and seeking to deter that activity with the nuclear threat.

The first statement of what became known as the massive retaliation doctrine was made by Dulles on January 12, 1954, before the prestigious Council on Foreign Relations. The heart of Dulles's statement, widely quoted since, was that the United States would deter communist aggression by depending "primarily upon a great capacity to retaliate, instantly by means and at places of our choosing."[8] Because the United States simultaneously cut back drastically its conventional ground forces, the only "means" available were strategic nuclear forces that could be delivered against the Soviet Union by air and sea.

The deterrent base of the doctrine was, in current jargon, "spectrum defense": the nuclear threat would be used to deter hostile activity across the range (or spectrum) of risk. The threat was purposefully ambiguous and did not bind the United States to act in any specific instance, because the administration would choose the "means" and "places" to employ nuclear weapons. Although the Soviet Union might assume that "an attack against the United States or its allies would be met with massive nuclear strikes,"[9] there was nothing automatic about such retaliation. Rather, the Soviets would have to assess the risk in any given situation, and the deterrent was effective when the Russians determined that the potential costs (devastation of their homeland) outweighed likely gains. Moreover, Dulles believed the policy gave the strategic initiative back to the United States by allowing us to "pick our own fights" rather than letting the Soviets decide competitive grounds. As Moulton puts it, Dulles "argued that the policy of massive retaliation would enable the President to take initiatives in foreign policy, whereas the Truman doctrine of containment had forced the United States to react constantly to enemy threats of aggression."[10]

Nuclear policy would thus become the means by which both deterrent purposes of armaments were carried out. The nuclear threat bolstered by indigenous forces in countries subject to communist incursion (those forces received considerable training and equipment support from the United States) were to convince aggressors that aggression could be stopped and thus was futile. At the same time, the threat to cause great suffering through massive attacks on the Soviet Union lurked in the not too distant background.

The key element (and problem) was the doctrine's credibility: it could serve to deter the Soviets only if they believed we would carry out the nuclear threat. Given the drastic action threatened against possible communist initiatives a great deal less violent, clearly something more than a fairly vague threat was needed to make the policy believable. Three basic means were devised to lend credibility to the policy: extension of alliance commitments; purposeful self-denial of other defensive means; and adoption of a stated policy enshrining the idea of preventive war.

The NATO alliance of 1949 had been America's first European peacetime alliance since its less formal agreement with France preceding the War of 1812 and was, from the European standpoint, necessary to convince the Soviets that the United States would indeed seek to repel a Russian attack on Western Europe. To make more credible the possibility that the United States would come to the assistance of beleaguered states, this concept was expanded to many endangered Third World (largely Asian) countries through either multilateral (such as SEATO, CENTO, ANZUS) or bilateral (as with South Korea and Nationalist China) mutual assistance pacts. These agreements were uniformly weaker than NATO in terms of what they required the United States to do in the event of an aggression, but the fact that they created a commitment was thought to add at least incrementally to basic U.S. credibility in making the nuclear threat.

The second credibility-enhancing technique was self-denial of other commitment-honoring capabilities. Kahan explains, "By denying itself meaningful nonnuclear capabilities and by publicly rejecting the premise that tactical nuclear war or any major conflict involving the United States and the USSR could remain limited, the Eisenhower administration apparently sought to increase the likelihood of escalation in the belief that its nuclear commitment would then become more credible."[11] This factor in combination with its NATO alliance commitment created the basis for invasion avoidance in the European theater: "In the United States and in Western Europe, possession of a wide margin of U.S. superiority over the Soviet Union in nuclear arms was widely believed to be the strategic foundation for NATO. The American nuclear threat kept Soviet armies from breaching the Iron Curtain and, in particular, kept Soviet tanks from rolling through the Brandenburg Gate into West Berlin."[12] Although there were other sufficient causes that could explain why the Soviets did not invade Western Europe during this period (such as internal problems in both the Soviet Union and Eastern Europe attending de-Stalinization), the nuclear threat was plausible and widely accepted.

To make the threat more strident and thus presumably more credible, an additional element was added. In carrying out massive retaliation, the United States might not wait to absorb a nuclear strike before unleashing an awful attack on the Soviet Union. Brodie explains the rationale for this position: "The case for preventive war, almost overwhelming in its simplicity, has rested primarily on two presumptions: first, that in strategic air war with nuclear weapons, hitting first is certainly a crucial advantage and, with reasonably good planning, almost surely a decisive one; and second, that total war is inevitable."[13] This assumption sounds remarkably like later arguments for "damage limitation" as a counterforce targeting objective or as a disarming goal in conjunction with a first-strike strategy and capability (Moulton denies this latter was intended[14]). Targeting accuracies with long-range bombers were, at the time, relatively low, meaning that a preventive attack "implies inevitably the unprovoked slaughter of millions of persons, mostly innocent of responsibility."[15] This countervalue targeting consequence, with the added requisite of the provocation of having absorbed a first strike and combined with belief in the inevitability of general nuclear war, became the major assumption of mutual assured destruction in the Kennedy-Johnson years. In massive retaliation, however, it had a somewhat different place.

In the contemporary world, massive retaliation has an unreal and incredible quality, largely because in a world of mutual second-strike capability, implementing the strategy is openly suicidal. The policy made a good deal more sense in the environmental context in which it was framed, and it must be assessed accordingly.

The most crucial environmental factor was external: in 1954, the Soviet Union did not possess the credible ability to initiate or retaliate against a nuclear attack on its homeland with a similar attack on the United States. Their strike force was composed of World War II-class bombers marginally capable of reaching U.S. territory unopposed. With the prospect of reasonable interdiction of an incoming force, marginal damage to the United States could be calculated. Combined with the preventive war doctrine that would further reduce Soviet retaliatory capability, the United States had an effective first-strike capability (whether intentionally designed as such or not) and was unconditionally viable, leaving the Soviets in a condition of insecure conditional viability. In this context, the threat of spectrum nuclear defense was much more believable then otherwise would have been the case.

At the same time, the doctrine was responsive to the Eisenhower administration's internal environmental perceptions. Replacing the burden of direct American troop involvement with native forces and definitionally limiting American participation to the nuclear level critically reduced the likelihood that American troops would get bogged down in another Korean-style conflict. Dependence on nuclear weapons was a good deal less expensive than maintenance of ground troops (because personnel costs were lower). Nuclear weapons produced "more bang for the buck," as then Secretary of Defense "Engine Charlie" Wilson was fond of saying, and thereby contributed to reducing defense spending and realizing the goal of a balanced budget.

Changes in the external environment, and most particularly those associated with technological development, evolved rapidly during the years following the enunciation of massive retaliation and formed the basic criticism of its adequacy. By the mid-1950s, the United States detected that the Soviet Union was involved in a program to develop and build jet-propelled bombers with an intercontinental range, clear evidence that the Russians were unwilling to accept strategic inferiority and consequently were vulnerable to nuclear blackmail. This discovery caused considerable (although premature) concern in the United States (it was premature because, as events later showed, the Soviets cut back the construction of aircraft drastically after the introduction of ICBMs).

The most important change was, of course, technological: the advent of intercontinental ballistic missilry. As discussed in Chapter 2, the major result of this innovation was the end of American unconditional viability. This event raised serious questions about the assumptions on which massive retaliation was based. The credibility of the spectrum application of nuclear threats was brought into question; and in the process conventional force denigration associated with the strategy came under critical scrutiny. In addition, it has been suggested that the doctrine of massive retaliation impeded understanding the dynamics of the nuclear system.

Doctrinal Credibility The introduction of an incontrovertible Soviet ability to attack the United States in a devastating manner with nuclear weapons raised serious questions about the ability to sustain the nuclear threat across the spectrum of risk posed by the Soviet Union and in the process presaged more recent debates about what nuclear weapons deter. A number of writers had raised the question of credibility in the context of situations that were less than vital to the United States. With Soviet capability secured, Brodie had a categorical answer to this question: "No responsible government will opt for massive retaliation except where it conceives its stake in the matter at issue to be absolutely vital."[16]

The massive retaliation doctrine had rested effectively on both American invulnerability to attack and a decisive strategic superiority that would ensure the decimation promised by implementation of the doctrine. According to Kahan, this latter doctrine was being questioned as early as 1955: "The rationale behind President Eisenhower's reluctance to accept the Air Force doctrine of superiority could be traced to a new strategic policy that emerged in 1955 and was given official standing early the following year—the policy of strategic 'adequacy' or 'sufficiency.'"[17] This shift in policy seemed to admit that "we can no longer effectively threaten general war as an initial response to anything other than a direct strategic attack on us."[18] This realization was further recognized and translated into policy in the latter Eisenhower years through programs to build relatively invulnerable (at the time) ICBMs

Evolution   57

and highly survivable SLBMs, all ongoing systems when John Kennedy came to office. These weapons, of course, are second-strike in orientation rather than having a primarily preemptive (or preventive) role.

Degeneration of Conventional Forces If there was question about whether American nuclear might could deter against the spectrum of possible communist challenge, then logically one had to ask what would take the deterrent place of those weapons. The loud response was that American conventional capability had been eroded to the point that there was no ready substitute. Moulton summarizes this criticism: "The most serious consequence of the Dulles manifesto of massive retaliation probably was the major deemphasis on adequate conventional ground forces to deal with conflict situations which would likely emerge once mutual deterrence became a recognized fact of international life."[19] Official policy had involved training and equipping local forces to provide the first line of defense, with American nuclear might the second line. When employing that second line entailed the potential destruction of the United States, the failure of native forces would leave this country with the potentially unpleasant options of nuclear holocaust or watching helplessly as a regime we supported fell. The need to interpose the American navy between the combatants during the Quemoy and Matsu shelling should have served as prelude to the ability to fight cold war battles with proxies, a fact brought home decisively in the abject failure of the South Vietnamese to ward off their enemies during the early 1960s.

Doctrinal Stultification It can, of course, be debated whether, in 1954, Eisenhower and Dulles should have been able to foresee the emergence of Soviet nuclear capability and strategic rocketry. In the administration's defense, projections about Russian capabilities to attain nuclear weapons had been consistently inaccurate, and skepticism prevailed about the strategic importance and potential of ballistic missiles. (This skepticism was largely associated with the air force, which saw missiles as a threat to the role of the manned bomber.) Regardless of capability or lack of foresight, the enunciation and sustenance of massive retaliation had a somewhat narcotic effect on strategic thinking. The doctrine's primacy and a general lull in Soviet-inspired aggressive activity coincided, and it was difficult for analysts to demonstrate that the relationship was spurious rather than causal. According to Tammen, ascendancy of the doctrine inhibited real study of emerging nuclear relationships: "The Dulles period of massive retaliation set back serious thinking about the Balance of Terror for at least five years. A new strategic relationship had been established, and the guidelines it instituted captivated the interests of most analysts."[20]

Flexible Response and the Doctrine of Mutual Assured Destruction If the Eisenhower years were the period of major innovation in weapons technology, the Kennedy-Johnson years were dominated by doctrinal adjustment and development. Much of the problem and many of the solutions with which the new administration had to work were legacies from the 1950s. As Wells points out, "America's sense of technological superiority was rudely jolted when the Soviet Union launched the first successful intercontinental ballistic missile (ICBM) in August 1957 and followed it in October with the orbiting of the first earth satellite, Sputnik I,"[21] but remedial action was well under way to reclaim the advantage. Using the National Defense Education Act of 1958 as a primary vehicle, a crash program to upgrade American space, and thus missile, capability was being translated into elements that would largely determine American force characteristics in the 1960s. Defense Secretary McNamara would, during his tenure, alter force compositions and add to the destructive capacities of those forces by authorizing technically innovative modifications of basic designs developed at the end of the 1950s, but basic characteristics had been established.

The Environment in the Early 1960s In the area of strategic doctrine, the picture was not so clear nor were the outcomes determined. There was arguable consensus on two basic points: the doctrine of massive retaliation was in disarray as the basis for American defense; and new Soviet capabilities (actual and projected) made increased attention to the deterrent purpose of nuclear weapons paramount. Spectrum defense including a war-fighting role across that spectrum had clearly gone by the boards. Moulton captures the essence of prevailing opinion on that point: "Once the Soviets were able to demonstrate that nuclear weapons could be successfully delivered against targets in the United States, any strategy this country might adopt which stressed the fighting of a nuclear war rather than the deterrence of such a war was unlikely to be credible to a nuclear opponent and very likely to be unacceptable to the American people."[22] There was, indeed, a kind of negative consensus about nuclear weapons, but, as Tammen pointed out (quoted earlier), little positive agreement on what should form the basis of deterrence. The theory-building and doctrinal fields were open to considerable suggestion, and the period was fertile for development of deterrence notions, many of which persist. Within the context of considerable change, the various environments can be overviewed to look for salient influences on the doctrines that would emerge.

The Internal Environment The Kennedy administration entered office in the midst of a public controversy over strategic doctrine that was at least partly of its own making and in a general public apprehension and horror about the consequences of nuclear weapons. The youngest president ever elected was faced with doctrinal disarray from the expert community, which agreed on the major problems but not on the solutions. Major new weapons systems were entering the strategic arsenals, and even more exotic variants would soon become available. From this melange a new strategic posture would have to be fashioned.

Campaigning as the Democratic presidential candidate, John F. Kennedy had raised public concern about American strategic standing by focusing on the so-called "missile gap." The basis of this issue arose from faulty intelligence estimates projecting the growth of the Soviet strategic arsenal and "demonstrating" that as the United States entered the 1960s, it was substantially behind the Russians in missile deployments. Because of limited intelligence-gathering capabilities, these projections had been made by estimating the peak capacity of the Soviet missile industry and assuming the Russians would build at the maximum level.

By the end of the 1960 campaign, U-2 overflights and the earliest satellite surveillance had revealed the estimates to be substantially in error: the Soviets had deployed very few missiles and presented no real threat to American nuclear superiority. That the "missile gap" was bogus is cataloged by Kahan: "Defense plans left by the Eisenhower administration called for a total of 250 Atlas and Titan ICBMs to be deployed by 1962 and included specific authorization for the procurement of 450 Minuteman missiles and 19 Polaris submarines. When combined with the force of more than 600 B-52 and nearly 1,400 B-47 bombers, the relative strategic position of the United States as it entered the new decade was one of overwhelming dominance."[23] President Kennedy maintained that he was not made aware of corrections in the estimates during the campaign. When he entered office, however, public perception, bolstered by vocal comment from the Congress, had raised concern about the adequacy of American forces. An action-reaction phenomenon, in which the administration would feel compelled to react to a threat it knew was not real but arose from political pressures it had helped create, seemed to impel force expansion. Administration reluctance substantially to alter the picture it had created in the campaign is best summarized by Secretary McNamara's oft-cited trepidation that he would get "clobbered" if he had to go to Congress and admit there was no missile gap.

A kind of fatalistic popular culture had also developed about nuclear weapons. Fueled by novels and motion pictures like *On The Beach* and *Doctor Strangelove*, an air of madness and inevitable doom surrounded the entire issue of strategic weapons and doctrine. Prophecies of nuclear Armageddon were commonplace and widely believed, and when the new administration suggested passive civil defense measures as a response to the Soviet threat, a small band of profiteers made their fortunes building bomb shelters in people's back yards.

At the same time, there was a rising belief within both the expert and lay communities that America's conventional forces also were inadequate. This question had been raised during the campaign and helped shape public acceptance of the administration's desire to expand these forces. Seeking to stimulate recovery from the recession of 1959 and not sharing its predecessor's depth of concern over balancing the budget, the new administration entered office committed to increased defense spending and with general public support for that position.

The External Environment Events and trends in the external environment played a major role in determining American strategic doctrine during the 1960s. The major factor in that environment, of course, was the changing Soviet arsenal: the emergence of a Soviet ICBM capability had required doctrinal reassessment in the first place, and the evolution of that arsenal would force alterations in doctrine during the period, ultimately triggering, as we shall see later in this chapter, a further profound debate that continues to the present. In October 1962, the Cuban missile crisis occurred, bringing home sharply the gravity of nuclear arms to both sides.

When the missile gap was debunked and more realistic estimates established, it became apparent that American deployed forces were substantially larger than their Soviet counterparts in the early 1960s. Estimates compiled by the International Institute of Strategic Studies show, for instance, that in 1962 the United States had almost a six-to-one advantage in ballistic missiles (438 ICBMs and SLBMs to 75) and a three-to-one lead in strategic bombers (600 to 190) for an overall strategic "launcher" superiority of nearly four to one (1,038 to 265). In absolute numbers, this advantage actually increased through 1967, when the United States possessed 1,510 more launchers than the Soviets.[24]

That there was indeed a missile gap, albeit in the opposite direction of that advertised in the 1960 campaign, was evident to the Russians. Apparently because of the Cuban missile crisis, the Soviets embarked on a massive development program that, allowing for the lead time from research to implementation, became manifest in the latter 1960s, as the Russians narrowed and closed the gap. After the United States reached its maximum launcher advantage in 1967, Soviet forces rapidly caught up, surpassing American levels of ballistic missiles in the 1971 estimates (1,950 to 1,710) and in total launchers in 1972 (2,227 to 2,165).[25]

The Cuban missile crisis was the most dramatic event of the nuclear age. As mentioned earlier, the Soviet-American confrontation over installation of intermediate and medium-range ballistic missiles on the island capable of striking American targets is generally agreed to have been the closest mankind has come to nuclear war. The vision of what might have been was traumatic to both sides and was the greatest spur to nuclear cooperation and arms control of any single event (see Chapter 6). Moreover, a major reason the Russians apparently perceived their need to back down was the marked inferiority of their strategic forces: if nuclear war had broken out, American ability to devastate the Soviet Union was certain, but the converse was not. The relationship between nuclear weapons and the attainment of other foreign policy goals in Soviet thought is discussed more thoroughly in Chapter 5, but it is clear that they found continued strategic inferiority intolerable. As a result, "The Cuban missile crisis of 1962 appears to have been the catalyst for a procurement program that generated the present force structure."[26]

Technology The level of technological development in the 1960s placed critical restraints on doctrinal opportunities. As mentioned, the technological innovation represented by workable ballistic missiles had already rendered true unconditional viability meaningless and had made a shambles of the massive retaliation doctrine. At the same time, technological accomplishment levels constrained some doctrinal choices, and emerging technologies during the 1960s further complicated strategic planning.

The major restraining influence was in targeting accuracy: the early ballistic missiles were indefensible, but the confidence that could be placed in their ability actually to hit a designated target was reasonably modest. Put in more technical contemporary terms, the "circular error probable" (or CEP, the radius around the target in which there is at least 50 percent expectation the missile will land) of early missiles had to be measured in miles. This fact had obvious implications for targeting doctrine.

In one sense, this limitation was not conceptually troubling, because most concepts of strategic targeting could still "be traced to the technical and conceptual limitations of strategic bombing in World War II."[27] The concepts developed in the saturation attacks on Germany and Japan were massive countervalue attacks, and the technological limitation of inaccuracy was well fitted to what emerged as MAD doctrine. William R. Van Cleave and Roger W. Barnett explain: "Strategically, assured destruction came to be measured in terms of destroying arbitrarily determined percentages of population and civil industry: (a) because of a lingering association of 'strategic' bombing with city bombing; (b) because during the formative years of the concept the combination of yield and inaccuracy then characteristic of strategic forces failed to lend itself to discriminate attacks avoiding major collateral damage, at least on targets in or near urban areas."[28] Retired Admiral Elmo R. Zumwalt stated bluntly: "For a cities-only attack, accuracy of the weapons used does not matter."[29]

Missile accuracy is, of course, crucial to the meaningful development of a counterforce targeting doctrine, and improvement of guidance mechanisms was a major thrust in the refinement process that largely defined R&D efforts in the 1960s. Much technological effort was placed on antiballistic missile (ABM) defenses and, in turn, means of overcoming ABMs. Although the ABM program has largely been abandoned since, developments begun as so-called "penaids" (penetration aids to ensure delivery to target) have continued. Combined with improving targeting accuracy, later versions of these devices (notably MIRV and its refinement MARV) would have dramatic effects on the nature of forces and thus doctrine by the end of the Kennedy-Johnson years and even greater implications in the 1970s and beyond.

Content of the Doctrine The basic defense doctrine of the United States during the Kennedy and Johnson presidencies was known as "flexible response." An explicit rejection of the notion of spectrum nuclear defense articulated in massive retaliation, "the key objective of flexible response was to maintain forces capable of meeting conventional threats so that the United States would not be faced with the choice of either using nuclear weapons or foregoing vital interests abroad."[30]

The basic doctrine emphasized developing sufficient conventional force levels to meet projected possible conflicts (for example, the so-called "two-and-a-half" war strategy—the ability simultaneously to fight two "major" and one "minor" conflicts) with forces appropriately specialized for foreseeable circumstances (such as the antiguerrilla warfare orientation of the Special Forces). Emphasis on the role of conventional armed forces gave strategic nuclear forces a more specialized role: deterrence of nuclear attack or massive conventional or nuclear attack on Western Europe.

Redefining the role of nuclear forces from a general to a more limited utility reflected the administration's assessment of its environment. Spectrum deterrent or war-fighting roles for strategic forces were either foolhardy or incredible or both with a Soviet ability to strike massively (more or less so, as the decade progressed and Soviet forces grew) at American cities. The decisive American edge in the early 1960s made visions of limiting the effects of nuclear war possible to conceive, but the enormous leaps in Soviet lethal potential during the decade erased those dreams.

America's strategic doctrine evolved as the environment to which it responded changed. Basically, doctrine went through two analytically distinct phases that more or less corresponded to the incumbency of one of the two presidents (remembering that Robert S. McNamara, the overarching influence on doctrinal development, served both men). During the early years, the emphasis was on a doctrine of "controlled response." After the assassination of John F. Kennedy, the strategic doctrine of Lyndon B. Johnson evolved to what is known as mutual assured destruction. The doctrines are distinctive enough to warrant individual consideration. Because much of the current doctrinal debate is presaged in the transition from one to the other, some examination of the reasons for that change is needed.

Controlled Response The idea of controlled nuclear response was a marriage of the basic tenet of massive retaliation with the flexibility of the overall defense posture. Flowing from these two basic thrusts, the doctrine ascribed two primary purposes to American forces: "First, in order to serve as a maximum deterrent to nuclear war United States strategic retaliatory forces must be visibly capable of fully destroying the Soviet society under all conditions of retaliation. Second, in the event that war was forced on the United States, its strategic offensive and defensive forces should have the power to limit the destruction to the nation's cities and population to the maximum extent possible."[31] The first purpose was the required destructive capability of massive retaliation, onto which Soviet capability to attack targets in the United States forced the addition of second-strike capability. As such, it was a modification in the face of environmental change rather than a fundamental conceptual alteration.

The second notion represented a philosophical break with the Eisenhower position that nuclear war automatically would be general and cataclysmic. Rather, the purpose of adding flexibility to projected force usage in war-fighting situations was to calculate how to minimize American loss in a nuclear war. As Kahan explains, "The purpose of the controlled response doctrine was to introduce flexibility into the use of strategic forces in order to reduce damage to the United States and its allies in the event strategic weapons were ever used."[32]

This basic notion became known later as "damage limitation." Tammen explains how the principle was to operate:

> It was the purpose of surviving U.S. forces to limit destruction of U.S. targets *such that projected losses were within a theoretical realm of acceptability,* thus allowing for the maintenance of a U.S. retaliatory capability. Defensively this meant ABM systems, air defense, and civil defense. Offensively the doctrine called for a portion of the U.S. retaliatory capability to be targeted against Soviet strategic systems—be they bombers, submarine pens, or missile sites. (Emphasis added.)[33]

The elements of these requirements warrant examination because some of these arguments and needs will recur. Since the policy maintained the retaliatory (thus second-strike) feature of force usage, calculating "acceptable" losses had two requirements.

First, effective active and passive defense systems were needed to interdict some portion of an incoming force and to protect portions of the population. As mentioned, abandonment of ABM is symbolic of the effectiveness of this option. Second, the Soviets would have to be somewhat cooperative in a nuclear attack: if there were to be any meaningful strategic targets to fire at (and thus limit damage), the Russians would have to refrain from firing everything they had at once to leave us something worth counterattacking against. In the heat of nuclear war, such thoughtfulness and cooperation surely would be problematical. At the same time, the doctrine has clear counterforce elements and thus requires counterforce-capable forces. On the one hand, it is arguable whether counterforce doctrine is meaningful (or believable) within a second-strike mode, for reasons discussed in Chapter 2. On the other hand, U.S. missile accuracies in the early 1960s, when the doctrine was unveiled, were at best marginal to carry out counterforce objectives in any but limited ways (such as against unprotected airfields).

Despite these objections, the possibility of limiting damage was at least arguable in 1962 given a small Soviet force of 75 highly inaccurate ICBMs and 190 marginal, mostly propeller-driven bombers, many of which probably could be intercepted. The doctrine, first publicly highlighted in McNamara's Commencement Address to the University of Michigan graduating class on June 16, 1962, in Ann Arbor, also was intended to bolster the credibility of the American nuclear umbrella to NATO allies, who were beginning to worry about whether the United States would risk societal destruction to protect Europe,[34] by showing that the United States might survive honoring our commitment. It was also an announcement of American nuclear strength and resolve, as Moulton points out: "The Ann Arbor speech was, among other things, a declaration by the Kennedy administration in favor of a strategy of controlled counterforce general war, and a rejection of a nuclear strategy for this nation based on 'minimum' or 'finite' deterrence."[35]

As he watched the inexorable growth of nuclear arsenals, Secretary McNamara came to question the ability to place limits on nuclear devastation and eventually concluded, along with others, that it was not possible. Abandoning the use of nuclear weapons for damage-limitation purposes left the massive retaliation role for American forces. This purpose became institutionalized in the form of mutual assured destruction. The flexible-use, damage-limiting concept, however, would return during the Nixon-Ford years to form an important part of the basis for the ongoing strategic debate, albeit with new variations.

Mutual Assured Destruction Although the Kennedy and later Johnson administrations came to question and implicitly reject the controlled response doctrine, it did not drop instantly from the strategic lexicon. The idea of damage limitation, whether feasible or not, had political value (a president trying to save American lives being more attractive than one who was not), and, according to Kahan, could be used "to justify expenditures for offensive forces, above the level necessary to provide a countercity deterrent, as a means of providing damage limitation options and contributing to a margin of U.S. superiority."[36] Quoting testimony by Secretary McNamara before the House Armed Services Committee on February 18, 1965, Kahan further shows that the dual purposes of forces survived into the early Johnson administration: "Secretary McNamara posited two strategic objectives: (1) 'to deter a deliberate nuclear attack upon the United States and its allies by maintaining a clear and convincing capability to inflict unacceptable damage on the attacker'; (2) in the event of war, 'to limit damage to our population and industrial capacities.' "[37] The move to MAD was thus primarily incremental, a matter of relative emphasis rather than dramatic replacement. Just as, in the *Animal Farm*, all animals were equal, but some were more equal than others, so it became with the dual goals of strategic nuclear doctrine.

The heart of MAD doctrine is "the maintenance of sufficient retaliatory forces capable of inflicting unacceptable losses upon an enemy which had attacked the United States."[38] In terms developed in the last chapter, this translates into possession of a secure second-strike capability and, in turn, claim to secure conditionally viable status. When this capability came to describe the arsenals of both superpowers, the ability to do unacceptable damage became mutual. Hence, the doctrine of mutual assured destruction is a description of a particular strategic situation as well as a statement of strategic or doctrinal preference.

Some of the ideas about deterrence systems developed in Chapter 2 should be reviewed briefly in light of a "MAD world." The mutuality of assured destruction is based upon possession of second-strike capabilities by both states, meaning highly survivable retaliatory forces with a high penetrability to target. The principal targeting imperative is a countervalue doctrine (although the Russians have always maintained a public preference for counterforce targeting). The dynamic that makes the system stable by removing any incentive for initiating a nuclear war is the so-called hostage effect: the ability in any circumstance to retaliate after any first strike and to do unacceptable damage (that is, to kill a prescribed number of people), such that the effect is to hold civilian populations as figurative hostages. In this condition, it is suicidal, and thus irrational, to contemplate use of nuclear weapons, since to do so is to guarantee your own personal and societal execution.

The logic of mutual assured destruction and the premises on which it is based are grisly and, to many, morally repugnant. Translating the doctrine to the operational level means calculating loss levels in unpleasant and repulsive ways. McNamara, for instance, operationalized "unacceptable damage" for the Soviet Union as destruction of 25 percent of the Soviet population and 70 percent of Soviet industries. In his 1979 *Annual Report,* Secretary of Defense Harold Brown (who was a secretary of the Air Force under McNamara) updated those figures: "It is essential that we retain the capability at all times to inflict an unacceptable level of damage on the Soviet Union, *including destruction of a minimum of 200 major Soviet cities*" (emphasis added). Using statistics on effects of cities attacks in the *Report,* this requirement translates into the ability to kill 34 percent of the Soviet population and destroy 62 percent of its industrial capacity.³⁹

Because operationalization requires calculating what are at best inhumane consequences of nuclear usage, it has been subject to criticism (reducing the doctrine to the acronym MAD was a euphemistic act by a detractor repelled by the doctrine) and to Strangelovian extrapolation. The moral objection is well summarized by Colin S. Gray: "A 'Carthaginian peace' would undoubtedly affront the moral sensibilities of Western publics today (Christian nations 'do not do that sort of thing'), but the ultima (ir)ratio of a kindly, homeloving and God-fearing American President is the guarantee of the assured destruction of the Soviet Union."⁴⁰ It also leads to some implications that appear almost bizarre. Representative is a conclusion reached by Richard Rosecrance: "From the standpoint of deterrent credibility, it might have been useful to put strategic missiles in urban areas. To attack these missiles an aggressor would have to hit our cities. The revenge motive would reinforce retaliation on the enemy's urban population."⁴¹ In other words, if MAD's consequences would be awful, they should be made as awful and as real to the average citizen as possible (how many Americans would tolerate a missile silo next door?).

Despite such criticisms, proponents find advantages to the doctrine. Among those advantages is that once the level of assured destruction is established, there are some clear guidelines for force requirements: "One of the advantages . . . was that the strategy could be implemented by a finite, limited number of strategic nuclear weapons."⁴² Further, if strategic forces are invulnerable to first strikes, incentives to continue the arms race are decreased: "Proponents of MAD argue that the relative size and capabilities of strategic forces are meaningless and that nuclear superiority is devoid of operational meaning."⁴³ Thus, once one has a sufficient force incapable of being preempted, the size of the opposing force is essentially irrelevant. As we shall see in Chapter 4, the present configuration of American forces resulted from this kind of calculation, although technological improvements in the forces have endowed them with lethal capabilities wildly in excess of those needed to carry out the basic mission.

One of the major reasons forces grew disproportionately to the assured destruction requirement was another artifact of the McNamara Defense Department: the greater-than-expected threat. John Newhouse explicates this phenomenon: "Another novelty of 1965 was the entry into the strategic lexicon of the Greater-Than-Expected Threat (GET). The term, or concept, signifies an enemy capability that exceeds the 'high end' of the range of threat in the National Intelligence Estimates (NIE's). Acceptance of the concept meant laying down an orderly process of planning how to hedge against it."[44] The "orderly process" became the conservative method of "worst-case analysis" described in Chapter 1. McNamara integrates this method with the doctrine it served: "Security depends upon assuming a worst plausible case, and having the ability to cope with it. In that eventuality we must be able to absorb the total weight of nuclear attack on our country—on our retaliatory forces, on our command and control apparatus, on our industrial capacity, on our cities, and on our population—and still be *capable of damaging the aggressor to the point that his society would be simply no longer viable in twentieth-century terms* (emphasis added).[45] Given that the Soviet Union's future capabilities are not always predictable and also the length of time necessary to prepare new systems, this conservative bias created force planning in excess of the basic needs of assured destruction. Many of the systems planned contingently during the period came or are coming into the strategic arsenal of the 1980s, often with dramatic effect.

The appropriateness of strategic doctrine during the Kennedy-Johnson years must, of course, be measured against the environmental context in which it arose. It was an adaptation to perceived environmental change and evolved as its environment continued to alter. At the beginning of the period, two considerations in that environment were, in retrospect, paramount: Soviet nuclear forces were capable of attacking the United States but were markedly inferior to American forces; and the accuracy levels with which strategic weapons could be delivered were minimal. Each of these factors led to strategic decisions.

The sheer fact of Soviet ability to attack targets in the United States obviated the spectrum defense role nuclear weapons were assigned under the massive retaliation doctrine. In other than protection of an absolutely vital interest of the United States, it is undoubtedly true that the American public would have viewed any loss as unacceptable, and those charged with security policy, as well as those who might challenge the United States, had to be cognizant of that reality.

United States strategic superiority, however, meant that nuclear massive retaliation threats remained viable in at least some circumstances. Even when John Kennedy entered office, the American force, if employed massively, possessed a capability approaching assured destruction. Since the Soviet Union did not possess such capabilities, nuclear threats by the United States were believable and were apparently decisive in removing offensive missiles from Cuba. At the same time, the marginal Soviet arsenal size and delivery capability made ideas like controlled response and damage limitation at least intellectually defensible.

More than anything else, enunciation of mutual assured destruction was an admission of and tribute to the growth of Soviet capabilities.

Implicitly, this description of the strategic environment admitted that the goals of damage limitation were no longer reasonable. By definition, mutuality of assured destructive capacity said that not only could we, under any circumstances, destroy the Soviet Union, but that the relationship was reciprocal. If the Soviets held us in the same hostage position we held them, then talk of limiting damage to some "acceptable" levels was a mere "will-of-the wisp." Thus, the only way damage limitation could be accomplished was through effective deterrence.

The deterrent base, of course, was premised on second-strike capability, and technological levels reinforced policy. The ballistic missile had established the hostage relationship, and the lack of targeting accuracy in existing systems reinforced it in two ways: missiles could be delivered accurately enough to devastate large targets like urban areas; but they could not hit small targets like retaliatory forces, thereby helping to guarantee their availability to carry out second-strike, hostage-executing missions. As a result, the balance of terror was stable and symmetrical.

Particularly by the end of the Kennedy-Johnson years, the balance was also uncomfortable to many observers. Just as Soviet achievement of the ability to attack the United States with strategic weapons made most observers feel that massive retaliation had lost its meaning, a number of observers wondered if the loss of American strategic superiority did not put further strictures on the utility of a doctrine the bottom line of which remained massive retaliation. This concern resulted in a doctrinal debate that raged throughout the Nixon-Ford years and beyond.

The Years of Doctrinal Debate: Strategic Sufficiency, Flexible Nuclear Response, Limited Nuclear Options, and Essential Equivalence That there would be a reassessment of strategic doctrine with the election of a Republican president was probably predictable on political grounds alone, if for no other reason than that Democrats and Republicans use different experts with consequently varying approaches and opinions on defense matters. In addition, the environment in which strategic doctrine is made had changed substantially since the inauguration of John F. Kennedy and visibly since the enunciation of mutual assured destruction as the American doctrinal base.

Environmental Factors in the 1970s These new and continuing environmental factors were and are highly interactive. The most obvious environmental changes facing the new Nixon administration were the attainment of strategic parity by the Soviet Union and the impending Strategic Arms Limitation Talks, factors interlocking because any arms control outcome of SALT would legitimize some form of balance in weaponry. Americans accustomed to superiority in arms found this a difficult adjustment to make, and it became the source of controversy.

Evolution									69

Technological forces were beginning to produce a whole new array of deadly weapons promising geometric increases in lethality and significantly improved targeting accuracy. With American public attention riveted on the rending Vietnamese experience, the most important internal environmental factor was the strategic debate itself, contested vigorously and continuously among defense intellectuals propounding the adequacy of MAD or its inadequacy and thus the need for new strategies.

The Internal Environment The basic question in the new strategic debate was the same one that was posed within the Kennedy administration: the credibility of the inherited doctrine in the face of new realities. Those who argued that MAD no longer comported with strategic realities focused on the irrationality of MAD should deterrence fail and thus sought alternative means to use nuclear weapons and alternative scenarios for nuclear use. To MAD proponents such arguments were heresy threatening to raise the likelihood of nuclear war. William H. Kincade summarizes the views of the opposing camps on the basic issue: "The concern of the deterrence advocates is that a war-fighting doctrine or posture will be self-realizing and therefore undermine deterrence. It may permit the Russians to believe, albeit erroneously, that a nuclear attack is being planned, encouraging them to 'preempt' or strike first. The opposing view is that increased war-fighting capacity enhances deterrence ('deterrence plus') and expands strategic options in the event deterrence fails."[46] Analytically, the debate can be segmented into three aspects that will be helpful in understanding doctrinal evolution: the ex post–ex ante debate; the likely ways a nuclear war would be fought; and the effects of the balance of forces both on deterrent credibility and arms control prospects.

The Ex Post–Ex Ante Debate: This debate about MAD revolves around the dilemma of massive retaliation as a means to deter aggression and as a strategy for actually conducting a nuclear war should the MAD threat fail. Richard Rosecrance, who coined the terminology, explains the basic problem: "But there is still a major distinction between deterrence *ex ante* and revenge *ex post*. . . . In the case of *ex ante* strategy the intent is to make an opponent believe that if he attacked, he would receive a swift and catastrophic retaliatory blow. But, as Herman Kahn was the first to recognize, the carrying out of any such massively punitive blow *ex post* was largely irrational."[47] In another work, Rosecrance explains the basis of this "irrationality" somewhat, saying, "ex post, one may or may not want to make a devastating response if hostile action does take place. There is thus a major difference between ex ante and ex post incentives."[48]

The point at which the gap in incentives is most apparent is in the case of a measured and limited first strike by the Russians, perhaps against

ICBM fields. As pointed out in Chapter 1, this position is most strongly associated with former Defense Secretary James R. Schlesinger, who argued in the 1975 Defense Department *Annual Report:* "Today, such a massive retaliation against cities, in response to anything less than an all-out attack on the U.S. and its cities, appears less and less credible. Yet . . . deterrence can fail in many ways. What we need is a series of measured responses which bear some relation to the provocation, have prospects of terminating hostilities before general nuclear war breaks out, and leave some possibility for restoring deterrence."[49] From the arms control end of the doctrinal spectrum, George Quester raises the question of whether it would ever make sense to carry out the MAD threat: "One could ask rhetorically whether there would ever be a need for retaliation after a nuclear strike, much less speedy retaliation. Even after a full Soviet strike on all the cities of the United States, would it not be far more rational to sue for peace, rather than destroying the one probable source of relief supplies?"[50] Green puts the irrationality of a decisive counterstrike more forcefully: "It is after all, simply impossible to imagine circumstances in which an annihilatory counterstrike makes any sense at all, by any standard of 'rationality' that is not equivalent to sheer vengefulness."[51]

This criticism of MAD, of course, has "straw man" elements that should be apparent from the last section. These arguments are strikingly reminiscent of the debate about the doctrine of controlled response in the Kennedy administration, although the reason for the debate was somewhat different. In the Kennedy-Johnson years, the concern was more on damage-limiting possibilities, although control of fighting levels was considered and debated, notably in the McNamara Ann Arbor speech (we will see, especially in Chapter 7, that technological advances in targeting accuracy have given a rebirth to damage-limitation notions as well). Schlesinger emphasized the likelihood of limited strikes, but even this emphasis has roots in the MAD era. Alain C. Enthoven, a McNamara Defense Department official at the time, wrote in 1965: "But, as time goes by and the size and destructive power of nuclear arsenals increase, total war between nuclear powers will, more and more, mean total destruction. It is my own opinion that with the widespread realization of this fact will come the general belief that all wars should be limited."[52] Not only are the arguments not new, but some of the solutions have been available as well. As Schlesinger admits in the same section of the FY 1975 *Report,* "We have had some large-scale preplanned options other than attacking cities for many years, despite the rhetoric of assured destruction."[53] The argument is thus a straw man to the extent that it attacks only the most prominent aspect of MAD while ignoring aspects that mitigate its inadequacy as doctrine.

Evolution

That the ex ante–ex post problem is real, however, is undeniable and is a concern with which strategic planners must grapple. The key issue is whether planning for limited wars makes them more or less likely. To argue that appropriate and proportionate responses to varying nuclear initiatives improve deterrence is to maintain a sort of options spectrum to cover a spectrum of nuclear threats. Whether that in and of itself is stabilizing or destabilizing depends on how one envisages war-fighting in a nuclear context.

War-Fighting Scenarios: The position one takes on the ex ante–ex post controversy largely determines positions taken on possible war-fighting scenarios and how to plan for them. As the quotation from Kincade indicates, adherents of the "deterrence-plus" school maintain that, given the obviously catastrophic nature of all-out nuclear war, such conflicts are the least likely form nuclear war would take. Finding that more limited engagements are the likely form of any Russian aggression, they maintain that deterrence is best served by having clearly developed contingency plans to deny the Soviets attainment of whatever goals limited attacks seek (basically the first, and traditional, goal of deterrence identified in Chapter 2). "Deterrence-only" theorists, of course, deny that nuclear weapons can be used for more traditional deterrence and war-fighting purposes because of uncertainty about the escalatory process (see Chapter 2) and that only pure deterrence (the second role of military force as a deterrent) is a legitimate purpose for nuclear weapons.

Former Secretary Schlesinger, as chief architect (or reviver, as the case may be) of the deterrence-plus position, explains why force use planning associated with assured destruction is doctrinally inadequate:

> Misunderstanding regarding the necessity of a credible response—or the desire to rationalize the avoidance of necessary costs—underlies the misplaced confidence too frequently placed in suicidal or quasi-suicidal responses based upon *the threat of the mass use of nuclear weapons, wholly disproportionate to the hostile action*. The threat of all-out city destruction, when one's potential opponent can respond in a similar manner, is one such example . . . The underlying problem for all such strategies is that their credibility is inherently questionable. (Emphasis added.)[54]

If less than massive Soviet attacks are most likely (or least unlikely), and assured destruction is incredible as a deterrent or as a response, what is the option? Polmar provides the answer of the deterrence-plus school: "US national leaders should have the means—both weapons and strategy—at least to consider a scale of options between all-out nuclear war and absolute capitulation."[55] Paul Nitze, a leading MAD critic and defense official during the Nixon-Ford years, agrees, adding his fear that failure to plan for a variety of exigencies would give the Soviet

Union an advantage should nuclear war occur: "The first question to be resolved is whether to concentrate on the countervalue level, or the theory that no sane Russian would think of risking the damage that even a limited retaliatory strike on his cities would produce, or whether to concentrate on denying the Soviets a superior intercontinental warfighting capability. It is my view that the latter alternative is the correct one."[56]

The deterrence-only school of thought finds such calculations threatening to the mutual hostage effect that defines stable deterrence. Taking direct aim at Schlesinger and the position he developed, Sidney D. Drell maintains: "By giving added emphasis to preparations for fighting and 'winning' limited nuclear wars aimed at one another's strategic arsenals, the Schlesinger doctrine threatens to undermine the stability of the strategic balance."[57] Wolfgang K. H. Panofsky agrees, adding that such planning increases rather than subtracts from potential confrontation situations: "Ill-founded attempts to 'sanitize' nuclear war are a disservice to the maintenance of stability, as well as the efforts to reduce areas of risk."[58] These statements, representative of the criticisms of deterrence-plus doctrine, point to two additional concerns in the debate: the effects on force comparisons and on arms control efforts.

Implications for Forces and Arms Control: As mentioned earlier in this chapter, a primary advantage of assured destruction doctrine is that force requirements arising from it are highly determinate. Once one has figured out what is to be destroyed, has protected the force from preemption, and has made adjustments for interdiction and malfunction, the size of force needed can be determined. These calculations do not, within a fairly broad range of confidence, require comparison with the adversary's force; and, as long as the outcomes do not threaten the weapons numbers one had deemed necessary for mission accomplishment, do not impede arms control efforts.

The growth of Soviet offensive and defensive capability, combined with concern for the kinds and amounts of force necessary to carry out limited responses to (supposedly) counterforce attacks, has raised serious questions about the adequacy of force planning in this mode. Because the Soviets have passed the United States by some measures of nuclear capability (the subject of nuclear "bean counting," as exercises in force comparison are known in the trade, is treated in Chapter 4) and are projected to surpass American levels in others, serious questions about force adequacy arise: "The band of tolerable inequality is not, however, infinitely broad. If probable casualties and damage to one side would be three, five or ten times the probable casualties and damage to the other, and if the absolute number of casualties on the stronger side would be a small percentage of the total population, it is not clear that the weaker side should or would meaningfully respond to a counterforce attack."[59] In a separate piece, Nitze summarizes his concern about the drift in relative U.S.–U.S.S.R. force levels: "In sum, the ability of U.S. nuclear power to destroy without question the bulk of Soviet industry and a large portion of the Soviet population is by no means as clear as it once was."[60]

The legitimacy of these concerns is hotly contested. First, although the number of strategic nuclear launch vehicles possessed by the United States has not grown since 1967 (it has actually declined somewhat as worn-out B-52 bombers have been retired), the number of warheads, their accuracy, and consequently their lethal capacity have increased dramatically. Second, American and Soviet forces are quite different (they "possess asymmetries," in the literature), leading many observers, including former chairman of the Joint Chiefs of Staff Maxwell D. Taylor, to conclude that comparisons are futile: "I have long since concluded that because of this inadequacy of weapons data and the inherent incomparability of such factors as missile numbers, accuracy, reliability, megatonnage and indestructibility, a direct comparison of our strategic forces with those of the Soviet Union can be only very approximate and can have only limited meaning."[61]

The concern (or lack thereof) one has about comparative force levels inevitably affects the way one views the arms limitation process. Deterrence-only advocates tend to be comfortable with the likely outcomes (although some would like to see cutbacks), because they feel that assured destruction capabilities are guaranteed within likely limits. Those who question the actual (or potential) adequacy of force comparisons likely to be largely frozen under probable agreed limits tend to view the process with some jaundice, particularly feeling that likely outcomes will institutionalize a superior Soviet war-fighting capability. Kincade warns of the potentially deleterious arms control effect of such consideration: "Most importantly, the experiment with war-fighting contributes to the frustration of strategic arms negotiations by weakening confidence in deterrent forces and by driving both sides to press for technological advantages and fixes. This course will continue to create new impediments to arms control and reduction."[62] Those who fear that current negotiations will place the United States in a position of permanent strategic inferiority would, of course, like nothing more than to see current negotiations frustrated.

The External Environment In retrospect, it is clear that the United States possessed a clear dominance over the Soviet Union in strategic arms for the first twenty years after World War II, but that the gap began to close rapidly between 1965 and 1970, when the Soviet Union achieved numerical launcher parity with the United States. The Soviets continued to build their forces beyond the parity level in the early 1970s, as Kissinger explained in his March 22, 1976, Dallas speech:

> The Soviet Union chose a different course. Because of its more limited technological capabilities, it emphasized missiles with greater throwweight compensated for their substantially poor accuracy. But—contrary to the expectations of American officials in the 1960s—the Soviets also chose to expand their number of launchers beyond what we had. Thus the Soviets passed our numerical levels by 1970 and continued to add an average of 200 missiles a year—until we succeeded in halting this buildup in the SALT . . . agreement of 1972.[63]

Although SALT curbed numerical deployment increases, it did not dampen Soviet defense spending. Viewing trends since 1958 as projected through the current (1976–80) Five-Year Plan, William T. Lee says:

> The essential points are:
> 1. Defense expenditures have increased as a share of the Soviet state budget, national income, and GNP since 1958. Similarly, weapons procurement has risen as a share of machinery production since 1958.
> 2. Soviet defense expenditures have grown at the annual rates of ten percent in the period 1968–1970, and eight to ten percent in the period 1971–1975.
> 3. The trend in the share of Soviet GNP allocated to defense has been eight to twelve percent in 1955, eight percent in 1958, 12 percent in 1970, and 14 to 15 percent in 1975.[64]

These figures compare to a current American defense commitment of 5–6 percent of GNP. The portion of those increases dedicated to strategic forces has raised questions about what the Soviets are spending their rubles on (capabilities), what they seek to accomplish with those forces (intentions), and what the United States needs to do to counter these efforts (action-reaction).

Since the Soviets have a large throw-weight advantage on the United States (they build larger missiles that carry heavier payloads) and since their investments under SALT cannot be made in more missiles, it follows that money must be spent in improving existing systems (allowed under SALT I). It is generally assumed that improvement translates into catching up with the United States in the related areas of mutiple warheads and missile accuracy. If this can be accomplished (and assuming that is what the Russians intend to do), their advantage in payload could give them a decisive strategic edge. Assuming that these projected capabilities would correspond with Soviet intentions, Christopher Lehman and Peter C. Hughes put forward a chilling scenario: "If the Soviet Union is allowed to maintain unilaterally its throw-weight advantage at the levels envisioned, it will be capable of launching a saturation attack against U.S. targets that would destroy a high percentage of this nation's military forces; the number of missiles still available to the Soviets would enable them to credibly threaten U.S. cities with certain destruction."[65] Tammen explains why a first strike might appear attractive in such circumstances: "Preemption theoretically becomes attractive when it can be calculated that a surprise first strike will blunt, disrupt, or significantly damage the other party."[66]

The Soviets have also invested heavily in recent years in a civil defense program presumably designed to minimize Soviet losses should a nuclear war occur. There is considerable disagreement about the effectiveness of the project (discussed in Chapter 5), but it raises an additional concern. To obviate these Soviet activities, Nitze proposes a dual response: "To restore stability and the effectiveness of the U.S. deterrent: (1) the survivability and capability of the U.S. strategic forces must be such that the Soviet Union could not foresee a military advantage in attacking our forces; and (2) we must eliminate or compensate for the one-sided instability caused by the Soviet civil defense program."[67] There is, of course, the inherent causal direction problem of the capabilities-intentions (or forces versus doctrine) relationship, and technological matters further complicate the issue.

Technology The major thrust of technological activity during the 1960s and into the 1970s has focused on improving systems begun in the 1950s and deployed in the early and mid-1960s. The most basic characteristic of these improvements has been in the area of targeting: giving increasingly pinpoint accuracy to formerly clumsy and inaccurate weapons. That these advances in technology carry important doctrinal implications is identified by Schlesinger in the FY 1975 *Annual Report* of the Department of Defense: "Much of what passes as current theory wears a somewhat dated air—with its origins in the strategic bombing campaigns of World War II and the nuclear technology of an earlier era when warheads were bigger and dirtier, delivery systems considerably less accurate, and forces much more vulnerable to surprise attack."[68]

The other major innovation of the 1960s was the multiple independently targetable reentry vehicle (MIRV, characteristics of which are discussed in Chapter 4), which, by allowing a number of warheads to be fired by a single missile at different targets, greatly increased nuclear capabilities for the United States and is doing the same for the Soviets as they deploy them (this is the basis of much of the concern raised in the previous section). The emergence of these two innovations is related, because MIRV is a highly accurate weapon. Moreover, improvements in computer capabilities and satellite guidance of missiles promise even greater accuracies (essentially, these innovations involve satellite computers reprogramming reentry vehicle programs to adjust course to targets; these are part of MARV and second-generation cruise missile technologies).

The major impact of these innovations has been to make possible realistic consideration of counterforce targeting. As Drell explains: "The resurgence of the doctrine of limited nuclear counterforce has been spurred by progress in weapons technology—in particular, the development of accurate and highly reliable MIRV's, which enable a single missile to attack several different targets with high accuracy."[69] This development has some obvious implications in terms of strategic firing doctrines and capabilities, as discussed in Chapter 2 and expanded upon in Chapters 4 and 7. Kissinger, however, offers a leavening note about the impact of innovations: "But today, when each side has thousands of launchers and many more warheads, a decisive or politically significant margin of superiority is out of reach. If one side expands or improves its forces, sooner or later the other side will balance the effort."[70]

Doctrinal Alternatives: Searching for Consensus in Chaos As the preceding discussion of environmental influences on strategic policy-making has sought to show, there is fundamental disagreement among experts in the field about the nature of the strategic environment in the 1970s and thus about appropriate doctrinal responses to that environment. Sampling the literature and positions taken, however, has provided a context within which doctrinal options have been raised, debated, modified, and in some cases rejected.

There is nothing in the current strategic doctrinal field that approximates the clarity and dominance attained by massive retaliation in the 1950s and mutual assured destruction in the 1960s. Rather, a number of proposed doctrines, some more formal and complete than others, have ebbed and flowed in the public debate. None have achieved doctrinal consensus or anything approaching it. The result has been a high level of doctrinal uncertainty that is the single most enduring and prominent feature of the doctrinal landscape.

Strategic Sufficiency Richard M. Nixon entered the White House faced with the prospect of opening the Strategic Arms Limitation Talks initiated by his predecessor (the scheduled opening sessions had been canceled as a response to the Soviet invasion of Czechoslovakia in 1968) and about which he had, according to Newhouse, personal reservations.[71] Realizing that his strategic posture and the SALT negotiating position had to be coordinated, he sought to move cautiously.

When he entered office, President Nixon was also faced with a series of commonly accepted views of the mutually deterring relationship between the United States and the U.S.S.R. that served as constraints on doctrinal choices he might make. Fred Ikle summarizes these views (which he calls "dogmas"):

One: our nuclear forces must be designed almost exclusively for "retaliation" in response to a Soviet nuclear attack—particularly an attempt to disarm us through a sudden strike.

Two: our forces must be designed and operated in such a way that this "retaliation" can be swift, inflicted through a single, massive and—above all—prompt strike. What would happen after this strike is of little concern for strategic planning.

Third: the threatened "retaliation" must be the killing of a major fraction of the Soviet population; moreover, the same ability must be guaranteed the Soviet population in order to eliminate its main incentive for increasing Soviet forces.[72]

Within this general consensus and his own interest in ABM systems (debate on which was controversial but moving toward its eventual negative conclusion), the president's first doctrinal delineation was cautious.

Some of this caution was evidenced by the fact that he chose as descriptor for his new policy a phrase developed during the Eisenhower years: as pointed out, the term "strategic sufficiency" surfaced first in 1955. The fact that the Soviets would catch up to the United States in deployed missile launchers in the year following Nixon's inauguration served to give the name a more plausible ring than it had in the mid-1950s.

Moulton summarizes neatly the four criteria that defined this initial position by the Nixon administration:

The sufficiency criteria enunciated by the Nixon National Security Council included:

Maintaining an adequate second-strike capability to deter an all-out surprise attack on our strategic forces.

Providing no incentive for the Soviet Union to strike the United States first in a crisis.

Preventing the Soviet Union from gaining the ability to cause considerably greater urban/industrial destruction than the United States could inflict on the Soviets in a nuclear war.

Defending against damage from small attacks or accidental launches.[73]

Except for the fourth criterion (which was offered as justification for the "light" Safeguard ABM system the administration was proposing), none of these policy goals was new. The first two were no more than a statement of the conditions for a secure second-strike capability that was central to MAD. The third recognized the growing destructive capability of the Soviet Union and suggested that damage capabilities should remain roughly the same for the two countries, a precursor to the essential equivalence idea discussed below.

Flexible Nuclear Response and Limited Nuclear Options The doctrine of flexible nuclear response or limited nuclear options (the two are conceptually similar enough to be used interchangeably) was formally announced in the waning months of the Nixon administration by then Defense Secretary Schlesinger, with whom it has since been associated (it is also known as the "Schlesinger doctrine"). Admitting that limited nuclear options had been part of nuclear planning for some time (Jeffrey Record, for instance, points out that the flexible or graduated use of theater nuclear weapons in Europe had been adopted by the Kennedy administration for defending Europe in 1962[74]), Schlesinger argued in the FY 1975 *Annual Report* that what he proposed was a change in emphasis: "Although several targeting options, including military only and military plus urban-industrial variations, have been a part of U.S. strategic doctrine for some time, the concept *that has dominated our rhetoric* . . . has been massive retaliation against cities" (emphasis added).[75] The purpose of thinking and planning for a variety of nuclear usages in numerous situations was to demonstrate American resolve to act in those situations. As he explained in testimony before a Senate Foreign Relations Committee subcommittee on March 4, 1974: "In my judgment, the effect of the emphasis on selectivity and flexibility, which I separate from any issue of sizing, is to improve deterrence across the spectrum of risk."[76] Elaborating on why enlarging "the spectrum of risk" encompassed by nuclear use was desirable, Schlesinger testified before an SFRC subcommittee in August 1975: "Deterrence, in our judgment, has certain characteristics: one, the opponent should see no vulnerabilities or asymmetries in the force balance that he can exploit; two, we should have the ability to clearly indicate the strength of our resolve and, three . . . if for some reason deterrence should fail, we should have the ability to terminate that conflict at the lowest possible level of violence."[77]

Lynn Ethridge Davis, who worked in the Schlesinger Pentagon and participated in framing the doctrine, lists four assumptions underlying limited nuclear options. First, deterrence would be improved if the United States had specific plans for a variety of contingencies (and the Russians knew it). Second, planning for the potential use of nuclear weapons would not make the decision to use them any easier. Third, planning would make political leaders more aware of the utility (as well, hopefully, of the disutility) of nuclear weapons. Fourth, deterrence is best served by denying the enemy a specific objective rather than punishing the enemy indiscriminately.[78] This last assumption is crucial to the doctrine. Davis describes it: "Proportionality was considered a general condition of deterrence. The new doctrine would promote deterrence across a broad spectrum from major aggression to the limited use of nuclear weapons in a crisis."[79] Assured destruction is implied as the ultimate threat should all else fail.

Edward J. Ohlert defines four major goals of this "flexible response" doctrine: "1. To deter a broad range of Soviet military options; 2. To provide incentive and demonstrate US capability to limit war should deterrence fail; 3. To permit continued negotiation even after the commencement of hostilities; 4. To terminate hostilities on terms acceptable to the United States and her allies."[80] The latter two goals could have been lifted almost verbatim from McNamara's Ann Arbor speech of 1962. They were, however, crucial to the doctrine; Davis explains that, along with the proportionality principle that seeks to improve deterrent credibility, the major purpose of limited nuclear options is "to end nuclear war quickly if it occurred and to re-establish deterrence as soon as possible."[81]

The doctrine has received some strong endorsements. C. Johnston Conover of the Rand Corporation, for instance, offers such support: "Limited nuclear options make deterrence threats credible *and* offer intrawar deterrence that is not possible with pure assured destruction forces. . . . The primary usefulness of limited nuclear options lies in demonstrating resolve—not in warfighting per se" (emphasis in original).[82] Rosecrance, though more restrained in his plaudits, agrees that limited nuclear options enhance deterrence credibility: "There seems little question that the Schlesinger strategy does improve the current position. The credibility of an all-out retaliatory attack on Soviet populations is so low that the combination of credibility and capability factors offers a very small deterrent product. The increased possibility of a much lighter attack seems to improve the net position."[83] As previous discussions have indicated, deterrence-only theorists disagree with nearly all of these assumptions and arguments. One who would hardly categorize himself as a MAD theorist, Klaus Knorr, reacting to the controlled response doctrine of the 1960s, offered a cautionary note in 1966 that summarizes current criticisms: "It is possible that a policy of firebreaks that encourages the initiation of limited military conflicts, or discourages them less than would be the case otherwise, may actually increase the probability of nuclear holocaust more than a policy that keeps the danger of escalation high and thus gives the powers the strongest incentive to shun all military conflict, at least as long as the balance of terror holds."[84]

Essential Equivalence What has become known as the "doctrine" of essential equivalence began, in the terms we have adopted here, as a statement about force composition and not as a strategic doctrine.

Arising from the so-called Jackson amendment to the legislation approving the SALT I Agreements (PL 92-448), it "stipulated that the United States should not have levels of intercontinental strategic forces inferior to those of the Soviet Union."[85] Essential equivalence was enshrined as part of strategic policy by Secretary Schlesinger in 1974, the same year limited nuclear options were first introduced.

Other than being a means to pacify Senator Henry Jackson's fears that the United States might be negotiating itself into strategic inferiority via the SALT process, essential equivalence was a policy means to accommodate Soviet strategic parity and to declare that the condition was acceptable but would not be allowed to degenerate. In this sense, the policy can be viewed as a force level corollary of the doctrine of strategic sufficiency.

Lehman and Hughes explain what Schlesinger meant when he used the term: "Although Secretary Schlesinger refrained from delineating a specific level and combination of force that would represent essential equivalence, he did cite three general criteria: (1) both sides should maintain survivable second-strike reserves; (2) there should be symmetry in the ability of each side to threaten the other; (3) there should be a perceived equality between the offensive forces of both sides."[86] The essence of this notion, of course, is in the perception of equal forces and hence equal ability to threaten. In turn, the need for the statement arises from the difficulty of equating asymmetries between American and Soviet forces in any given measure of effectiveness, while allowing a hedge for growth in forces if the Soviets attempt to gain strategic advantage. As Van Cleave and Barnett put it, "Essential equivalence . . . has to do with the sizing of forces and depends upon Soviet force developments."[87]

Essential equivalence has obvious psychological and perceptual implications. John C. Culver sees it as a way to adapt to strategic reality: "The reality of essential equivalence, and hence incomplete security, has confronted us with a difficult psychological adjustment."[88] In that light, enunciation of the doctrine can be seen as a way to convince ourselves, and our allies, that parity is not debilitating. As Secretary Donald H. Rumsfeld argued in the FY 1978 Defense Department *Annual Report,* this is necessary because, "if friends see the balance as favoring the Soviet Union rather than the United States, their independence and firmness may give way to adjustment, accommodation, and subordination."[89]

Parity as a stabilizing concept also receives some support at the theoretical level. As Tammen explains: "The 'parity' argument rests on the assumption that a state of near equality releases tensions, eases negotiations, and restores stability to the balance of terror. It also emphasizes the importance of the appearance of equality, regardless of the purely mathematical probabilities inherent in any force level."[90] Perception and appearance are, of course, highly subjective and interpretive matters that lend themselves to differing conclusions, thereby creating a climate of uncertainty in force calculation. Schlesinger, however, argued before an SFRC subcommittee in August 1975 that this uncertainty itself was a virtue: "To the extent that there are these uncertainties, and they are perceived by the leadership on both sides, they do impose restraint, and they contribute to deterrence."[91]

The essential equivalence concept has not been without criticism. The most obvious objection relates to definitional specificity: what does it mean? Archie L. Wood states this criticism: "Unfortunately, this term has defied definition in a generally acceptable way. Some wish to count only missiles; others want to include missile throw-weight; still others want to include bomber forces; and some want to count warheads. Still others would consider accuracy."[92] George W. Rathjens points to the problem operationalizing the concept in specific, and particularly categorical, ways would have for American policy makers: "U.S. demands for symmetry in certain areas of strategic arms (numbers of launchers and throw weight) are not likely to be matched by U.S. willingness to accept equality or symmetry in others where it has advantage (notably in access to foreign bases and in the technology relevant to strategic arms)."[93] On the other hand, Stefan H. Leader and Barry R. Schneider maintain that the concept, particularly left vague, has arms-race implications: "The current emphasis on 'essential equivalence' rather that simply building forces to accomplish 'assured destruction' permits and encourages the building of many additional strategic weapons and delivery systems."[94] Paul C. Warnke reaches the same conclusion: "Today, the danger is that hardliners both here and in the Soviet Union might use the concept of strategic parity as an excuse for endless continuation of an arms race, trying to match opposing strategic force structures across the entire spectrum of weaponry."[95]

The Carter Inheritance The new Carter administration inherited the strategic situation described in the preceding pages. To say that Jimmy Carter inherited a clear doctrinal statement or an environment the characteristics of which were agreed upon would obviously be a misstatement of the first order. Rather, the legacy of the Nixon-Ford years was a series of partial doctrines and force statements that reflected rehashing old ideas under new catch-phrases (strategic sufficiency, limited options) or political compromises (essential equivalence). The expert community was as divided on doctrinal propriety as it had been in 1969, and although everyone could agree that the growth of Soviet nuclear might constituted a threat, opinion varied widely about what and how much of a threat there was. Discussion focused emotionally on the outcome of

SALT II and what its impact would be on American security. At the same time, new weapons systems with the potential to alter the balance in dramatic ways lurked on drawing boards and in testing laboratories, straining to enter the strategic inventory despite inadequate attention to whether they were needed or what their impact would be on nuclear stability.

The strategic doctrine the Carter administration inherited can be reduced to three basic elements. First, the ultimate deterrent remained assured destruction of the Soviet Union, with a reciprocal Russian capability that produced deterrence against general nuclear exchange through the dynamics of secure second-strike capability and strategy. Second, since massive countervalue exchange was viewed as the least likely form of nuclear war, flexibility in nuclear response across a spectrum of nuclear risk was part of policy. This emphasis included planning for proportionate responses to limited attacks, including limited counterforce targeting, and was premised on the belief that such response would deter limited nuclear aggression and allow early cessation of nuclear hostilities should deterrence fail. Third, essential equivalence, loosely defined as the ability to respond proportionately to any Soviet action, was the basis of force doctrine. Collectively, Secretary Brown calls this melange the "countervailing strategy."

As the preceding discussion has sought to show, an ample stock of defense experts were willing vigorously to defend or attack any of these elements. Thus, any position the new administration might take would predictably be simultaneously applauded and condemned, as Carter's early forays in the strategic wilderness demonstrated. When Carter opted against the B-1 bomber, he was praised by the arms control community for his restraint in not adding yet another spiral to the arms race and condemned by others concerned with equivalence for giving away a potent weapons system without getting Soviet concessions in return (both the "bargaining chips" argument and the B-1 controversy are analyzed in subsequent chapters). His proposals to reduce launcher ceilings within SALT II below the limits agreed upon by President Gerald Ford and Secretary Leonid Brezhnev at Vladivostok in 1974 drew concomitant accolades and fire. Within that context, it is not entirely surprising that the new administration has proceeded cautiously in its assessment and that it has not produced a distinctive new policy, but instead an incorporation of the various strands available under the ambiguous countervailing strategy title.

The early activity of the Carter administration has not been to reject any of its doctrinal inheritance, but rather to adapt and shade the meanings of existing ideas. Van Cleave and Barnett would agree that this is an appropriate approach: "The issues are not either-or as frequently suggested, but more or less: more or less enhanced flexibility; more or less selectivity; more or fewer options; more or less emphasis in planning on limited strikes, on restraint, on precision, on military doctrine."[96] J. I. Coffey, overviewing the first year of the Carter administration in a paper at the 1978 International Studies Association convention, concludes that the early results of this reappraisal have not produced significant doctrinal adjustment: "If Mr. Harold Brown's statements reflect the views of the President, as presumably they do, an outsider can only conclude that the Carter Administration has made relatively little change in the requirements for deterrence—and consequently in the requirements for strategic nuclear forces."[97] Although changes have not been dramatic, some incremental adaptation is discernible.

The major spokesman for strategic policy in the Carter administration has been Defense Secretary Harold Brown (who, it will be remembered, had his formative training in the McNamara Defense Department), and the major statement of Carter strategic policy has been the FY 1979 and 1980 *Annual Report* of the Defense Department. Within the context of those documents one can gain some insight into Carter policy on the major issues: deterrence only (assured destruction) versus deterrence plus (assured destruction plus limited options); essential equivalency; and the nature of the Soviet threat.

Assured Destruction and Limited Nuclear Options The Carter administration has taken the basic position of its predecessors on this issue. According to Secretary Brown, "Our second-strike forces must have the capability to execute either a full-scale retaliatory strike or smaller-scale counter-attacks on selected targets while the rest of the force is withheld."[98] The reason for possessing this capability is that "we have to plan our forces on the basis of two assumptions: first, that deterrence might fail; and second, that our forces must be given the capability to frustrate any ambition that an enemy might attempt to realize with his strategic nuclear forces. In other words, we cannot afford to make a complete distinction between deterrent forces and what are so awkwardly called war-fighting forces."[99] Earlier in the *Report,* what is meant by "frustration" is defined largely in terms of the first traditional purpose of deterrence (punishing aggression by denying its success):

> Clearly we must have the military capabilities necessary to persuade an adversary that, regardless of the circumstances, he will:
> —either have to pay a price to achieve his objective that is more than the objective is worth;
> —or be frustrated in his effort to achieve it;
> —or suffer both high costs and frustration.[100]

As mentioned earlier, Brown defines assured destruction as the ability to destroy two hundred Soviet cities (and consequently 38 percent of population and 62 percent of industry). General George S. Brown, chairman of the Joint Chiefs of Staff, described the limited options role of American forces: "Should deterrence fail, these forces must be capable of flexible employment in order to allow conflict termination at the lowest feasible level. If such escalation control fails, then the forces must permit conflict termination on the most favorable terms possible."[101] Sensitive to the Schlesinger scenario of what we would do if the Soviets launched a selective attack against ICBM facilities, Secretary Brown states the specific criteria that "surviving strategic offensive forces" must be able to:

—implement a range of selective options to allow National Control Authorities (NCA) the choice of other than a full-scale retaliatory strike if needed; and

—hold a secure force in reserve to ensure that the enemy will not be able to coerce the United States after a U.S. retaliatory strike.[102]

All of this argumentation bears a strong resemblance to the deterrence-plus position of the Nixon-Ford years and appears to embrace the limited options policy. Probably reflecting his association with the McNamara Pentagon, Secretary Brown expresses serious reservations about the flexibility idea: "None of this potential flexibility changes my view that a full-scale thermonuclear exchange would be an unprecedented disaster for the Soviet Union as well as the United States. *Nor is it at all clear that an initial use of nuclear weapons—however selectively they might be targeted—could be kept from escalating to a full-scale thermonuclear exchange,* especially if command-control centers were brought under attack" (emphasis added).[103] This reservation, reflecting the prevailing view of assured destruction theorists, stands in some contrast to the bulk of the *Report*. It is too early to tell if it will reflect a subtle movement away from limited use doctrine, but it certainly leaves the door open for such an emphasis change.

Essential Equivalence Because it was politically necessary to do so, the Carter administration kept the essential equivalence language in its strategic lexicon (Senator Jackson being in the Senate and knowing future arms control agreements require Senate approval). In the FY 1979 *Annual Report,* Secretary Brown gave his own definition of the concept: "By essential equivalence, I mean a condition such that any advantages in force characteristics enjoyed by the Soviets are offset by other U.S. advantages. *Although we must avoid a resort to one-for-one matching of individual indices of capability,* our strategic nuclear posture must not be, and must not seem to be, inferior in performance to the capabilities of the Soviet Union" (emphasis added).[104] Brown rejects a "hard" and precise comparison of forces on grounds largely derived from assured destruction doctrine (possibly further indicating an intention to move strategic doctrine back in that direction), as shown by two statements in different parts of the *Report:* "In principle, if the conditions of deterrence are present, questions about relative power and influence should not arise as a consequence of comparing strategic forces. . . . Our primary measure of strategic capability is our ability to retaliate after a Soviet first-strike."[105] These statements could have come directly from the mouth of Brown's mentor, former Secretary McNamara, and thus breathe hope or despair, depending on which side of the doctrinal fence one sits. Taken in tandem with his caveat about limiting nuclear engagement, however, there is some indication that MAD remains the ultimate threat within the countervailing strategy structure.

The Nature of the Soviet Threat The new administration has had to face the familiar capabilities-intentions problem regarding Soviet nuclear forces. Brown raises the basis of administration concern regarding Soviet force characteristics: "If these forces are dedicated simply to pure deterrence, or even to large-scale, second-strike assured destruction—conservatively designed—we must still wonder whether they are not excessive in quantity and mismatched in characteristics to either of these purposes."[106] Ohlert, writing in the *Naval War College Review* in 1978, agrees, adding that "Soviet procurements are plainly guided by mission requirements other than those of mutual assured destruction."[107] That American capabilities are equally excessive for the dictates of assured destruction, making it equally appropriate for the Russians to ask the same questions about American intentions, will become clear after inventorying the American arsenal in the next chapter. At the same time, statements of this kind must be viewed in context of an adminstration committed to arms control and arms reduction (as pointed out, positions more compatible with assured destruction than deterrence-plus doctrine) and may have the primary purpose of establishing negotiating positions within the SALT process.

Chapter 4:
American Strategic Forces

The strategic nuclear forces that a nation possesses at any point in time define that nation's capability and thus its ability or inability to exercise strategic options. As has been pointed out, the technology involved in strategic nuclear weaponry has been one of the most dramatic aspects of the nuclear relationship between the United States and the Soviet Union, possessing a dynamism that has frequently and inexorably altered the nuclear balance. This pivotal role of weapons "hardware" thus makes understanding the American strategic arsenal (as well as its Russian counterpart discussed in Chapter 5) necessary to assessing strategic doctrine and its development.

The characteristics a nuclear force possesses result from the R,D,T,&E process, and we begin our investigation at that point. After describing the ways various weapons improvements enter the arsenal, the discussion will move to the actual composition of American forces as defined by the TRIAD concept, including debates about strengths and variations in each of the "legs" and how the various components combine to produce the American deterrent. Because force capabilities gain meaning only in the dyadic relationship between two nuclear parties, the methods by which force comparisons are made, the so-called "bean-counting" process, will be examined. The discussion will then move to two areas related to the computation of strategic balance; the forward-based, theater forces (emphasizing the role of theater nuclear weapons) and active and passive defenses against nuclear attack. The chapter will conclude with a discussion of weapons systems that could enter the strategic inventory in upcoming years.

The Force Development Process Strategic weapons systems enter the nuclear arsenal through a complex process that combines perceived military needs and the often serendipitous pace and direction of development efforts. Because actual force numbers and characteristics that define strategic capability reflect the technological "state of the art" at a particular time, and because current development activity will shape forces in the future, knowing something about how that process works is a precondition to understanding those forces. In turn, understanding the various components of weapons systems and how development efforts in one area reinforce efforts in the others assists in comprehending the entire process.

Components of Nuclear Weapons Systems There are three distinct components of nuclear weapons systems: warheads (explosives), delivery vehicles (launchers), and command and guidance capacities. These components interact with one another so that advancements in one technological element facilitate progress in the others. Nuclear warheads were the first technology developed, and the developmental trend has been toward making them smaller and lighter yet simultaneously more powerful and "cleaner" (reducing the amount of residual radiation emitted by each device). The ballistic missile was the most salient development in delivery vehicles, and efforts at improvement have been largely centered on increasing range on advanced propulsion systems. Command and guidance functions refer to the ability to control the launch of delivery systems and to hit a target. Major advances in this area have been in targeting accuracy and have been linked to burgeoning computer technology intertwined with the use of satellites.

Warheads As mentioned in Chapter 1, the original atomic warheads were primitive compared to present designs: to obtain a blast estimated at fifteen to twenty kilotons (KT, or the equivalent of fifteen to twenty thousand tons of TNT) required a device weighing nearly five tons (ten thousand pounds). By contrast, small warheads based on the deuterium-tritium reaction used in the enhanced-radiation (or neutron) warhead can produce a one-kiloton explosion by using "about six grammes of deuterium and about six grammes of tritium"[1] (or about thirty pounds of material). This progress, accompanied by cleaner warheads, was made possible by advancing knowledge in the area of nuclear physics.

There are three basic forms of nuclear reaction that can be applied to warhead construction: fission, fission-fusion, and fission-fusion-fission.[2] Fission is the simplest and most primitive form. In the fission reaction, energy is released by splitting atoms of unstable elements like U-235 or plutonium using a chemical explosive to achieve critical mass. The resulting reaction literally involves breaking apart these atoms (thereby creating blast effects). In the process, a number of highly radioactive particles are dispersed and find their way back into the ecosystem with long-range effects (residual radiation), which is why these bombs are referred to as "dirty." Fission was the reaction form of the original "atomic bombs," and, because it is a relatively inefficient reaction form, requires larger (thus heavier) amounts of material than other forms.

Fission-fusion is the second reaction form. Energy production is accomplished by fusing together atoms of deuterium and tritium under very high temperatures. The result is to create helium atoms (energy is released when the new atom is created) and the emission of a "fast" neutron. Albert Legault and George Lindsey describe the process: "The main fusible nuclei are the heavy isotopes of hydrogen: deuterium (H^2) and tritium (H^3).... At temperatures of tens of millions of degrees, H^2 and H^3 will fuse, liberating a very fast neutron and a great amount of energy."[3] The process gets the name fission-fusion because a small fission detonation must be created to produce the heat necessary to initiate the fusion reaction (fission serves as a "trigger"). Because the elements created by fusion are not themselves radioactive, fusion produces no residual radiation. The fast neutrons released in the process do create enormous doses of highly lethal initial neutron and gamma radiation, which dissipates rapidly as the process ends, and small amounts of residual radiation are created by the fission trigger. The fission-fusion principle was the basis for the first "hydrogen" or "thermonuclear" bombs and, in sophisticated form, the controversial enhanced-radiation or neutron warhead. Although there are no theoretical limits to the size of a fission-fusion reaction, neutrons are released randomly in it, meaning that not all possible fusions occur in the milliseconds the process takes. As a result, practical warhead designs (small and light enough to be placed on ballistic missiles) are limited to the one-megaton (MT or equivalent of one million tons of TNT) range.

Fission-fusion-fission reactions are necessary for multiple megaton explosions. As the name implies, the nuclear process consists of three stages. First, a fission trigger is detonated. This event triggers a fusion reaction. The device is "coated" with an additional layer of fissionable material (U-236 or plutonium normally) which, in turn, is triggered by the heat and neutron emission of the fusion reaction. The result is a chain reaction producing very spectacular results: the Russians have been reported, for instance, to have detonated an eighty-five-megaton device using this process. The ultimate link in the chain, of course, is fission, meaning that residual radiation is inevitably created in the process.

These sophistications on the nuclear reaction process have created the dual effects of making warheads cleaner (emitting less residual radiation) and more efficient. Fission-fusion bombs are, in a sense, the cleanest variation (the controversy surrounding the prompt radiation effects of the neutron bomb will be discussed in the section on emerging nuclear weapons), and advanced fission-fusion-fission bombs produce considerably less residual radiation than earlier models. Greater efficiency has meant more explosive power from less material. As Polmar explains, "Possibly more significant on a long-term basis, the advent of small nuclear warheads meant that the delivery of weapons of mass destruction would become feasible with unmanned missiles."[4]

Delivery systems The importance of ballistic missiles to deterrence theorizing has already been emphasized and need not be reasserted here. Although relatively primitive in capability compared to contemporary models, the early successes had an enormous impact on destructive capabilities. Since the Sputnik days, there have been improvements in propulsion systems of the rockets themselves and also in the reentry vehicles (RVs) that the missiles carry.

The original ballistic missiles operated on what are now seemingly elementary principles. The basic idea is to propel a missile into the outer reaches of the atmosphere and then to let the forces of gravity bring it back to earth on a course determined by the original trajectory of the firing. Since the reentry course is determined by the arc of the original trajectory, it is possible, by knowing that trajectory, to know at what a given missile is aimed and to plot its course to target (this kind of calculation formed the basis of ABM interception planning).

The early missiles were, by and large, fired with highly volatile liquid propellants that were both bulky and unstable. These characteristics were undesirable because they made handling difficult and dangerous and very large missiles were required to fly long, intercontinental-range missions. At the same time, missiles using nonstorable liquid propellants cannot be fired rapidly, making them vulnerable to preemptive attack. Much developmental work has thus been devoted to producing solid propellant fuels that are composed of relatively inert physical materials and that can be stored safely for long periods on the missile and allow comparatively rapid firing of the missiles.

A related problem occurred when dictates of invulnerability triggered missile storage in underground silos and submarine tubes. Because enormous amounts of energy are required to allow the missiles to take off, launching could badly damage the silos or tubes, rendering them inoperative for extended periods of time (such that they could not be "reloaded" and thus reused). In order to combat the problems associated with what are called "hot launches," a new propulsion technique, the so-called "cold launch," was devised. Basically, this technique involves using tremendous gas pressure literally to "pop" the missile free of the silo or tube, after which the engines are detonated and the missile begins its mission. The result is that virtually no damage is done to the launcher.

The realization that missiles followed fixed and predictable trajectories inevitably led to thinking about how this knowledge could be applied to plotting ways to intercept an incoming missile force. This emphasis on ballistic missile defense (BMD) in turn triggered interest in how to negate the effectiveness of such efforts and major development activity in reentry vehicle technology that has produced some of the most spectacular innovations in nuclear technology.

There were two basic solutions to the potential problems created by successful BMD: multiple-warhead missiles and evasive tactics allowing incoming missiles to deviate from fixed trajectories. The result was to transform a weapon whose basic principle was that of the simple bullet into a sophisticated weapon capable of changing course in midflight and spitting out numerous warheads like a machine gun. The vehicle for launching multiple warheads was the so-called "MIRV bus," and evasion has been accomplished with the MARV system.

Because of its tremendous impact on quantitative arms levels, the story of MIRV is highlighted later in the chapter. Basically, however, the MIRV bus is not unlike the last stage of a space launch. After the missile's engines have projected the rocket to the zenith of its trajectory and the rocket has detached itself, what is left is a device containing several (how many depends on the model) warheads, each with its own small rocket engines. The rockets, in turn, can be fired in various combinations to alter the trajectory of the bus, thereby evading interceptor ABM missiles. The warheads can be fired simultaneously at a target (thereby overwhelming defenses by sheer numbers) or selectively at different targets within a defined range (known as the "footprint"). A later version, the maneuverable reentry vehicle (MARV) essentially adds one more sophistication by putting what amount to tail flaps on the warhead, which allow its trajectory to be altered as it descends toward its final destination.

MIRV's technical sophistication also makes it a highly accurate weapon. As such, it stands as a bridge between delivery vehicle and guidance technologies and as a symbol of the major thrust in weapons systems improvements in the 1970s. As George Rathjens predicts, "Whatever changes in nuclear capabilities we can look forward to are much more likely to be a consequence of improvements in delivery vehicles and in command and control than in warheads."[5] The bottom line of this activity has been increased targeting accuracy.

Guidance and Control Improvements in guidance technology have been made possible basically by two phenomena: through improvements (especially miniaturization) of computer technology that allow small computer-based devices to be placed in RVs; and through interface between computer programs on satellites orbiting the earth and RVs. In turn, sophistication in warhead and delivery system design has facilitated this process: as warheads become smaller and propellant space needs decreased, more space was available for guidance equipment on a strategic vehicle.

The original ballistic missile targeting form, as suggested in the last section, amounted conceptually to little more than the application of artillery firing principles over much longer distances and was thus subject to the vicissitudes of aiming attached to those more conventional means of launch. Early in the nuclear period, the most accurate delivery form was the strategic bomber, originally equipped with simple gravity bombs and later with air-to-surface missiles using radar-based and other homing devices. As pointed out in Chapter 3, early ballistic missiles were highly inaccurate, with CEPs of several miles. From that modest beginning, weapons are becoming available that can be launched several thousand miles to within, in some cases, one hundred feet or less of their destination.

MIRVed weapons are equipped with a computer program designed to guide warheads to a predetermined set of targets. As mission requirements dictate (for instance, what has and has not already been destroyed in previous missile firings), these programs can be changed at the launch site, and, in the case of moving systems like SLBMs, adjustments are made based on the position of the submarine at the time missiles will be launched. This process allows great control over where individual warheads will be fired.

Coming available in the 1980s will be guidance techniques that make this technological level seem almost primitive. These techniques involve using satellites as intermediary devices between offensive systems and command-and-control centers to monitor progress to target and to transmit back to the RV's guidance system course corrections. John L. McLucas describes how this process could be applied to future generations of cruise missiles: "One can conceive of a future fleet of cruise missiles that is tied together through data links at a control center, which keeps track of their position and performance. The center could direct evasive action, keep track of the actions carried out by each member of the fleet, and reschedule another cruise missile for a target when one is shot down or otherwise incapacitated."[6] As we shall see, perfection of the TERCOM guidance system made cruise missiles, the basic design work for which was begun in the 1950s, practical, and it is not too far-fetched to imagine extension of this principle of guidance and control to adjusting incoming trajectories of weapons with MARV capability during their final approaches to target.

Command and control has a second meaning that will be encountered in discussing the relative merits of various weapons systems. That meaning relates to the degree of positive control the National Command Authority (NCA) has over the actual employment of weaponry, particularly after nuclear conflict has begun. This problem is greatest regarding communication with submerged submarines seeking to evade enemy detection, but would affect other systems as well depending on the damage levels that had been sustained at any point in time.

Although these descriptions have a certain science fiction quality, they all point to progress in our capabilities. In a society where scientific progress has a very positive value, it is natural to assume that these advancements equate to improvements in the strategic equation (more precise weapons, after all, being "better" than less precise ones). It is not, however, so obvious nor automatically true that all scientific and technological advance is positive in its impact. It is, for instance, self-evident that one would not have to worry about preventing nuclear holocaust had nuclear fission not been discovered. It is not clear that innovations in targeting accuracy will not fundamentally alter the basis of nuclear deterrence as we know it, at worst, or more minimally undermine arms control efforts as currently defined. Because the pace of technological development often moves toward unpredictable ends with unanticipated impacts and in ways not always entirely controllable, it is necessary to amplify remarks made in Chapter 1 about the technological process.

The R,D,T,&E Process The key to understanding why it is difficult to control with precision the outcomes of technological developments in weaponry is to realize that such efforts are creative. Like most scientific inquiry, it may be possible to identify the area in which one proposes to seek new knowledge; but, by definition, if one needs to investigate a subject in the first place, it is because one does not know what there is in the subject to find out about. As a result, the outcomes of scientific inquiry are necessarily unpredictable, and the ways in which new knowledge can be applied to solving particular problems are largely a matter of the limits of human ingenuity.

In an insightful article, Harry G. Gelber classifies the component processes that are contained in what he calls the "innovatory process":

> I propose to try to discuss it in terms of the following classification: discovery, invention, innovation, development, technological drift, and testing. By discovery I mean the discovery of new ideas about the physical universe or its organizing principles. By invention I mean the creation of new things, which may or may not be based on new discoveries. By innovation, I mean the creation of new systems from previous inventions or known components, or improvements in products or processes whose essential characteristics are known and remain unchanged.[7]

To this impressive list of ways in which new ideas may enter the strategic inventory might be added the notion of technological transfer: the process by which the results of any ideational process in an independent, and possibly unrelated, area are applied to solving a weapons problem. The MIRV bus, for instance, was originally designed as part of the space program to allow multiple satellite launches at different orbits from a single rocket, and the idea was borrowed and applied to solving the weapons systems problem posed by the Soviet ABM program.[8]

The notion of technological drift, a concept akin to the concept of technological determinism identified in Chapter 1, is basic to understanding why it is difficult to regulate and control technological outputs.

As Gelber describes it: "Technological drift is the useful term, coined by J. P. Ruina, to denote processes which do not, or need not, result from the decisions of higher authority or the formal R&D machinery at all. It involves minor improvements in systems and components, to cope with minor snags which have appeared during development, marketing, deployment, or servicing, but whose cumulative impact can amount to or make possible substantial system changes."[9] This means that basic scientists and engineers are inveterate tinkerers dedicated to producing the best product they can. Thus, if something does not work precisely the way it is supposed to function (for example, a warhead will not land as close to a target as it is supposed to with adequate consistency), then the answer to the problem is to make whatever modifications are necessary to produce the desired outcomes.

The basic nature of the technological process provides a momentum of its own, as Gelber explains: "Research and development programs are increasingly expensive and produce even more complicated and costly types of equipment. Worst of all, at both the scientific and economic levels, the R&D process has a built-in tendency to proliferate."[10] The more one inquires into a problem area, in other words, the more one discovers and the more one produces. In turn, discovery leads to new directions in research, and inventions beg to be made more sophisticated. This self-generating process is reinforced because the whole R,D,T,&E process contains its own prisoner's dilemma: "Fear of an enemy technological breakthrough is one of the driving forces which justifies arms research . . . levels."[11] As we shall see in Chapter 6, this prisoner's dilemma is very difficult to overcome, because R&D efforts do not become "public" (that is, capable of being observed, normally by some form of spying) until they reach the testing stage. By that point in the development process, if one side has sole possession of a particular innovation, it will have an advantage for some years.

The results of the R,D,T,&E process enter the strategic arsenal in the form of specific and individual weapons systems, and it is for those specific purposes that they are designed. As was pointed out in Chapter 1 and as Colonel Richard G. Head reiterates, these innovations have a cumulative effect on strategic doctrine: "One of the results of the doctrine of quality and technological substitution has been a certain tendency for technology to drive both strategy and doctrine."[12] To see how weapons inventories cumulatively define capability and thus form parameters around doctrinal choice requires looking at the individual components of the American strategic arsenal and how those components orchestrate in overall strategic posture.

TRIAD: The Shape of American Forces American strategic capability is defined by the concept "TRIAD": the possession of three separate strategic systems, each of which is capable of carrying out an assured destruction retaliatory attack on the Soviet Union following any conceivable Russian preemptive attack. The concept, as Newhouse puts it, "came about more by accident than by design"[13] in the sense that the development of three separate systems was the result more of technological availability than careful preplanning. The strategic bomber force was in place as nuclear weapons were being developed; the ICBM, when it became available, was a natural delivery vehicle; and the SLBM flowed from application of missilry research to naval purposes.

Just as force configuration was an evolutionary process, so has been the process by which the concept has been defended. The virtues of TRIAD have been gradually realized as experience with the concept has accumulated. John F. McCarthy summarizes the prevailing view that has emerged about TRIAD over time: "There is nothing magic about the number three, but the three present strategic systems work in concert to minimize their individual weaknesses and maximize their individual strengths. This synergistic effect is one of the bases for having more than one type of system."[14] As experience and conceptual comfort with the configuration have grown, the TRIAD concept has become chiseled in granite as official policy. The 1979 Department of Defense *Annual Report* formalizes the virtue of a three-legged strategic force:

> The TRIAD gives us the necessary diversity. No potential enemy could expect to destroy the ICBM's, alert bombers, and on-station SLBM's in a simultaneous attack. In most circumstances, at least a large fraction in two out of the three components of the TRIAD would survive. The enemy's defenses would then have to deal with weapons approaching him from different directions, at varying speeds, and along a variety of trajectories. There would be no way for him to escape without unacceptable damage.[15]

Secretary Brown's statement neatly summarizes the second-strike capability that TRIAD provides: the combination of systems is highly invulnerable and thus survivable; and the multiple defensive problems it creates for a potential enemy guarantee its penetrability to target. Defenses of TRIAD stress these two aspects.

James L. Garwin states the basic case regarding the invulnerability of forces configured in the TRIAD: "The most obvious benefit of the interaction of these three types of weapons in the Triad is the dilemma they present to an attacking enemy."[16] The reason for this "dilemma" is inherent in the nature of the relationship between the three components and the ways one would have to attack each. To be effectively disarming, a simultaneous attack would have to be launched and completed, and, as Polmar says, "simultaneous attacks against all three would be impossible."[17]

Without going into systems detail reserved for succeeding sections, the basic attack problem can be examined. To disarm the strategic bomber and ICBM legs of TRIAD would require an attack coordinated such that incoming forces arrived on target at very close to the same time. This is an exceedingly difficult task, because the Soviet ICBMs aimed at U.S. missile fields would require long enough to reach their destinations to allow alert bombers to take off and thus avoid destruction on the ground. The ability to destroy the ICBM fields is today problematical in the first place, since they are stored in concrete-reinforced silos that require almost a direct hit (which the Soviets do not have the current capability to make) to disarm, leaving, at worst, a residue of missiles for retaliatory purposes, along with the alerted bombers (to destroy them on the ground requires firing "depressed trajectory" SLBMs that can reach their targets faster than the planes can take off, another capability the Soviets have not perfected).

Even if a successful attack on the two legs can be projected by waving the worst-case planner's pen, the SLBM retaliatory force would remain. This third leg of TRIAD is universally agreed to be the most invulnerable force component, with no short-term prospect of developing an effective preemptive capability against it in the foreseeable future on either side. Since first-strike capability requires a high degree of certainty about the results, the TRIAD configuration is maximally deterring, as Brodie explains: "If the enemy is obliged to think of not one but two or more kinds of hard-core forces which he would have to eliminate in a surprise attack, his uncertainty of success is disproportionately enhanced."[18]

The characteristics of the weapons in the TRIAD are also justified because of their penetrability in a retaliatory strike. The heart of the argument is that each would attack in a different way: the ICBMs along fixed trajectories; the SLBMs with fixed trajectories but launched from locations the enemy could not identify in advance and enhanced by MIRV characteristics; and bombers attacking from a variety of directions along nonfixed routes at varying altitudes. Defending against an attack by any single component presents staggering problems, and the three force components in combination greatly exacerbate the situation. As McLucas explains, "The existence of the triad compels our adversaries to defend against all three components, thus reducing the defenses which each one would otherwise encounter."[19]

As should be obvious from the above, much of the TRIAD's virtue is its planned redundancy: any one of the force components has the capacity alone to carry out the assured destruction mission. Such configuration and capability are justified on the basis of the offensive and defensive problems the combination creates, as noted above, but an additional reason for having all three is as a hedge against one component becoming vulnerable or ineffective. Archie L. Wood summarizes this justification: "This diversity gives added assurance that United States retaliatory capabilities will not be undermined by some unforeseen development in opposing forces or by some unanticipated weakness in our own."[20]

It is within this context of mutual reinforcement that the three force components must be examined. In the case of each strategic system—the ICBMs, the SLBMs, and the "air breathing" forces—the primary criterion for assessment is the degree to which the weapons category contributes to the overall TRIAD second-strike capability. Thus, in addition to describing each system's basic characteristics, each must be assessed in terms of the dual second-strike criteria of vulnerability and penetrability.

The Intercontinental Ballistic Missile (ICBM) Force The land-based ICBM force consists of 1,054 missiles in hardened missile silos in the western United States. Pending any alterations in force composition arising from a SALT II agreement, this force consists of 54 Titan liquid-fuel missiles, 450 Minuteman II, and 550 Minuteman III missiles. Titan is the oldest and largest component, carrying the only multiple-megaton warhead in the American strategic arsenal. It is also the least effective missile in terms of accuracy, reliability, and vulnerability (because of its liquid propellant) and is a likely victim of launcher reductions included in SALT II. The Minuteman II is also a single-warhead missile (carrying a weapon in the megaton range) that uses a solid propellant. The heart of the ICBM force is the Minuteman IIIs. Launching a reentry system containing up to three MK12 reentry vehicles, it is the most accurate missile in the U.S. inventory, with a CEP in the 0.2 mile range. The Minuteman III force has recently been "modified with a command data buffer system which permits retargeting of a single mission in about thirty-six minutes."[21] Projected improvements in warhead design (the Mark 12A) would give these missiles a targeting accuracy approaching "hard-kill" capability.

Once the pride of the strategic nuclear forces, the ICBM's glamour has come under question in recent years as targeting accuracies increase on both sides. The projection that missile accuracies will approach true counterforce proportions in the coming years has raised the question of ICBM vulnerability. There is general agreement that fixed-site, land-based missiles are potentially the most vulnerable American strategic system to preemptive attack should counterforce capability be achieved (or once it is achieved). At the same time, there is considerable disagreement about how much of a problem this projected vulnerability creates and even more disagreement regarding what should be done about it.

Polmar states the basic problem facing the ICBM force: "The principal threat to the Minuteman-Titan ICBMs is the potential threat of a preemptive attack by Soviet land-based ICBMs against the US weapons while the latter are still in their underground silos."[22] Joseph Kruzel compares the vulnerability of the various systems and explains why

ICBMs are the most vulnerable: "Of the three types of strategic offensive forces—ICBM's, SLBM's and heavy bombers—fixed land-based ICBM's are potentially the most destabilizing. With satellite photography, each side knows the precise location of the other's ICBM's. This knowledge may under certain conditions provide the incentive for a pre-emptive strike."[23] In simple terms, a nonmoving target the location of which is known is easier to hit than a moving target the location of which is not known. Polmar cites an additional problem: "ICBMs are susceptible to failures. For an ICBM to function perfectly, hundreds of different parts must work within limited tolerances; although the percentage possibilities of individual failures are small, in the aggregate they significantly reduce the effectiveness of an ICBM force."[24]

ICBM vulnerability is a serious matter, because invulnerability is one of the two basic requirements of an effective second-strike weapon. To the extent that the viability of the TRIAD concept is dependent on the effectiveness of each part, it is a matter of still greater concern. There is agreement that the ICBMs are potentially the least invulnerable TRIAD member, but there is not universal agreement that this means they are becoming dangerously vulnerable and thus an Achilles' heel rather than a strong TRIAD leg. Among those who most strongly argue that the vulnerability hue and cry is a bogus issue are John D. Steinbruner and Thomas M. Garwin: "The strategic forces of the Soviet Union, even if very aggressively modernized, will not be sufficient to threaten with true credibility the decisive destruction of the United States Minuteman force."[25] Basing their analysis on sophisticated computer simulations of possible Soviet attack scenarios against American missile fields, they reach their conclusion largely on the basis of the so-called "fratricide effect": "Since the land-based missiles of the United States are concentrated on a few bases, the Soviet Union could not conduct an approximately simultaneous attack with several warheads per silo without having some of the attacking warheads either deflected or effectively destroyed by previous explosions in the attack sequence—explosions resulting from earlier attacks on the same silo or on nearby silos."[26] They further argue that such an attack would require precision in timing well beyond reasonable expectations and would deplete the Soviet arsenal badly, leaving the United States with an overwhelming advantage in surviving nuclear power. Secretary Brown also muses on the extent of the problem in the 1979 *Annual Report:* "In recognizing that the MINUTEMAN vulnerability problem is a serious concern for us, we also realize that the Soviets would face great uncertainties in assessing whether they would have the capability we fear—and still greater uncertainty as to its military or political value."[27]

Although there is less than general agreement about the degree of ICBM vulnerability and the extent to which that vulnerability is or will become a strategic problem, there is consensus that, at a minimum, the

Minuteman force is the least invulnerable TRIAD leg. To deal with this situation, a number of solutions have been proposed. Newhouse summarizes the most common remedies: "If Minuteman is the least stable system in the arsenal, what, if anything, can be done about the Minuteman force? The options have been four: (1) dismantling all or part of it; (2) superhardening the silos; (3) defending it; (4) putting it on mobile launchers."[28] Without attempting to exhaust the argumentation supporting these positions (some of which will recur in other contexts), each can be examined briefly.

The idea of doing away with the ICBM force has most frequently been associated with supporters of the SLBM force. As Kruzel points out, "Neither side appears anxious to build additional fixed land-based ICBM silos,"[29] because, as targeting accuracies improve, they become an increasingly dubious investment. In the SALT I negotiations, this position was formalized in U.S. proposals for the so-called "one-way freedom to mix": the notion that, in configuring total launcher numbers agreed to, each side would have the option to dismantle ICBMs in favor of SLBMs, but not vice versa. This position arises from the general agreement that SLBMs are the most invulnerable force and that heavy investment in submarine-based missiles maximizes survivability and discourages preemption, whereas the ability to calculate destruction of ICBMs encourages first-strike notions. Opponents of this option point to the distinctive advantages of the TRIAD configuration discussed earlier, arguing that the absence of ICBMs would make Soviet defense plans easier and would make technological breakthroughs in other areas such as antisubmarine warfare (ASW) much more traumatic.

The second option is "superhardening" the missile silos. The U.S. force has already been protected by reinforcement of silos with concrete to withstand blast overpressure of approximately one thousand pounds per square inch (or psi, which is one thousand times normal atmospheric pressure), but proponents of this position (largely in the air force, which is responsible for the ICBM force) favor dramatic increases that would require an incoming missile virtually to land on the silo cover to disable it. A number of methods have been proposed, including siloing in deep caves, but the position has not met with great support largely on the basis of cost and the ability to overcome any improvements by increasing accuracy. As Colin S. Gray puts it: "The arithmetic is not promising for its effectiveness (e.g. an increase in nominal silo blast resistance from 2,000 to 3,000 psi . . . can be offset by an improvement in missile CEP of 60 feet)."[30]

The third option is to provide some form of active defenses for the ICBM force. The Nixon ABM proposals to place defensive missiles at the ICBM fields in Grand Forks, North Dakota (an installation that was built and later dismantled), were the most prominent attempt to provide this kind of protection, but foundered because of expense and the belief that such defenses could be overcome by increases in the sophistication of offensive forces (such as MIRVs). Garwin, however, maintains that the technology is currently available for effective ICBM defense through "pebble-fan" devices. An an example of possibilities in the R&D area, he explains: "Both airburst and groundburst low-drag RVs attacking silos can be countered by a pebble-fan projection—an east-west line 300 meters north of each Minuteman silo and 300 meters long, consisting of propellant emplaced in the ground to project a curtain or fan of pebbles up to 300 meters in the air."[31] Garwin contends that the effect of the contact between the pebbles and the incoming reentry vehicles would be to cause premature detonation of the warheads, thereby avoiding silo destruction.

The fourth option is to deploy mobile land-based missiles and arises because the main source of ICBM vulnerability is that they are fixed targets at which to aim. Gray lists some of the suggested ways mobility could be accomplished: "The major options include the following: dispersed shelters; garage mobility; buried trench mobility; off-road random crawling; road and rail mobility; canal-deep mobility; lake bottom mobility."[32] The more "public" forms of mobility, such as placing missiles on railway cars or river barges, have largely been discounted because of likely negative public reaction to their visibility. Development work continues on a mobile missile employing the so-called racetrack technique, the controversial Missile X or MX. As will be pointed out in Chapter 7, any movement toward mobility has enormous arms control implications.

A far more serious potential problem of ICBM vulnerability is its effect on firing doctrine. As Fred C. Ikle states: "Those branches of American and Soviet military services that believe they must continue to press the case for land-based missile forces will—because of the increasing vulnerability of these forces—be even more tempted to stress launch-on-warning as an option."[33] The prospect of launch-on-warning (firing missiles when an incoming attack force is detected to ensure they are not destroyed in their silos) is chilling: it provides an incentive to fire before having absorbed an attack, thus undercutting second-strike doctrine. This, in turn, can cause an "itchy finger on the nuclear button" that could lead to premature launching of a missile force to ensure that it gets fired at all (see arguments in Chapter 2).

The Submarine-Launched Ballistic Missile (SLBM) Force The United States launched the first nuclear-powered ballistic-missile submarine, the USS *George Washington*, in November 1960. Since that time, the early Polaris-class submarines have been improved with the introduction of the Poseidon system, and warhead design has been upgraded by adding

MIRV capabilities in the Poseidon C-3 warhead, which "has a reported range of 2,500 nautical miles and can deliver up to 14 RVs, each with a yield of about 50 KT."[34] The formidable SLBM force of forty-one submarines will be augmented by the new Trident submarine, which will, according to the Defense Department, achieve initial operating capability (IOC) in 1981[35] and will gradually replace earlier Polaris models.

The great advantage of the nuclear submarine force is its invulnerability to preemptive attack. Powered by quiet (and thus very difficult to detect with sonar) nuclear engines, these vessels can remain submerged in the millions of square miles of the earth's oceans for extended periods and can fire their missiles from under water at targets thousands of miles away. Although submarines experience greater command-and-control problems then the other TRIAD legs (the watery medium that protects them can also make communication difficult) and historically have had greater CEPs than other systems (which, of course, is not critical in countervalue targeting), their survivability makes them the ideal second-strike weapon. As such, they form the "backbone" of TRIAD and support the other legs. As Stefan H. Leader and Barry R. Schneider put it: "The notion that the Soviets would risk a surprise nuclear attack on U.S. land-based missiles, leaving the formidable (4,500 nuclear weapons) U.S. sea-based forces untouched, makes little or no sense."[36]

The introduction of SLBMs has had a dramatic effect on the nuclear balance. According to the French strategist Pierre Gallois, "But, today, the submarine armed with Polaris missiles has deprived thermonuclear war of its last semblance of rationality."[37] Congressman Thomas J. Downey agrees that, particularly as the range of sea-launched missiles increases, there is a stabilizing effect on the nuclear system: "Long SLBM range is stabilizing for both sides, since it not only increases submarine survivability but also increases the warning time if these missiles are used."[38] Moreover, submarine forces are particularly well suited for the United States, as Newhouse notes: "The sea for the United States is a congenial strategic environment. Its boats have easy access to their operating areas and can remain on station for extended periods. It has maintenance facilities at Holy Loch, Scotland, and Rota, Spain, and these permit as many as thirty boats on station at any given time."[39] Although the Rota base was closed in 1979, the increased range and capability of Trident will compensate for the loss, leaving Newhouse's point unblemished.

The already formidable U.S. SLBM force will be greatly enhanced with the addition of Trident/ULMS (underwater long-range missile system), which comprises both a new submarine and advanced missiles. Phil Stanford describes the Trident submarine: "It will be 535 feet long—almost as long as two football fields; 43 feet high—as tall as a four-story office building; and will weigh 18,000 tons. It will carry 24 missiles, each with a range of 4,500 miles. These could eventually be replaced by a later version with a range of 6,000 miles."[40] Based in Washington (despite the objections of some local residents), the new submarines will have increased speed, longer periods of operation before maintenance is necessary, and more silent operation than the Polaris/Poseidon fleet, all enhancing their second-strike capability.

The increase in missile range is the most dramatic feature of the new system. The Trident I missile, "now in production after a successful test flight program,"[41] will have a 4,500-mile (7,400-kilometer) range, as noted, thereby greatly enlarging the range from which the missile can be fired and thus adding literally millions of square miles to the effective "on station" area in which the submarines can operate. If authorized for testing and production (President Carter declared a moratorium in 1978), Trident II's 6,000-mile range would greatly add to that operating area and in the process even more greatly complicate defensive measures. Moreover, Trident missiles are compatible with Poseidon submarines, and, according to Secretary Brown in the FY 1979 Defense Department *Annual Report,* "TRIDENT I missiles will be backfitted into twelve POSEIDON submarines to support a deployed level of up to ten Trident submarines,"[42] thereby enhancing the Polaris/Poseidon force as well.

Although "most authorities appear to agree that Soviet ASW is not now, nor for the near future, capable of destroying a significant portion of the Polaris/Poseidon submarine force,"[43] it is the development of effective antisubmarine warfare devices that most threatens the SLBM force. Kosta Tsipis points out the enormous logistical difficulty that ASW represents: "To threaten the Polaris fleet as a deterrent force, an opponent's ASW forces must be capable of destroying within a few minutes all the Polaris submarines while his offensive missiles attempt to wipe out the land-based Minuteman ICBMs, the SAC bombers, . . . and all the aircraft-carrier-based bombers capable of delivering nuclear weapons."[44] Of all these complex problems, submarine destruction is the most difficult. There are essentially two barriers to ASW—tracking and actual attacks on the vessels. Those who seek to keep the SLBMs as invulnerable as possible have consequently favored limitations on development in each of these areas.

To track nuclear submarines requires monitoring the location of the boats, an extremely difficult task given the speed, high level of silence, range, and length of time they can stay submerged. One possible method of accomplishing this task (which neither the United States nor U.S.S.R. employs) is installation of a complex network of electronic monitoring devices in the world's oceans that could detect the presence of submarines. The practical limitations on such a system arise from diffraction of sound and electronic waves in water and the vast areas in which the boats operate (with Trident II missiles, for instance, the submarine would be on station virtually anywhere it was). Hedging against the possibility of a breakthrough in this area, however, Tsipis suggests, "The first measure would be to forbid the installation in the oceans of the world of large acoustical active or passive arrays capable of tracking missile-carrying submarines at large distances."[45]

Even if the submarines can be tracked, destroying them presents additional difficulties. They operate too deep for traditional antisubmarine devices to be effective against them and are too fast to be pursued by conventional diesel-powered submarines. Thus, the only effective weapon against them would be specially designed nuclear-powered "hunter killer" submarines. George W. Rathjens proposes a ban on such weapons: "One could envisage the negotiation of limits on number of highspeed, nuclear-powered attack submarines—the only ones that could trail the missile launchers—without limits being imposed on conventionally-powered, or lower-speed, nuclear-powered attack submarines or on other kinds of equipment useful for ASW in a long war, such as aircraft, helicopters, and mines."[46] Kruzel agrees with this distinction between conventional (tactical) and anti-SLBM (strategic) ASW: "For arms control purposes an ideal curb on ASW would restrict the deployment of strategic ASW forces but not affect research, and at the same time leave tactical ASW completely unlimited."[47] Some observers maintain that conversion of all submarines to nuclear power will render the strategic-tactical distinction meaningless.

Therefore, unless radical breakthroughs in ASW occur, the SLBM force will remain the most invulnerable component of American strategic forces. R&D in the ASW area is not highly developed, and the movement to ban ASW deployment acts as a further barrier. At the same time, the new Trident system promises to compound the difficulties of overcoming SLBM invulnerability. As a result, this combination of invulnerability and penetrability (as pointed out earlier, SLBMs are difficult to intercept because one does not know in advance the direction from which they are coming) makes the SLBM force an ideal weapon for implementing a second-strike strategy and should continue to play that role for the foreseeable future.

The "Air-Breathing" Forces With the introduction of the cruise missile into the American strategic arsenal (the first weapons may be delivered in 1980), what formerly was known as the strategic bomber leg of TRIAD has been officially given the cumbersome designation "air-breathing" forces to cover both weapons encompassing the nonballistic component of American nuclear might (the name refers to the fact that both have engines that use oxygen from the air). The present force is composed of an aging B-52 bomber fleet that is declining in numbers and some FB-

111 fighter-bombers capable of reaching the Soviet Union from forward bases in Europe and will be augmented by the cruise missile.

The evolving air-breathing force has been controversial in recent years. The B-52 force reached its maximum size (630) in 1963 and since 1967 has physically declined as attrition has forced retirement of some planes while others were lost in Vietnam War combat (the Department of Defense *Annual Report* for 1980 listed the current force at 316 aircraft[48]). Declining numbers and projections about when the B-52s would "wear out" have sparked an intense debate about the need for a new strategic bomber temporarily quelled by President Carter's decision to build cruise missiles rather than the B-1 bomber, but the issue is likely to resurface. Although the exact configuration of air-breathing forces is in an evolutionary state, "it is clear that in the future the United States will have a mixed force of cruise missiles plus short-range attack missiles and gravity bombs carried on some penetrating aircraft."[49] Each aspect of this evolution merits individual attention.

The Strategic Bomber That the strategic bomber force is an important part of American strategic forces is undeniable. As General Brown points out, "Fully generated, the bomber force carries approximately one-third of all deliverable US strategic nuclear weapons and half the total megatonnage."[50] Bombers also have advantages over other delivery systems. As McCarthy points out: "Bombers fulfill missions which missiles cannot. Their viability, controllability, versatility, and effectiveness are necessary to our strategic deterrent forces. The United States can only achieve its policy of essential equivalence if bomber payloads are included in our strategic forces."[51] Robert Berman agrees, adding that bombers contribute to the complicating role of the TRIAD concept for the Russians: "Manned bombers increase the number of targets the Soviets must aim at and they increase the size and complexity of an attack against which the Soviets must defend."[52]

The obvious advantage of the manned bomber is command-and-control: it is the only strategic vehicle that can be called back or retargeted after it has been launched (although, as pointed out, other systems may have this latter capability in the future). As a weapons system carrying out American strategic policy, however, the bomber force must be assessed in terms of the major criteria for second-strike weapons—survivability and penetrability.

The major problem involved in bomber survivability is the danger that the planes could be destroyed on the ground by a Soviet preemptive attack. To deal with this problem, 30 percent of the B-52 force is kept on ground alert that would allow it to take off between the time incoming missiles were first observed on advanced radar and when they landed. The greater danger, however, is Soviet perfection of so-called "depressed trajectory" techniques for their SLBMs, allowing incoming missiles to reach their targets in five or fewer minutes (as opposed to twenty-five to thirty minutes for ICBMs fired from the Soviet Union). The United States has responded to the danger of preemption partially by dispersing B-52 squadrons to airfields around the country, and there have been various proposals, such as installing quick-start pyrotechnic cartridges in the engines, to decrease takeoff time.

The problem of penetrability arises because the Soviet Union has the most sophisticated air defense system ever devised. Although U.S. attack plans include clearing "corridors" of Soviet air space with ICBMs and SLBMs to minimize those obstacles, the ability, particularly of the aging B-52s, to reach targets is debatable. Defenders of the bomber concept point to advantages of aircraft that assist penetration. Comparing bombers to other force components, McCarthy observes that "ICBMs have known, fixed trajectories, while the other two systems can approach from different directions, the bombers having the advantage of the least predictable approach."[53] A second advantage is that defense is complicated by the variety of ways that bombs are delivered: "Some B-52s and the FB-111s will penetrate to targets using SRAM [Short Range Attack Missile] and gravity weapons, while other B-52s will launch cruise missiles."[54] It has also been pointed out that the B-52s had a remarkable 97 percent survival rate in missions during the Vietnam War against extremely sophisticated defense.

The basic problem many observers see in the air-breathing leg is the age of the B-52s. As Polmar says, "The existing B-52 force is predicted to 'wear out' in the 1980s, when the newest of the aircraft will be over 18 years old."[55] The B-52 is a subsonic aircraft, and although the "newest" models (the B-52 G and H series) can fly at very low altitudes, it is difficult for the aircraft to evade detection. Moreover, this inability to "hide" from Soviet radar will be decreased as the Soviets deploy "lookdown, shoot-down" capability: the radar capacity to track an incoming plane from a high altitude (by satellite or observation aircraft) that is flying below the effective profile of conventional radar (about two hundred feet). The United States has this capability, but to date the Soviets have only recently demonstrated it over land masses. Secretary Brown, however, projects that Russians eventually will perfect the technology: "The main long-term effort is likely to go into the development of a true look-down radar and the shoot-down capability to go with it. Such a combined capability could become operational as early as the early 1980s, although it is more likely to take place later."[56] By the time the Soviets achieve this capacity, not only will the B-52s be old, but their penetrability will be dubious.

These concerns raised the question of the need for a replacement aircraft for the B-52 and centered on the B-1 bomber, a supersonic aircraft capable of flying at very high speeds below normal radar detection and equipped with very sophisticated penetration aids. Extensively developed and tested, the aircraft was scrapped in 1977 by the

Carter administration on cost-effectiveness grounds: the cost of each plane was estimated at over $100 million, giving the projected deployed force of 244 a cost of nearly $25 billion. Combined with fears that Soviet look-down–shoot-down capacity would eventually render the B-1 vulnerable as well, the decision was reached not to build it. As Secretary Brown explained in the FY 1979 *Annual Report*: "Given assumptions as to scenario, the task to be done, costing ground rules, and other factors, coupled with assumptions regarding Soviet defenses that, if anything, favor the B-1 over the cruise missile, a B-1 force that would have had a capability equal to B-52s with cruise missiles would have been about 40 percent more expensive."[57] Although these costs figures are hotly contested (possibly most eloquently by McCarthy), the B-1 appears to be a dead letter for the present time, if for no other reason than that the start-up time to resume production (reassembling the old production crew or training a new one) would literally take years. Most attention among supporters of a new bomber has shifted to the FB-111H, a "stretched" version of the FB-111 equipped with B-1 engines and capable of flying missions of around three thousand miles (approximately the same as the new Soviet "Backfire" bomber discussed in Chapter 5).

The Cruise Missile The concept of the cruise missile is not particularly new: the prototype was the V-1 "buzz bomb" developed by Nazi Germany in World War II and, as pointed out in Chapter 3, development work on the weapon began in the United States in the 1950s. Major breakthroughs in guidance technology brought this weapons system to the forefront, although its emergence "was largely a historical accident, reflecting the technological advances and bureaucratic interests of each of the armed services."[58] As Robert S. Metzger points out, its juxtaposition as an alternative to a new manned bomber and its potential effects on arms control have made it a controversial addition to the American arsenal: "Originally justified as a dispensable bargaining chip of little military value, the small, pilotless, low-flying winged aircraft . . . has emerged in recent years as one of the most controversial weapons systems under development by the United States."[59]

As Metzger indicates, the cruise missile is a small (the air force version, for instance, is about twenty-five feet long) missile that flies at subsonic speeds (five to six hundred miles per hour) toward its target at extremely low altitudes (less than two hundred feet, thus below the effective level of ground-based radar). As General Brown describes it, "Powered by a small turbofan engine, the missile will incorporate an inertial guidance system updated by Terrain Correlation Matching."[60] This terrain matching system allows the missile to fly at very low altitudes and to adjust to topographical changes (it does not run into the sides of hills). Culver explains the effect this innovation has had on the missile's attractiveness: "Terrain-comparison guidance allows pinpoint accuracy, sufficient to destroy many hard targets. Miniaturization and a highly efficient engine permit a long range in this hard-to-detect missile. These achievements surprised those who had cancelled U.S. cruise missile problems in the early 1960s."[61]

Although the air force's air-launched cruise missile (ALCM) has attracted the most attention, the navy has developed a sea-launched model (SLCM) and the army a shorter-range tactical version. The ALCM and the SLCM are of strategic significance, and Secretary Brown explains that the decision on which model to deploy has not been reached: "Since we must be certain of its success, I believe we must, as a matter of prudence, maintain both the Air Force air-to-ground cruise missile AGM-86B (ALCM-B) and the air-launched version of the Navy TOMAHAWK cruise missile . . . until a competitive flyoff determines which missile can best be employed in the air-launched mission."[62] Because cruise missiles are relatively small and light, they can be carried on any number of delivery platforms: SLCMs fit on standard attack submarines[63] and the decks of surface ships; the ALCM can be fitted onto existing bombers (Culver estimates that "a force of 70–120 modified bombers could carry 1,400–2,400 long-range cruise missiles; specially designed wide-bodied cruise missile carriers would probably hold more than twice that number"[64]) or on 747-type aircraft or C-135 transports or on a specially designed STOL-type (short takeoff and landing) aircraft. Because cruise missiles are relatively cheap by strategic standards (about $1 million per missile), it is possible to produce large numbers comparatively economically.

Cruise missiles have a number of virtues. Their basic strength is that "nuclear-armed cruise missiles . . . will be capable of attacking strategic targets in the Soviet Union."[65] Current models have a range of about 2,700 kilometers, meaning they can be launched by aircraft in a so-called "standoff" mode: the plane can launch the weapon outside Soviet air space. Because they relieve B-52s of the need to penetrate the Soviet Union, it is argued that cruise missiles lengthen the effective life of the B-52 (it does not have to meet as exacting performance requirements if it does not have to penetrate), thereby obviating the need for a new bomber. Moreover, the guidance system makes it extremely accurate. Ohlert, for instance, says, "Accuracy for the cruise missile over a 2,000 km range has been estimated at under 100 feet."[66] Garwin estimates a CEP of less than 100 meters over a 2,700–3,600 km range.[67] As we shall see, such accuracy adds greatly to the lethality of these weapons.

Cruise missiles, it is also argued, are highly supportive of American strategic doctrine on two grounds. First, they are alleged to complicate defensive problems for the Soviets:

> To defeat a mixed force of penetrating bombers and ALCM-carrying aircraft, the Soviet Union would have to spend an estimated $10 to $50 billion to develop new weapons, including new long-range interceptor

aircraft to attack ALCM carriers, a sophisticated look-down radar to distinguish low-flying cruise missiles and bombers from ground clutter, and a vast array of terminal defenses using numerous ground-control radar, low-altitude surface-to-air missiles (SAMs), and radar-directed rapid firing anti-aircraft artillery.[68]

Moreover, the ALCM's small size gives it a small radar "profile," and, according to Ohlert, the weapon "provides better coverage of certain elements of the target structure"[69] than other weapons systems. Second, the weapons are suitable only for second-strike use, making them strategically stabilizing. As Ohlert explains: "The cruise missile represents the ideal in offensive weaponry for a second-strike-oriented nation. Its slow flight speeds preclude its use as a first-strike weapon, while its high prelaunch survivability deters an opponent's first-fire decision."[70]

The second-strike advantages of cruise missiles have a Damoclean arms control impact, however. Their survivability largely results from the fact that they are small and thus can be concealed, meaning they cannot be destroyed because the enemy does not know where they are (in retaliatory strikes, this has the added virtue that the enemy does not know where they are coming from, thus exacerbating defense problems). These virtues become vices in any arms control agreement that involves monitoring numbers of weapons. As Richard Burt points out: "Possibly more intractable is the difficulty of monitoring an agreement that includes cruise missiles; their potential for deployment in large numbers aboard stand-off aircraft or attack submarines would seem to make adequate verification impossible. The adaptability of cruise missiles to both strategic and tactical roles makes it increasingly difficult to differentiate between weapons suited for one or the other."[71] Leader and Schneider point to an additional problem posed by SLCMs: "The sea-launched cruise missile could significantly complicate arms control efforts because of the difficulty of identifying delivery vehicles that might carry it."[72] These problems are raised here to indicate that cruise missiles do complicate strategic concerns. In terms of impact on the stability of the strategic system, these difficulties will be assessed in Chapter 7.

Strategic Bean Counting Force capabilities are, of course, principally meaningful in comparison with the strategic arsenal of a potential adversary, thus creating a legitimate basis for trying to equate force levels. This exercise, known as bean counting, has gained particular prominence in an era of Soviet-American parity, arms control, and essential equivalence.

Comparing Strategic Arsenals As argued in Chapter 3, the difficulty in comparing American and Soviet strategic forces is that in many ways they are not comparable. By some measures, the United States has advantages over the Soviet Union, and by others the relationship is reversed. Polmar summarizes the contemporary balance: "In the mid-1970s, the United States had advantages with respect to strategic weapons in (1) MIRVs and reentry vehicle technology, (2) guidance technology, and (3) nuclear weapons technology, that is, smaller weapons with a better yield-to-weight ratio. At the same time, Soviet advantages were in (1) numbers of launchers, (2) missile payloads, and (3) ongoing missile development programs."[73] Since all things are rarely equal in importance, the question of what these advantages mean to an overall comparison of force capabilities has spawned a series of means to create comparability. Stephen A. Garrett offers a cautionary note about the exercise: "Data on the Soviet-American military balance is so complex and comprises so many categories and non-comparable elements that it is child's play to pick and choose among the facts to prove almost anything one wishes to."[74] With that caveat in mind, we can examine the five most common static measures of force comparability: strategic launchers; numbers of warheads; raw megatonnage; equivalent megatonnage; and lethality.

Strategic Launchers The simplest and most straightforward means of counting capabilities is simply to add up the number of strategic delivery vehicles each side has and compare the results. This means involves computing the number of bombers capable of carrying out strategic (homeland-attacking) missions, the number of ICBM silos, and the number of SLBM tubes on the strategic submarines. This has been the method used in making comparisons within SALT, largely because it is a comparative method amenable to so-called "national technical means of verification" (see Chapter 6). Using figures from *The Military Balance, 1975–1976*, the Soviet Union had an advantage by this measure of 2,537 to 2,147 (in conformance with SALT I limits affirmed by the 1974 Vladivostok Accords, the Soviet launcher number is to be reduced to the agreed limit of 2,400 by retiring outdated ICBMs).[75]

Although it is the simplest form of comparison, counting launchers can also be the most deceiving. Because of the ability of a single launcher to carry more than one warhead and because the count is not sensitive to the size or accuracy of warheads, this simple count does not yield much precision about the deadlines of an arsenal. Bombers, after all, carry numerous warheads, a single launcher (airplane) can carry a very large number of cruise missiles, and MIRVed missiles carry numerous warheads of varying yield. Such factors led Congressman Downey to conclude that "aggregate numbers of missiles is such a poor measure of capability that we must recognize this as political window dressing necessary to impress the impressionable."[76]

Numbers of Warheads A second bean-counting method that takes into account multiple warheads and thus reflects the number of explosive

American Forces

devices that can be delivered to targets involves, quite simply, counting the number of warheads in each arsenal. Using 1975 data as illustration, Paul Walker presents the data given in Table 4.1 to show this comparison.[77]

Table 4.1: Comparative Warhead Numbers

	U.S.			U.S.S.R.	
Launchers	Warheads per Launcher	Number of Warheads	Launchers	Warheads per Launcher	Number of Warheads
54 Titan	1	54	80 SS-8	1	80
450 MM II	1	450	288 SS-9	1	288
550 MM III	3	1,650	970 SS-11 (I)	1	970
96 Polaris A-2	1	96	60 SS-13	1	60
176 Polaris A-3	3	528	40 SS-11 (III)	3	120
384 Poseidon C-3	12	4,608	10 SS-17	4	40
421 B-52	8	3,368	10 SS-18	1	10
		10,754	52 SS-19	6	312
			528 SSN 6	1	528
			322 SSN 8	1	322
			40 Bison bombers	4	160
			100 Bear bombers	6	600
			40 Backfire bombers	6	240
					3,730

This measure, which the Defense Department refers to as "force loadings," reflects the considerable American superiority in multiple warhead technology and shows the United States with about a 3 to 1 advantage. This measurement is quite sensitive to deployments of new weapons: if one were to replace the ten Polaris submarines carrying A-2 missiles with Trident submarines (which carry 24 tubes) equipped with the Trident II missile (16 MIRVed warheads each) and convert 100 B-52s to carrying 50 cruise missiles each, there would be a net warhead increase of 6,508, for a total U.S. inventory of 17,162 warheads.

Although Leader and Schneider deem force loadings "perhaps the best measure of strategic power,"[78] it has limits. Warhead counts omit any reference to the size of warhead(s) on each launcher and thus their destructive capability. This is an important consideration, particularly since the Soviets tend to build much larger missiles than does the United States.

Raw Megatonnage The most basic means of calculating the amount of destructive capacity a nation possesses is through the gross (or "raw") megatonnage of its weapons. This calcuation is done by multiplying the megatonnage of each type of warhead in each arsenal by the number of warheads in the arsenal and summing the products. Walker[79] provides this comparison (see Table 4.2).

Table 4.2: Megatons Comparison

	U.S.	U.S.S.R.
ICBMs	1,715	8,028
SLBMs	374	850
Bombers	2,889	950
	4,978	9,828

The reason for this nearly 2 to 1 Soviet advantage is their preference for very large missiles that consequently have a larger payload or throw-weight (defined as "the weight of that part of a missile above the last boost stage"[80]) than their American counterparts. Simply put, larger throw-weight means that more explosive can be put in each missile, resulting in larger yields as measured in megatonnage. Although the U.S. lead in warheads largely compensates for this advantage currently, a number of observers are concerned with the consequences for the strategic balance at the point Soviet MIRV technology equals that of the United States. Paul Nitze states this problem: "If we do not add new strategic programs to those which are now programmed, the U.S. will end up the ten-year program with a half to a third of the throw-weight devoted to multiple warheads, or MIRV."[81]

Megaton Equivalents Raw megatonnage figures do not provide an entirely satisfactory measure of destructive capability, however, because "beyond a certain point, larger nuclear bombs do proportionately less damage than smaller weapons," and "beyond a certain yield, no more explosive yield is needed to destroy a given target."[82] The result is the need for a measure to standardize and thus make comparable the blast effects of different weapons. That unit is the megaton equivalent: "The unit of area destruction—the megaton equivalent (MTE)—is defined as the area destructive power of a one-megaton weapon. The megaton equivalent of a given weapon is derived by taking the two-thirds power of its yield in megatons."[83] Total MTE for a country is derived by applying the basic formula $MTE = NY^{2/3}$, where N is the number of weapons in a class and $Y^{2/3}$ is the derivation of yield expressed in megatons (or fractions thereof) and adding the products for each weapons class. Leader and Schneider, using 1974 data, show the United States with a 4,456 to 3,847 MTE advantage;[84] the Walker data[85] result in a 4,350 to 4,150 Soviet advantage. Recent MIRVing of additional Soviet ICBMs has increased Soviet MTE advantage.

Megaton equivalencies add sophistication to calculations by compensating for inefficiencies in yield for large weapons, but they do not capture the entirety of destructive power. What is missing is the accuracy of the weapons to target: it matters little how large or small the yield of a weapon is if it misses its target.

Lethality To standardize the effects of accuracy (or inaccuracy) on the likelihood an incoming RV will destroy the target at which it is aimed, a measurement of lethality (sometimes referred to as the "Kill" or "K-factor") has been introduced into the strategic lexicon. The K-factor is the most sophisticated measure of nuclear effect and is derived from the relationship of MTE to accuracy.[86] So defined, the K-factor is expressed in the formula

$$K = \frac{NY^{2/3}}{(CEP)^2}$$

The K-factor, as examination of the formula reveals, places enormous emphasis on accuracy (because CEPs are squared as the divisor). Since the divisor is measured in fractions of a mile and becomes smaller when squared, the result is that a much smaller warhead more accurately delivered can have the impact of a much larger missile less accurately fired. Walker applies these calculations to his data and concludes with an American lethality advantage of 25,000 to 7,000.[87]

There is an old saying that "figures can lie, and liars can figure" that is not an inappropriate way to leave the discussion of strategic bean counting. As should be evident from the foregoing, there are measures of the strategic balance available to support any position one wants to take: for those who find the balance unfavorable, throw-weight and launcher statistics are available; if one finds the balance equitable or favorable to the United States, warhead counts, megaton equivalents, and lethality factors can be confidently cited to reinforce the position.

The precision that reducing these measures to numerical representations appears to create is also deceiving. The bottom line in all these calculations is how much devastation using these weapons would produce: how many people would be killed; how many cities leveled; how many missile silos destroyed. Nuclear weapons carry out their deadly missions in a variety of ways: through heat from the fireball, enormous atmospheric pressure caused by the explosion (blast overpressure), and initial (prompt) and residual radiation. Each of these effects works in different ways and maximizes different targeted objectives, and how these occur is also dependent on where and how the weapons are detonated. The following quotation gives some idea of the interaction of factors:

> If the attack strategy were to emphasize immediate or "prompt" casualties and damage to the industrial facilities of a local target area, then air blasts

would probably be used, since blast and thermal radiation effects would be great. However, if the purpose were to threaten large numbers of total casualties over time, then surface blasts would probably be used since they would maximize local fallout, which can be very lethal to both human population and food and livestock supplies.[88]

These concerns are not reflected in force comparisons, and estimates of weapons effects are also subject to a great many unpredictable factors such as wind and rain at the time of detonation.[89]

Thus, the bean-counting exercise has an elegance that is to some extent undeserved, and arguments based on one measure or another should be viewed with some caution. The exercise, however, does clearly show the dramatic impact multiple warhead missiles and increasing missile accuracy have on destructive arsenals. Since MIRV is a highly accurate weapon in addition to carrying multiple warheads, it has especially contributed to the lethal spiral, thus warranting special attention.

The Multiple Independently Targetable Reentry Vehicle (MIRV) The advent of MIRV is important for at least three reasons, as suggested in various places in the discussion. First, it is the single most obvious and prominent example of "technological determinism" or "technological drift" of any weapons system, as those concerns have been raised in Chapter 1 and earlier in this chapter. Second, MIRV has done more to increase the destructive spiral of the arms race than any other single weapons innovation, as described directly above. Third, the accuracy improvement process begun by MIRV, and particularly sophistications of its basic design, promise the potential radically to alter the basis of nuclear deterrence and the nuclear balance we have known. For these reasons it is worthwhile to examine MIRV by asking three questions: why is there MIRV? how does it work? and what has been its impact?

The Origins and Evolution of MIRV The basic answer to the question of why there is MIRV is that the technology was available to develop and build it, so we did. Tammen, whose work is the most thorough and comprehensive treatment of the subject, summarizes this genesis: "MIRV was a logical offshoot from a broad technological base accumulated in the penetration aid and space programs. It is a classic example of technology shaping a strategic decision."[90] MIRV was thus a result not of a policy decision to create such a system, but an extension of other technological activity that probably was impossible to avoid without concerted effort. As Herbert F. York describes it: "All the technologies needed for MIRV had other reasons underlying their development, and so MIRV would very likely have emerged at about the same time if the need for ABM penetration had not been perceived until much later, and possibly even if it had not arisen at all."[91]

American Forces

Once there was a MIRV, there was the concomitant need to justify its deployment. Happily for those who had designed the system, its emergence happened to coincide with developmental ABM programs on both sides. Not knowing what Soviet intentions or capabilities were (or were likely to become) in the ABM field, worst-case analysis dictated a crash program of countermeasures for ABMs, collectively known as penetration aids. Some of these were designed simply to confuse enemy defenses, but MIRV offered an alternative method: "Multiple warheads . . . penetrate defences simply by saturating or exhausting them."[92]

The Soviet ABM threat turned out to be abortive, as Tammen explains: "It must be concluded that MIRV development was stimulated by an intelligence coup (the U-2), justified by intelligence failures (Tallinn Line and Leningrad ABM), and developed contrary to intelligence estimates (no nationwide system and Moscow ABM slowdown)."[93] The U-2 coup, of course, was the discovery by overflight that the Russians were building an ABM system around Moscow (ABM systems are distinctive largely because of the size and configuration of their accompanying radar). The intelligence failure was the interpretation that the Tallinn Line and Leningrad systems were ABM precursors to a nationwide ABM system (they proved to be air defense systems against conventional attacks), and even when this error was corrected, MIRV deployments continued.

It is possible that the U.S. MIRV program had something to do with the failure of Soviet ABM systems to develop, because, as Malcolm W. Hoag points out, MIRV deployments "make a 'thick' ABM defense of Soviet cities distinctly less effective, and therefore less likely to be deployed."[94] Tammen rejects the ABM justification more as ex post facto rationalization. He says: "MIRV was developed out of a complex relationship of factors, the vast majority of which were domestic. The few inputs that were related to Soviet weapons programs were so muted in influence as to be of little importance."[95] Rathjens agrees with this assessment, ascribing the decision more to the momentum of bureaucratic influence: "It has become increasingly clear that strategic weapons programs have the bases of their support in a multiplicity of interests and that, once underway, expedient and changing rationales will likely be used to sell them."[96] These conclusions are difficult to argue with because MIRV programs have continued and expanded in the absence (indeed the treaty-based prohibition) of ABMs.

The Mechanics of MIRV Without going into great technical detail, the heart of the MIRV system is the so-called MIRV "bus," that part of the missile remaining after the last booster engine has been expended and decoupled. Equipped with power sources that allow repositioning and carrying multiple warheads that can be fired independently of one another, MIRV warheads can be fired at a number of targets within a prescribed range of territory (the so-called "footprint," for most systems an area about seventy by thirty miles).

The basis for MIRV technology comes from the space industry's attempts to develop the means to launch multiple satellites at different orbits, as mentioned earlier. The solution to this problem (first applied when the Transit II A and NRL Solar Radiation Satellite were launched from the same rocket on June 22, 1960[97]) required development of the Agena restartable engine which, as Tammen explains, had direct applicability to MIRV, and specifically MIRVing the SLBM force: "The significance of the Agena restartable engine lies in its conceptual similarity to the MIRV bus. As a second stage, Agena essentially was a dual purpose carrying vehicle used not only for escape velocity, but for orbital changes. As such, it bore marked similarities to the MIRV carrying bus with its own self-contained propulsion system. Lockheed's Agena experience paid dividends in their Poseidon MK-3 program."[98] The "self-contained propulsion system" on MIRV in turn is based on the use of vernier engines configured in a triangular pattern on the cylindrical bus. By firing one or more of the small rocket engines for a predetermined period, the pitch and yaw of the vehicle can be altered, allowing the bus to align itself properly to fire the projectile to its target. According to Tammen: "If the vernier engine could realign the reentry vehicle after dropping off the third stage, it could just as easily, with a little more energy, computers, and guidance, control the trajectory of a bus carrying multiple reentry vehicles. Each alteration in bus trajectory would make warheads individually guided and thereby increase target coverage."[99] In addition to increasing the number of targets that can be brought under attack with a single booster, the ability to make in-course adjustments has markedly increased the accuracy with which warheads can be delivered to target. Therefore, it is no coincidence that the Soviets lag behind the United States in both MIRV and targeting accuracy and that those fearful of Soviet strategic advantage when they have MIRVed to allowable limits under SALT (given the Russian throw-weight advantage) phrase those doubts in terms of counterforce targeting possibilities.

The Effects of MIRV MIRV has been an intensely controversial addition to strategic arsenals. Polmar states what is probably the prevailing sentiment that "multiple warheads have been a severely destabilizing influence on strategic weapons development."[100] The weapon has also been defended, as Tammen summarizes: "MIRV can also be interpreted as a stabilizing influence on U.S. defense policy. First, MIRV relaxes the pressure for additional deployment of new launchers which look provocative. Second, it is said that MIRV redresses the Soviet advantage in throw-weight or megatonnage. Third, MIRV reinforces the logic of deterrence by making it senseless to contemplate attack."[101] The basis of controversy has arisen from the two major contributions attributable to MIRV: warhead proliferation and improved targeting accuracy.

Arguments about warhead proliferation point to the vast quantitative leaps in destructive capacity MIRVs have allowed. Increases could be justified as necessary to saturate a thick ABM defense, but when this defense failed to materialize, the United States found itself deploying systems grossly in excess of assured destruction requirements. For instance, even were it necessary to target five warheads from each leg of TRIAD to guarantee the assured destruction of the two hundred cities cited by Secretary Brown in the 1979 *Annual Report* of the Defense Department, the United States would need only three thousand warheads, or less than 30 percent of the inventory figures cited earlier.

Given this growth in weapons systems well beyond stated doctrinal needs and the fact that the Soviets knew the Tallinn and Leningrad systems were not ABMs (making that justification false), it was unavoidable that they would question why the United States was deploying these systems. Although the simple reason probably was that the United States deployed MIRV because we had it, Moulton states, "It is difficult to imagine how Soviet leaders could avoid interpreting the MIRV decision as a deliberate American escalation of the strategic arms race."[102] The result was an action-reaction phenomenon producing a spiral in the arms race. "Virtually any technological development by one side can be matched by the other, and so it has been with multiple warhead technology."[103] Moreover, this spiral was possibly avoidable through the SALT process had the United States made different decisions about testing MIRVs, as will be shown in Chapter 6.

Warhead redundancy combined with increased missile accuracy to create a second MIRV impact. Georgi Arbatov, director of the USA Institute in Moscow, offers this observation: "Military planners immediately began to look for a rationale to justify the 'redundant' warheads—in other words, to look for new targets for them. That was how the idea of 'counterforce' use of strategic weapons was revived."[104] This interpretation comports at least temporally with the facts: conclusive MIRV testing was completed in 1968, and MIRVed warheads began to enter the arsenal in 1970. "A reduction in megatonnage and accuracy . . . inhibited MIRV from becoming an efficient hard-target system in its early stages,"[105] but by the time then Secretary Schlesinger announced the limited option doctrine in 1974, with its counterforce implications (see Chapter 3), many of these difficulties had been overcome.

The counterforce aspect of MIRV creates conceptual problems. "MIRV technology is destabilizing in that it offers the possibility of one ICBM force destroying another of similar size in a preemptive strike," and, as a consequence, "in a MIRV world, sea-based systems or mobile land-based systems appear to be the only alternative available to insure a stable assured destruction capability."[106] The first point speaks to the ICBM vulnerability problem and thus the continuing viability of the TRIAD. The mobile missile alternative poses the same kind of arms control problems associated with the cruise missile.

Other Force Concerns: Nuclear Defense and Theater Nuclear Weapons In describing the array of programs and force configurations that contribute to the American deterrent posture, two additional, if basically unrelated, matters need to be considered. The first is defensive measures that would be available to the American population, productive base, and retaliatory forces in the event nuclear war broke out. The second is the theater nuclear weapons (TNW) force arrayed in Europe for defense of the NATO alliance in the event of a massive Soviet conventional or nuclear (or combined) attack.

Defenses against Nuclear Attack At the outset it should be pointed out that the possibility of effective defense against nuclear weapons has not been a prominent feature in American strategic thinking. Tied closely to beliefs about the ability to limit nuclear war below the cataclysmic level of general homeland exchange, the ability (and thus the meaningfulness of the attempt) to try to save some particular portion of the civilian population and urban industrial base has received attention when limited war notions have been raised (notably during the early Kennedy days of controlled response and recently), but have generally suffered from disrepute otherwise. Skepticism about defensive possibilities arises because "the doctrine of the primacy of offense over defense has been the cornerstone of U.S. strategic policy in the nuclear age."[107] This belief arises both from convictions regarding the qualitative changes in warfare occasioned by thermonuclear weapons (as discussed in Chapter 1) and the observation that, in nuclear weapons technology, offensive developments invariably outstrip defensive innovations (for example MIRV and ABM systems).

With this somewhat somber introduction, the nature of American defense efforts can briefly be described (as we shall see in Chapter 5, nuclear defense occupies a more prominent place in Soviet planning). Basically, defenses against nuclear (or conventional) attack are classified in two ways: active and passive measures. As Brodie explains: "Defense against hostile weapons in all forms of warfare . . . has always basically consisted of two things: first, measures to reduce the number of enemy weapons dropped or thrown or to spoil their aim by hitting the enemy as he attacks (i.e. active defenses); and second, preparation to absorb those weapons that actually strike home (i.e. passive defenses)."[108] The most prominent active defense efforts of the United States have been air

American Forces

defenses against conventional attack (interceptor aircraft and associated equipment such as radar) and the ABM program. The damage-limitation, counterforce aspects of the limited options doctrine also, in a convoluted way, can be argued as an active defense measure. Passive defense has included civil defense efforts to protect the civilian population and efforts to protect retaliatory forces.

Active Defenses The United States does not possess impressive active defenses, particularly when compared to Soviet air defense. American active air defenses are minimal in size and capability, reflecting to some degree a recognition that the Soviets have not invested heavily in conventional delivery systems (the deployment of the new Backfire bomber, discussed in Chapter 5, has raised some question about U.S. air defenses). At the same time, U.S. ABM programs beyond the R&D stage were ended when the Grand Forks facility was dismantled in 1976.

ABMs have been the most controversial active defense measure. Essentially, there are two types of ABM systems: so-called "thick" and "thin" configurations. Thick systems (the original Johnson Sentinel proposal falls into this category) involve two types of missile: large payload interceptors capable of interdicting an incoming force high above the atmosphere (the U.S. Spartan, for instance, used a five-megaton weapon that was exploded to create such intense blast effects as to destroy an incoming force); and smaller interceptors capable of attacking missiles that penetrate the deep shield (the American Sprint missile being capable of cold launch within fifteen seconds and carrying a warhead in the one-kiloton range).[109] Arrayed in large numbers, the thick system was to intercept a massive Soviet attack. Thin systems, on the other hand, consist only of short-range interceptors that protect specific targets against small attacks, as opposed to the area defense aspect of thick systems (the Nixon Safeguard system was justified as a thin or "light" ABM shield).

The effectiveness of ABM systems has been the most contentious issue about them, and the debate revolves around what kind of system one is proposing. Thick systems have always been the more controversial, largely because of the deep interdiction requirement (which is based on computer calculation of known, fixed ICBM trajectories that can be evaded) and because American and Soviet arsenals are so large that saturation is almost assured. Thin systems are not so controversial: they are designed against accidental launches or launches by minor nuclear powers (who do not possess sophisticated penetration aids as components of their arsenals). Against small attacks, thin defenses stand a reasonable chance of success. In the light of potential nuclear proliferation, Donald G. Brennan, a leading advocate of defensive measures, sees a place for thin systems: "This examination of multipolar strategic problems does reinforce my belief that active defense—missile and air defense—is important for the United States. The greater the number of nuclear countries that could attack the United States in a crisis, the more it seems to me we should have defenses capable of making a genuine difference."[110] Hoag believes a thin system would improve deterrence against the Soviets by forcing them into massive, and suicidal, attacks rather than more limited attacks: "The first area function of a thin U.S. ABM defense is to force the Soviets to consider so high a level of nuclear counterdemonstration against U.S. territory, using high-confidence penetration aids in a heavy attack, that they will be deterred from such attacks even if there are theater nuclear attacks abroad."[111] The contrary view is that expenditures on even a thin system are extraordinarily high and hardly justified given likely effectiveness.

There is a related theoretical concern in terms of the impact of ABM on the balance of terror. Knorr states the problem: "The balance would gain in stability if ABMs were deployed only for protecting the retaliatory forces, thus rendering them less vulnerable than they might be otherwise, and adding assurance to each antagonist's ability to devastate the other. Yet if they were installed to afford substantial protection to populations, then they would—even were they far from perfect—introduce new uncertainties, and thus tend to destabilize the condition of nuclear terror."[112] This argument relates back to the hostage effect central to MAD doctrine: anything that contributes to retaliatory invulnerability (and thus assuredness of destruction) is stabilizing to the system; anything that allows contemplation of surviving a nuclear attack weakens that most basic disincentive of knowing one is a hostage.

Whether ABMs or any active defenses are effective is also a matter of disagreement and concern. Given the preponderance of offensive weaponry, Brodie cautions that "one must have extraordinary faith in technology, or a despair of alternatives, to depend mainly on active defenses."[113] Whether one could, or should, have that kind of faith is debatable, since technological efforts in the area of missile defense have been modest compared to the emphasis on offensive weaponry. In the future, there is the possibility that breakthroughs in "exotic" defense systems "like lasers, charged particles, or electromagnet waves"[114] could make defense attainable (although these development at least implicitly are banned by SALT I). The uncertainty regarding ABM efficacy leads Jack Ruina to conclude, somewhat wearily. "It would be I think a tremendous step . . . if we could just eliminate ABM from the nuclear equation entirely, so that nobody could even confuse the issue about what small amount . . . ABM might mean or might not mean."[115]

Passive Defenses Like their active counterparts, passive defense systems can be designed to protect either counterforce or countervalue targets. Brodie lists these objects: "Passive defense falls into one or more of three patterns: first, measures involving concealment of the target (including devices to deceive or confuse enemy missiles or bombers); second, measures involving some form of armoring or hardening of the target (e.g. provision of underground shelters); and third, measures involving . . . dispersion or mobility of targets."[116] American passive defense efforts have largely focused on protecting retaliatory forces and can be viewed in Brodie's categories.

Concealment of countervalue targets is problematical at best (how can one hide a factory?) and probably impossible in an open society. Concealment of retaliatory forces is prohibited within SALT. Cruise missile and mobile missile options have aspects of this form of passive defense, however. Hardening has been applied to ICBM missile silos, and the United States does have a limited civil defense program including shelter provision in urban areas. These latter programs have been largely unsuccessful, and "public apathy or skepticism which for years has effectively inhibited shelter programs of any kind would first have to be overcome"[117] before such programs could be effective. Dispersion of B-52 bombers has been undertaken to reduce their vulnerability, but population dispersal (such as evacuation and industry relocation) plans have never been implemented. Given the great speed with which a nuclear attack could occur and the enormous devastation attendant to it, there is considerable disagreement about whether attempts at passive defense, particularly of civilian populations, is worth the effort.

Theater Nuclear Weapons (TNWs) The most controversial part of the American nuclear arsenal is the nearly 7,000 theater or tactical nuclear weapons deployed in Western Europe. The controversy surrounding these forces arises from three basic sources: the effect their employment would have on Europe itself; the danger that their use would inevitably entail an escalation to general strategic war; and whether those parts of the arsenal capable of reaching Soviet territory should be counted against American strategic armaments totals in the SALT process.

Nature and Purpose The United States maintains at least 7,000 nuclear warheads in the European theater, which are opposed by Russian tactical nuclear forces estimated at around 3,500 warheads. Jeffrey Record describes the characteristics of these weapons: "The range (and for aircraft, the operational radius) of U.S. TNW varies from eight nautical miles for the 155-MM howitzer to about 1,500 miles for the F-111. Warhead yields extend from approximately 0.1 kiloton, deliverable by both artillery and aircraft, to over one megaton deliverable only by aircraft."[118] These weapons are dispersed throughout the NATO area, with the largest concentrations (particularly of the battlefield nuclear artillery shells) in West Germany, which would be the focal point of a Warsaw Pact invasion of Western Europe.

The purpose of these forces is generally to compensate for NATO's conventional force inferiority compared to the general purpose forces of the Soviet Union and its Warsaw Pact allies. For a variety of reasons that do not concern us here, the NATO alliance forces have always been markedly smaller than their Warsaw Pact counterpart (by any comparative force measure), meaning that deterrence of a Soviet invasion and thwarting such an invasion should deterrence fail have dictated integration of TNWs into NATO defensive plans. According to Secretary Brown, TNWs "are intended to deter theater nuclear attacks, to deter conventional attacks in conjunction with conventional forces, and, if necessary, to respond as appropriate in the event of attack."[119]

In the FY 1979 *Annual Report,* Secretary Brown elaborates on the ways these forces might be employed:

> We believe that three types of theater nuclear options are necessary. . . . They are:
> —Limited nuclear options designed to destroy selectively a number of fixed enemy military or industrial targets and, in so doing, to demonstrate a determination to resist attack by whatever necessary means.
> —Regional nuclear options intended, as one example, to destroy the spearheads of an attacking enemy force before they could disrupt the front and achieve a major breakthrough.
> —Theaterwide nuclear options directed at counter-air and counter-missile targets, lines of communication, and troop concentrations in the first and follow-on echelons of an enemy attack.[120]

Record describes the two types of TNWs in the force that would carry out these missions:

> 1. Battlefield or short-range theater nuclear weapons are . . . designed to influence directly the outcome of combat by destroying engaged enemy military forces. . . .
> 2. Long-range theater nuclear weapons are . . . designed to influence indirectly the outcome of combat by interdicting the movement of enemy troops and supplies to and from the battle area or by destroying rear area installations—airfields, marshaling yards, supply depots—vital to the enemy's prosecution of hostilities.[121]

Because of conventional inferiority, TNWs are considered vital to a credible NATO deterrent; and because NATO forces might well be defeated rapidly in strictly non-nuclear combat, they would be used early in a European conflict. Record argues that the willingness of the United States to use these weapons has deterrence-enhancing value: "TNWs represent for most Europeans, particularly West Germans, visible evidence of . . . commitment as well as symbol of U.S. willingness to cross the atomic threshold relatively early on, with all the risks of escalation such a decision would entail."[122] Rosecrance, however, points out a basic enigma between the deterrent value of TNWs and overall strategic stability: "In pure strategic terms, Western conventional weakness required superiority at the nuclear level if deterrence in Europe were to succeed.... But this superiority, so necessary to deterrence in Europe, meant that there was no stable bilateral deterrent relationship between the United States and the Soviet Union."[123] Rosecrance's observation, relating to the difference in credibility of nuclear threats in situations of American strategic superiority and parity, is but one source of controversy about TNWs.

The Effect on Europe Europeans and Americans tend to have different perspectives on TNWs, and Europeans themselves are divided about whether having nuclear weapons stationed on their soil for their defense is a virtue or not. The basis of disagreement and concern arises from differing perspectives on the strategic-tactical weapons use dichotomy and the probable consequences for Europe of nuclear defense.

The use of nuclear weapons in Europe may be tactical for the United States, but it would certainly be strategic for Europeans. As Wayland Young (Lord Kennet) observes, "A Londoner is perfectly certain that the immolation of London is not a matter of tactics, but entirely 'strategic,' whatever the range of the attacking missile."[124] It is also widely recognized that nuclear defense of Europe would in all likelihood be Pyrrhic. As Brodie puts it: "Whatever their ignorance about the specific effects of nuclear weapons, the Europeans have learned quite enough about war generally; they also know enough about how much worse it would be with wide-scale use of nuclear weapons to be clear that they cannot afford flammable irredentas."[125] The problem is analogous to the ex post–ex ante dilemma identified in Chapter 3: although the nuclear threat may be necessary to make deterrence credible, actual nuclear use might be entirely inappropriate. As Barry R. Schneider summarizes, "It would seem to make little sense to use theater nuclear weapons extensively to defend an ally if, in the process, the use of such weapons destroyed the people, the cities, and the countryside of that ally."[126]

The size, nature, and distribution of American TNWs adds fuel to these concerns. The sheer number of warheads guarantees that, if they are used, large-scale damage would occur, and this perception is exacerbated by the fact that "present yields are too large to permit their actual use without risking the wholesale destruction of the very Europe that NATO was founded to protect."[127] Moreover, this force is stationed widely across Europe, and it is not unreasonable to believe that, prefatory to an invasion, the Soviets might launch a nuclear strike against them to minimize the amount of damage they would do, with consequent widespread collateral damage. Under these combined circumstances, it is not surprising that the nuclear disarmament movement is particularly strong in Europe.

The Escalatory Ladder Possibly the most basic controversy about TNWs is whether their use could be limited to the theater level, or whether their use would lead to an escalatory process eventuating in general exchange between the United States and the U.S.S.R. As was pointed out in Chapter 2, there is no empirical evidence that directly supports either side of this argument. Given the tremendous uncertainty about the ability to limit weapons use, there is considerable debate, very similar to that surrounding the limited options doctrinal debate, about whether plans to use these weapons lowers the firebreak (with its potential escalatory possibilities). This problem also relates to the definition of "strategic" adopted in this work: if one concludes that TNWs fall in the category "use of weapons which significantly threaten homeland exchange," then the distinction between theater and strategic weapons is largely meaningless (see the definitional discussion in Chapter 2).

There is a considerable body of opinion supporting the notion that tactical weapons use would significantly threaten escalation. Gallois, for instance, argues: "No one agreed—or yet agrees—on the eventual role of these tactical nuclear weapons. *But nearly all the experts agree that their wartime use would lead to immediate escalation.* Once engaged in an exchange of short-range nuclear missiles, the belligerents would probably delve into their arsenals for increasingly more powerful weapons rather than accept defeat" (emphasis added).[128] Because both the United States and U.S.S.R. define Europe as vital to their national interests, of course, the ability to accept defeat short of escalation is problematical.

There are two other issues that relate to the escalatory process: use of so-called "mininukes" and the effect of strategic parity on the European theater. Controversy over the mininuke (very small, low-yield nuclear devices, in some cases with a yield smaller than large conventional bombs) centers on whether they lower the nuclear threshold. Record summarizes this concern: "A . . . major argument against miniaturization is that the deployment of mini-nukes . . . would make nuclear conflict more likely by lowering the nuclear threshold to the point where a conventional and a nuclear response would become indistinguishable. The very 'usability' of the weapons, particularly if they were governed by a doctrine of immediate employment, would obliterate the political constraints on crossing the threshold."[129] Schneider agrees, adding that the small size (and thus effect) of these weapons "may create the illusion that a limited nuclear war can be fought."[130] Recently, the controversial W70-3 Lance enhanced-radiation warhead and the new Pershing missile have served as the focal point of debate on this issue.

Finally, there has been conjecture about the effect Soviet strategic parity has had on European nuclear use. As mentioned in Chapter 2, some Europeans (notably the French) concluded at the time the Soviets attained the capacity to attack the United States with nuclear weapons that the United States would no longer honor alliance commitments (thereby justifying an independent nuclear force). This conclusion has been attacked, and there is disagreement on the effect of parity. On the one hand, Burt maintains that parity increases the likelihood of escalation: "In a time of strategic parity, and when fixed-site ICBM's are increasingly vulnerable to attack, the danger of escalation to full-scale nuclear war is implicit in the link between nuclear use in the European theatre and the American strategic arsenal."[131] Record, however, disagrees: "To Europeans and to many Americans, parity has served to reduce the credibility of the strategic deterrent and, by extension, of ideas that the use of TNW might pave the way for its employment."[132] The inability to demonstrate the truth of either assertion, of course, simply adds to the uncertainty about the effect.

Forward-Based Systems (FBS) The existence of "a 'fourth nuclear force' consisting of forward-based fighter-bombers . . . capable of nuclear attacks on the Soviet Union"[133] in Europe has long been an irritant to the Russians and has been a recurring issue within SALT. The Soviets' position is described by Brennan: "The Soviet negotiating position has been that these are American forces capable of attacking targets in the Soviet homeland, and are therefore 'strategic' for purposes of the Strategic Arms Limitations Talks, while the Soviet IRBM's deployed in European Russia, which are obviously capable of attacking strategic targets in NATO Europe, cannot attack targets in the American homeland, and are therefore not 'strategic.' "[134] The United States maintains that the FBS are not strategic weapons "since the primary mission of the forces involved is not to deliver nuclear weapons against the U.S.S.R., and since the interests and, in some cases, the forces of other nations are very directly involved."[135] The United States maintains that these systems are more appropriately considered in the context of Mutual and Balanced Force Reductions (MBFR) negotiations, a position which, we shall see in Chapter 6, the Russians have apparently accepted in SALT II. Kruzel, however, maintains that these forces were the most vexing issue in SALT I and that the inability to agree on FBS "more than anything else, is why there is an interim agreement rather than a permanent treaty on offensive forces."[136]

Future Weapons Systems As has been stated repeatedly, nuclear weapons technology has been extremely dynamic. Despite the dampening effect of the SALT process on launcher numbers, technological development has continued through replacement and improvement of existing weapons systems. The most prominent technological trend has been in the area of targeting, as Steinbruner and Garwin point out: "At the frontiers of the American weapons program, technologies are now being projected which promise drastic increases in the most sensitive variable—accuracy."[137] This factor has already been manifested in the form of the ICBM vulnerability debate and will undoubtedly have an even more dramatic influence in the future.

The burgeoning of weapons technology has, curiously, been stimulated by an artifact of the SALT process: the bargaining chip. Defined as a projected or actual weapons system which one is willing to trade away for a concession from the Soviets or to improve a bargaining position within SALT, the R&D for a number of new weapons systems has been authorized on the basis that the program would be a good bargaining chip. Raymond Garthoff, a senior member of the SALT negotiating team, explains the dynamics of the bargaining chips process:

> Defense Department weapons analysts and programmers in particular, but also senior executive and legislative figures, concluded that it was possible to add crucial support for particular military programs by calling them "bargaining chips" in SALT. Former Secretary Kissinger, shortly before leaving office, ruefully confessed that he had initially supported strategic cruise missiles for this reason, and then later found Pentagon adherents unwilling to give up the systems in bargaining with the Soviets.[138]

The problem with bargaining chips, in other words, is keeping them from becoming deployed weapons systems. Paul Warnke explains why exercising this restraint is often difficult: "New weapons starts acquire a bureaucratic momentum which makes them extraordinarily difficult to stop. This momentum is increased by commercial interests in the continuation of weapons contracts."[139]

New weapons programs, directions, and systems are constantly being proposed and added to the American strategic inventory, whether they emerge from careful needs planning and assessment, technological determinism or drift, or from bargaining chip-induced momentum. Some of these are incremental and have relatively little impact on the strategic balance. Others have dramatic effects, sometimes unrealized at the time of inception and development. In the remainder of this chapter, we will examine three development efforts on the horizon that have potentially great implications: the mobile missile (MX); the movement toward hard-kill targeting accuracy (including MARV); and enhanced-radiation warheads.

The Mobile Missile The most prominently discussed hedge against ICBM vulnerability by those who support the TRIAD concept is the mobile Missile Experimental or MX. In the late developmental stages, MX is an extremely powerful, expensive, and controversial system that can be deployed in existing ICBM silos or in a mobile mode. According to the FY 1979 *Annual Report*, both options are under consideration, with the missile scheduled to reach "the point of full-scale development before the end of FY 1979."[140]

MX is a very potent weapon, according the published accounts. Quoting a *New York Times* report on August 6, 1977, Coffey describes the missile's characteristics and capabilities: "The programmed M-X is twice the size of MN-3 (Minuteman III), has more than twice the thrust (enabling it to carry 7–14 RVs, each more powerful than the 170-KT warhead now carried by MN-3) and is potentially capable of being delivered within 100 yards of the target, compared to the 300 yard CEP of Minuteman III."[141] Downey reports a doubled targeting accuracy and fivefold throw-weight advantage over Minuteman III,[142] and McLucas states flatly that "each M-X is equivalent to about ten Minuteman missiles."[143] The cost of M-X is equally impressive, however, as Brodie reports: "The cost per missile in a 200 to 250 missile force would be some $100 million."[144] The total force cost, in other words, would be $20–25 billion, about the same as the projected cost for the scuttled B-1 bomber.

One of the most commonly discussed mobile deployment schemes for MX is the slit trench option. In this mode, shallow trenches would be constructed, with the missile riding horizontally along a track in the trench. To ensure invulnerability, the trench would be covered with a thin layer of dirt to conceal the missile's location at any point in time. To fire MX, a cowling over the missile would uncover it and the MX would then be tilted to its upright launching position (recent tests have indicated this option may be faulty, because detonations near the trench render it inoperable for considerable distances). More recently, official discussion has centered around the Multiple Protective Structures (MPS) system, whereby missiles would be shuttled between numerous hardened silos.

This sophisticated weapon has been criticized on two basic grounds. Justified as "a major hedge against projected ICBM vulnerability in the late 1980's,"[145] MX has been questioned on the basis of its cost given the strategic gains it purchases (arguments not unlike those against the B-1). As Metzger says, "With two survivable retaliatory forces—bombers and submarines—able to maintain target and penetration capability, the justification for proceeding with . . . the mobile MX is sharply reduced."[146] Proponents point to the weapon's enormous counterforce capabilities and the virtues of the TRIAD configuration in rebuttal.

The second major criticism relates to MX's arms control implications. On the one hand, and analogous to the cruise missile, the problem of deployment verification arises. The survivability of MX is based on its concealment, which in turn makes it difficult to count the number of missiles actually deployed. For example, trench rotation, with multiple (usually described as three) trenches per missile, is often discussed as an

MX option. The verification difficulty is that the Soviets would have no effective way to tell if there was one missile in every three covered trenches, one in each trench, or more than one per trench. On the other hand, a move to mobility by the United States might, in an action-reaction manner, force the U.S.S.R. to do the same thing. The Soviets already have an intermediate range mobile missile for use in a European conflict (the SS-20), and it could be converted to strategic, intercontinental range with the addition of another booster stage (the SS-16).

Hard-Kill Capability Improvements in missile accuracy to the point that an ICBM can be destroyed in a hardened silo have raised the entire ICBM vulnerability, and thus mobile missile, issue. Although there is considerable debate about the necessity or wisdom of obtaining this capacity that relates to the theoretical impact of its attainment (discussed at length in Chapter 7), the United States (and presumably the Soviet Union) have ongoing programs that could produce true hard-kill capability.

Downey explains that targeting effectiveness can be increased in three basic ways: "One can increase the number of warheads (equivalent to increasing the number of missiles); increase the yield of each re-entry vehicle (RV); and/or decrease inaccuracy. Of the three, reduction of inaccuracy is by far the most effective."[147] This effectiveness, in turn, arises from gains in lethality associated with greater accuracy: "Hard-target capability is proportional to the square of accuracy (doubling accuracy increases counterforce capability fourfold); directly proportional to the number of warheads (double the number of warheads doubles the counterforce capability); and proportional to the two-thirds power of yield."[148] Three programs that have significant lethality, and thus counterforce, implications have already been discussed: the Trident II missile, MX, and the cruise missile (although the slow delivery speed of cruise limits its counterforce application). Two other improvement programs, MARV and the MARK 12A warhead, offer the possibility of accuracy approximating hard-kill as well.

In a very real sense, MARV is the "son of MIRV," and its development and purposes closely parallel those of MIRV. The warhead was originally designed as the final "stage" of MIRV for the purpose of evading ABM defenses, but, like MIRV, it has continued despite the absence of a counter-ABM mission. Barry Carter explains how MARV works:

> By definition after the MARV has separated from the "bus" . . . it can maneuver almost up to impact in order to correct its flight path. . . . The most likely development is the homing MARV, what some call the true MARV. A sensor in the warhead would acquire an image or images of the target or of prominent terrain features nearby (or perhaps would simply acquire an "altitude profile" of the terrain along its flight path). An

onboard matching device would match this information with a map stored in its memory. The warhead's flight path would then be corrected either by gas jets or by aerodynamic vanes.[149]

MARV is thus the projectile fired from the MIRV missile except, unlike earlier models, it contains its own propulsion and guidance systems, rather than being a "simple" bomb. It is not inconceivable, in the future, to imagine the computer system in MARV interacting with a central command and control computer in the United States via satellite to make even more precise in-course correction or for retargeting to different locations. Even the currently conceived MARV, however, "will presumably have an accuracy of a few hundred feet" and would "give even warheads the size of the Poseidon's very effective hard-target capability."[150]

The Mark 12A is the successor warhead to the Mark 12 currently deployed on the Minuteman III missiles and is thus viewed by some as an alternative to MX deployment. Although the warhead, as currently conceived, would not allow mobile deployment, it will greatly enhance the capability of Minuteman III. Downey offers a comparison of the new warhead and its predecessor:

> The Mark 12A RV ... will have the same weight, size, and shape as the presently deployed Mark 12, but twice the yield. Other technological improvements will give the 1980's-generation Mark 12A perhaps half the inaccuracy of the 1970's-generation Mark 12, again without increasing weight. The yield upgrade will improve lethality 59 percent; the accuracy upgrading will improve it by a factor of four, giving an overall hard target capability 6.3 times its predecessor.[151]

The net effect of these programs is to give meaning to the possibility of counterforce targeting and thus the realistic ability to consider counterforce strategies, at least against fixed-site and unhardened targets, both options that were not truly available with earlier delivery systems. In turn, the possibility of threatening the destruction of retaliatory forces is implied, with concomitant implications for incentives to preempt or fire first in a nuclear conflict. As we shall see in the next chapter, these technologically driven opportunities are particularly important for the Soviets, who have invested much more heavily in land-based ICBMs as a component of their overall strategic forces than has the United States.

The Enhanced-Radiation Warhead The so-called neutron bomb "burst" into public attention on April 22, 1977, when the House Appropriations Committee released testimony from the Energy Research and Development Agency regarding the intention to produce the weapon as a battlefield artillery shell in conjunction with the Lance W 70-3 missile. The stated purpose of the warhead, a fusion bomb using the deuterium-tritium thermonuclear reaction as its basis but "improved" to enhance initial radiation effects, was to counter Soviet armor concentrations by penetrating the armor of tanks, killing the crews with intense neutron radiation.

The principles on which the enhanced-radiation warhead is based are not new. Robert Gillette has reported that development work on the technique, investigated originally for short-range ABM interceptors, began in December 1958 at Lawrence Livermore Laboratories in California and that the first experimental neutron bomb was detonated in the Nevada desert in 1963.[152] As a battlefield weapon, the device has been praised and condemned. As an advance in warhead technology with strategic weapons implications, it has been largely ignored.

The principle of the enhanced-radiation warhead is to rearrange the order of importance of nuclear blast effects. As Harold M. Agnew explains, "The *fusion* process produces neutrons, heat, blast and fallout but produces many more neutrons and, specifically, more high-energy neutrons in relation to the other products than does the fission process."[153] As was explained earlier in this chapter, very little residual radiation (only that created by the fission trigger) is created. At the same time, blast and heat effects are reduced, and, with detonation in the atmosphere, can be virtually removed.

In the enhanced-radiation variety, the objective is to achieve something approaching a mono-energetic effect of neutron radiation, so that "ER warheads release about 80 percent of their yield in the form of very high doses of neutrons and gamma radiation."[154] The warhead is very deadly against humans because "certain of the radiations such as neutrons have what the medical profession call a high LET (linear transfer). This means they interact with living tissue in a strong manner."[155] The principle of cell destruction is thus the basis of killing (it is the same principle that is applied to radiation treatment for cancer), but neutrons do not harm other media such as buildings. Combined with the suppression of heat and blast effects, they thus produce far less collateral damage than fission-based warheads.

Neutron warheads are thus very potent weapons, particularly against basic countervalue targets (people). They can be made to produce very small yields, but at the same time, "since the matter of critical size does not arise, there is no limit to the amount of fusible material that can be placed in the bomb."[156] The weapons thus would seem amenable to strategic application, possibly as a means to bolster the hostage effect in the face of Soviet civil defense programs (see Chapter 5). At the time the weapons were publicly announced, however, the Arms Control Impact Statement was mute on strategic plans: "The ACIS does not intimate whether it expects anyone will perceive a strategic application for ER weapons. . . . It does not comment on whether the United States has

American Forces

plans for applying the ER concept to strategic weaponry."[157] The effectiveness and efficiency ("a workable neutron bomb would probably have the same radiation (neutron)-killing capability, at a given range, as a 'normal' nuclear weapon of five times the explosive power"[158]) of ER warheads makes one suspicious that consideration of their strategic application must be occurring somewhere in official circles.

The ER warhead in its Lance configuration has received two criticisms. First, because it can be produced as a very small weapon, it is opposed as a mininuke: "Therein lies the real objection to the neutron bomb. By contributing to the illusion that nuclear weapons are useable . . . the development of neutron bombs could greatly increase the chances of . . . nuclear war."[159] Second, it is opposed on humanitarian grounds, because "the high lethality of these weapons, and their potential for causing unnecessary human suffering, are sufficient reasons for banning them."[160] Pat Towell quotes Senator John Heinz II (R-Pa.) more strongly on this point: "To perpetrate death by neutron radiation smacks of the sort of chemical and biological warfare that has historically outraged civilized nations."[161] The first objection is largely irrelevant in the strategic context; the second may be relevant, if it is meaningful to talk about the humanitarian aspect of nuclear warfare.

Chapter 5:
Soviet Strategic Doctrine and Forces

The most obvious and strongest external environmental factor shaping American strategic doctrine and forces has been and will continue to be the Soviet Union. Just as American concepts and force characteristics have evolved during the nuclear age, so have their Soviet counterparts and American perceptions of Russian capabilities and intentions. In this context, the latter 1970s have given a particular urgency to understanding how the Soviet Union views the nuclear system. The Russians have achieved effective strategic parity with the United States, and overall Russo-American relations totter through a critical period, marked by an increasingly expansionist Soviet foreign policy and a likely leadership transition with unpredictable outcomes.

Within these considerations, further conditioned by arms control discussions that are becoming even more controversial in the light of apparent Soviet truculence in other foreign policy areas, an examination of Soviet views of the nuclear system becomes necessary. There is an increasing realization in the United States that Soviet and American views of nuclear balance and nuclear weapons utilities are not mirror images of one another. At the same time, the enormous Soviet nuclear arsenal, and the even greater potential it possesses, have created grave concern among American analysts and policy makers. These dual concerns justify and form the organizational basis for detailed analysis.

Soviet Views on the Nuclear System The way Americans view the Russians generally and by extension the nuclear relationship between the two countries has undergone considerable alteration as the nuclear age has evolved and the strategic balance has changed. During the first twenty years after World War II, when the United States possessed first a monopoly and then a decisive superiority in strategic armament, that view evolved from the perception of nearly absolute intractability and interest incompatibility marking the apex of the Cold War to the belief that the Soviets were more rational beings who looked upon nuclear weapons in basically the same way that Americans do (which, of course, made them rational). This prevailing view of the "mirror image" effect was looked upon by many as a cardinal accomplishment of American policy (I recall a State Department official describing it as "the greatest accomplishment" in American strategic policy) and was summarized well by Brodie in 1966: "Now that they have had [nuclear weapons] for more than a decade and had opportunity to test them and reflect on what they imply, Soviet ideas on what can be accomplished, for example by surprise nuclear attack, seem to be developing along lines familiar in the United States."[1] These comforting and comfortable views flourished in the spirit of nuclear cooperation that followed the Cuban missile crisis and reached their pinnacle with the announcement of the spirit of detente by President Nixon and the commencement of SALT in 1969.

In the midst of this trend was another, and more unsettling, development. As detente became the watchword of Soviet-American relations, the Russians were catching up with, and by some measures surpassing, the United States in strategic capabilities. This development was, in many ways, incompatible with American notions of the intellectual symmetry in American and Soviet nuclear thought: if they accepted the mutual balance of terror as the basis of stable deterrence, for instance, why were they building forces obviously excessive for assured destruction dictates? (That our forces were also clearly larger than those needed for a BOT world was less intellectually troubling.) Inability to answer such questions within conventional frameworks fueled a movement to reexamine Soviet intentions and capabilities.

A primary thrust in this reappraisal has been to analyze more carefully what the Soviets say about nuclear relationships. Implicit in this form of analysis was at least a partial rejection of the mirror-image notion. As Raymond L. Garthoff says, "It is fallacious to assume that the Soviet leaders' ways of thinking and objectives, and therefore their strategic perceptions and their political-military intentions, are the same as our own."[2] One way to determine where those differences exist is to analyze the emerging body of Soviet literature and policy pronouncements on strategic concerns, recognizing that interpretive limits must be placed on such endeavors. Jack L. Snyder warns: "At first glance, doctrinal statements by Soviet military writers and political figures seem to be of limited value on two counts. On one hand, such material often seems hackneyed, unsophisticated, propagandistic, and devoid of serious analysis. . . . Beyond this the Soviets may not only be inscrutable, but also inveterate liars."[3] Recognizing the rejoinder that public pronouncement often has political as well as expository purposes, the Soviets have been reasonably consistent in their descriptions of the nuclear system. That the views they espouse are in some juxtaposition to American thinking is summarily stated by Colin S. Gray: "With respect to the Soviet Union, one is competing with a state which (a) views war (at all levels) as an instrument of policy, (b) views a good defense as a good deterrent; and (c) views Western interest in detente processes as a fairly direct consequence of the rise in the relative military strength of the Soviet Union."[4] It is, of course, possible to overstate the differences in American and Soviet strategic thinking and thus to overextrapolate the consequences flowing from those differences, just as one can dismiss Russian statements as mere propaganda and underestimate their effect. A more balanced view is necessary. I will attempt to examine Soviet doctrine looking, in turn, at the basic Soviet public view of the nuclear balance, their perceptions of deterrence, and how a nuclear war might be fought should deterrence fail.

Soviet Views of the Nuclear Relationship Soviet and American ideas about nuclear weapons, and particularly their utility as a tool of foreign policy, are divergent. Although the Russians reject nuclear war between the two superpowers as an inadmissible forum for the continuing struggle between capitalist and communist societies, they also believe that possession of a massive nuclear arsenal provides foreign policy opportunities (or removes constraints) not otherwise available. For the Soviets, the apparent lesson of the Cuban experience was that the absence of strategic parity or superiority could be critical in prosecuting successfully their interests in less direct Soviet-American conflicts. These basic differences in outlook are summarized in the conflicting meaning of the Soviet description of the strategic relationship under the rubric "peaceful coexistence" and the American description "detente" and differing notions about nuclear war fighting.

Nuclear Weapons and Foreign Policy Goals American strategic thought has not emphasized the role of strategic nuclear power as a foreign policy tool beyond deterring nuclear war since the doctrine of massive retaliation was discredited in the latter 1950s and early 1960s. Rather, the utility of nuclear weapons has been viewed as narrow and specific, and the contrary view that such weapons can be used for political leverage has been discredited as adventuristic and dangerously threatening to the nuclear threshold.

The Soviets, on the other hand, look upon nuclear weapons possession as a complementary component of their overall foreign policy, with indirect applicability in the broader context. This position is described by the term "world balance of forces" and is summarized by Leon Goure and his associates: "The development of Soviet nuclear forces has transformed the Soviet Union, for the first time, into a global military power, thus intensifying the basic trend toward globalism in Soviet foreign policy and providing new opportunities for an expansion of Soviet political and military influence on a world scale."[5] Although some analysts also emphasize the role of the expanded Soviet navy in making the U.S.S.R. a global power, strategic nuclear weapons thus possess utility in attaining other goals to the Russians that is not found in American thought and has further ramifications in their own thought on the subject. As Snyder suggests: "Having reached a position of strategic parity, they do not have to bow to anybody's attempts to force 'alien doctrines' upon them. Their public statements on strategy suggest that they are primarily attuned to such issues as the immediate political utility of a favorable balance of forces rather than to the conceptual esoterica of intrawar conflict theory."[6]

To some extent, these calculations of utility reflect a kind of realpolitik, power politics view of nuclear weapons. Influenced more heavily by civilian deterrence theorists, American doctrine has not embraced this nuclear worldview, but "the current Soviet leaders apparently agree with the military that increases in military strength do render tangible diplomatic and security benefits."[7] As was pointed out in Chapter 3, Soviet military (and particularly strategic) expenditures have risen dramatically in the past twenty years without the apparent opposition to defense costs that has occurred in the United States. Richard Pipes suggests that this may be attributable to Soviet perceptions of benefits that such weapons bring: "Ultimately, it is political rather than strictly strategic or fiscal considerations that may be said to have determined Soviet reactions to nuclear weapons and shaped the content of Soviet nuclear strategy."[8]

The Cuban Lesson The considerable impact of the professional military on strategic policy formulation in the Soviet Union is a recurring theme that goes back to original Russian thought on the subject. Pipes maintains, "The guidelines of Soviet nuclear strategy, still in force today, were formulated, during the first two years of Khrushchev's tenure (1955–1957), under the leadership of [Field Marshall] Zhukov himself. They resulted in the unequivocal rejection of the notion of the 'absolute weapon' and all the theories that U.S. strategists deduced from it."[9] Nuclear weapons, from this military vantage point, were another component of the military arsenal, and although American thought posits the sharp qualitative change brought about by the thermonuclear revolution, Snyder rejoins, "It is wrong to dismiss non-Western conceptions of the dynamics of deterrence simply because they seem unsophisticated by Western standards."[10]

The notion that nuclear weapons have definite and strong geopolitical implications was brought home strongly to the Soviet leadership by the humiliation of the Cuban missile crisis. As Moulton describes the general impact of the situation, "In retrospect, it seems clear that the decisive nuclear superiority possessed by this nation combined with overwhelming conventional superiority in the immediate conflict area converged to forestall the boldest and most reckless gambit of Premier Khrushchev's career." [11] Congressman Downey states the point more bluntly in terms of the strategic capabilities of the United States not reciprocally held by the Russians: "The Soviets were not forced to back down in 1962 because our numbers were higher than theirs. They were forced to back down because we had a countervalue force sufficient to destroy their society and they lacked one sufficient to destroy ours."[12] The Cuban crisis, of course, helped lead to Khrushchev's fall from power. It also confirmed in Soviet minds what the military had been saying about the role of nuclear forces and gave support for those favoring strategic expansion. As Jerome H. Kahan says: "The U.S. 'victory' in Cuba both vindicated Moscow's view that strategic strength conferred political power and raised the prospect of other circumstances in which U.S. strategic preponderance would again become decisive."[13]

Peaceful Coexistence and Detente What emerges from these considerations is a Soviet view that nuclear inferiority has a deleterious, if indirect, effect on the prosecution of other foreign policy goals. The possession of an equivalent nuclear arsenal is viewed as an important psychological element in being able to act in decisive ways, because the absence of such a nuclear presence creates a climate in which it is possible to be "bullied" out of attaining goals.

In a sense, then, nuclear parity removes crucial restraints on foreign policy. That it does not create unconstrained opportunities nor make the dangers of nuclear war greater, however, derives from the policy of peaceful coexistence. Goure et al. posit: "According to Moscow's specifications the prime purpose of peaceful coexistence is to reduce the chances that the continuing struggle between systems will lead to war between the great powers, or, as Brezhnev has said, 'will make possible a shift of the historically inevitable struggle (between systems) onto a path that will not threaten wars, dangerous conflicts and unrestricted arms races.' "[14] The "continuing struggle" between the two systems, of course, is an axiomatic feature of Marxist-Leninist thought. The concept of peaceful coexistence seeks to create a "ground rule" for that competition saying that direct superpower competition that could devolve into nuclear conflagration is unacceptable. This position is compatible with the general Soviet position because, "Even though Soviet spokesmen predict a Soviet victory in the event of a nuclear war, they consistently acknowledge the enormous damage that would result from such a war and hasten to deny that it can serve as a desirable midwife of revolution."[15]

Peaceful coexistence thus emerges as a limited term describing and prescribing Soviet-American nuclear relationships but not extending to the totality of superpower interaction. As such, the concept stands at some odds with the American description of the superpower relationship as "detente." That term, which means literally "relaxation of tensions," was, at the time it entered the foreign policy lexicon, thought to be a generalizable phenomenon, a spirit "of a radical normalization of the international situation, of the limitation of arms, of disarmament, of . . . cooperation to their full extent in the relations between states."[16] Such euphoric descriptions go well beyond what the Soviets mean by peaceful coexistence and, if taken literally as Soviet policy, would require explicit rejection of some of the most basic tenets of communist ideology. It is, however, gradually being recognized in the United States that there are sharp limits to the extension of cooperation arising from nuclear accord. That this realization is wrenching, however, is attested to by the continuing controversy over "linkages" between Soviet-backed conflicts in Africa and progress within SALT (discussed later in this chapter) and rumblings of a "new Cold War" from the Carter White House. It is possible that much of the disillusionment surrounding Soviet behavior has resulted not so much from their actions as from our own extrapolations of peaceful coexistence into detente.

Soviet Views of the Basis of Deterrence Just as the doctrine of peaceful coexistence flows from ideological definitions, Soviet notions about the basis of nuclear deterrence are partly the result of Marxist-Leninist dogma that has contributed to "the development of separate and distinct strategic cultures in the two countries."[17] The Soviets disdain mutual assured destruction as sufficient for a stable deterrence system. Instead, they maintain that the success of deterrence requires a Soviet capability to defeat the United States in a nuclear war, a position arising from the communist precept of "capitalist encirclement." Each of these positions merits some attention.

MAD as Inadequate Deterrent Base As has been pointed out, the American doctrine of mutual assured destruction at least implicitly defines nuclear war as almost necessarily devastating and, as a result, no one could expect to gain from such a conflict. Because the Soviets have, in their strategic thinking, been more influenced by traditional military concepts viewing weapons as tools of policy, they tend to reject these notions. As William T. Lee puts it:

> The concept of deterrence as the assured second-strike capability to kill one fifth to one quarter of the Soviet population while destroying up to three quarters of Soviet industry is a U.S. (and European) concept. The concept that deterrence so defined is the essence of strategy in the nuclear age—that is, that nuclear war cannot be conducted for traditional political-military objectives, and that there is no such thing as victory in a nuclear war—has no Soviet counterparts.[18]

This rejection, in turn, arises from the view of Soviet strategic thinkers that MAD as doctrine is "second-rate. In their view, U.S. strategic doctrine is obsessed with a single weapon which it 'absolutizes' at the expense of everything else that military experience teaches soldiers to take into account."[19]

The Soviets maintain as well that mutual assured destruction is a temporal accident of sorts, "a distasteful, temporary state of affairs necessitated by the peculiarities of the technological limitations of the day."[20] This position, of course, has some truth value, as has been noted: the countervalue, strategic bombing basis of the MAD threat was about the only targeting doctrine possible given the limits on targeting accuracies at the time the doctrine was formulated. Because of the pace of technological innovation in the nuclear weapons field, Goure et al. attribute to Soviet strategic thinkers the conclusion that the doctrine is "inherently unstable in view of the possibility of new breakthroughs in weapons technology as well as for political reasons."[21] This conclusion finds support in the dramatic increases in targeting accuracy coming available. Snyder applies this reasoning specifically to Soviet projections about the strategic future: "At best, the Soviets see MAD as a transitory phenomenon that must be accepted until it can be nullified by civil defense measures or by improvements in Soviet ICBM capabilities that would make possible the destruction of American land-based retaliatory forces in a surprise first strike."[22]

American doctrine is also criticized on political grounds. Goure et al. note "the impossibility of compromise of the class objectives of the two systems because of a situation" where, "while the danger of a general war has been reduced, it continues to exist and will as long as capitalism survives."[23] The Soviet Union maintains that its basic interest is in avoiding nuclear war (peaceful coexistence). Ideologically, however, the doctrine of capitalist encirclement and intersystem intractability by definition makes the Soviet Union the defensive force in the strategic dyad. Although this construction of the relationship seems incongruous to the Western reader, it allow the Soviets to conclude that, whatever else its shortcomings may be, MAD is a kind of backhanded compliment to Soviet nuclear capability. According to A. Trofimenko: "But since, in reality, it was a question all these years not of restraining the USSR from attacking the United States but, on the contrary, of the USSR restraining U.S. imperialism from unleashing a big thermonuclear war, the *U.S. concept of deterrence signifies,* in fact, *a recognition that the Soviet Union possesses such strategic strength which effectively restrains the United States from direct use of its nuclear missile arsenal against the USSR*" (emphasis added).[24] Although Americans would contest the initial premise of American aggressive nuclear intent, it is also true that the "mutuality" of assured destruction is the direct result of the growth of the Soviet strategic arsenal.

The Basis of Deterrence Stability As suggested above, a basic ideological driving force in Soviet deterrence formulation is the concept of capitalist encirclement: the notion that the Soviet Union is ringed by a series of hostile capitalist nations harboring the aggressive intention to destroy the Soviet state. Developed in the wake of Allied intervention against the Red Armies at the end of World War I and supporting perceived needs to maintain a closed, authoritarian society, the doctrine redefines (at least from an American vantage point) who is the potential aggressor and who is the potential victim of aggression in a nuclear conflict. American strategic formulations begin from the position that the United States would retaliate only after a Soviet preemptive blow; the Soviets turn the table. Arbatov argues: "In all the postwar period the military might of the USSR has never presented a threat to the security of the United States excluding only one situation—*a retaliatory blow in the event of America's attack on the Soviet Union*" (emphasis added).[25] The Soviets have been consistent in maintaining this position (the Trofimenko quotation in the last section, for instance, starts from the same premise). The result is that, by definition of the situation, "their position is that the Soviet Union never initiates war but only fights wars imposed by a foreign enemy."[26]

It is, of course, dangerous to take ideologically based positions at their face value, because the statements arising from them normally have multiple, including domestic propagandistic, purposes. To the extent that these positions do represent Soviet perceptions of strategic balance dynamics, however, they present perceptual difficulties for Americans attempting to assess Soviet capabilities and intentions, since we begin from the perspective of being the reactive, rather than the aggressive, actor in the system. Garthoff brings this perceptual problem into focus: "We attribute modest deterrence requirements to the Soviets because we know we would never attack first, and that the Soviets do not need, for example, to absorb a first strike. After postulating that the Soviets have more than we feel they should need for deterrence, we imply that they have aggressive interests beyond deterrence."[27] A Soviet worst-case defense planner would, of course, be as irresponsible as his American counterpart if he believed the United States posed no first-strike threat to the Soviet Union and planned strategy and forces based on that assumption.

The reversal of aggressive roles thus places the Soviet Union in the defensive position and gives their nuclear forces a predominantly defensive role. Having rejected massive countervalue assured destruction as an adequate conceptual basis for deterrence, however, they posit a different means for preventing American nuclear aggression: possession of a capability, and American knowledge of that capability, to defeat the United States in a nuclear war should deterrence fail. Paul Nitze summarizes the Russian position succinctly: "Soviet spokesmen see deterrence and a war-winning capability as complementary, not opposing, concepts. To their mind, a Soviet force capable of defeating an attacker and going on to win the resulting war is *clearly better able to deter the other side than any lesser force*" (emphasis added).[28]

This Soviet position has been the basis of considerable concern among American strategic thinkers, but it can be (and has been) overstated and overemphasized. A balanced view of war-winning capacity as the deterrent base must include reference to the companion concept of peaceful coexistence and the resulting inclusion of deterrence as the major stated goal of Soviet strategic policy and forces. Losing a nuclear war is not, in some sense, a much more gruesome prospect than that of having a sizable part of your civilian population incinerated and most of your productive base leveled, which the United States posits as the deterrent base. If the United States feels comfortable in the belief that it deters a Soviet nuclear attack (which they might or might not be interested in launching) by the threat of massive destruction, and the Soviet Union feels it dampens American nuclear aggressive intentions (held or not held) by offering the prospect of sure defeat, the end result, mutual deterrence, may be what counts most. It is legitimate and necessary to contemplate and plan contingencies based on Soviet stated policy, but becoming fixated with those policies may be overreactive. As a result, it is possible to conclude, as does former U.S. Senator James L. Buckley, that Soviet "military thinking is not bound by the essentially defensive concept of deterrence, but rather seeks to ensure a favorable outcome vis-a-vis the United States should war occur."[29] At the same time, however, "Existence of this capability, *as distinguished from any intentions regarding its use*, is in Soviet thinking an essential prerequisite for the effectiveness of the Soviet deterrence posture" (emphasis added).[30]

Soviet View of Nuclear War-Fighting Unifying deterrence with war-fighting capability creates a major conceptual advantage for Soviet strategic thinkers as opposed to their American counterparts. As Nitze summarizes, "In essence, Americans think in terms of deterring a war almost exclusively. The Soviet leaders think much more of what might happen in such a war." The ex post–ex ante dilemma and the fierce debate about limited nuclear options that has cleaved the American strategic community (see Chapter 3) are not problems for the Russians. Rather, "Soviet doctrine and military posture do not distinguish between deterrent and war-fighting nuclear capabilities, but appear to view them as 'fused together' in dialectical unity."[31] This difference in perspective in part arises from the organizational and historicocultural setting in which Soviet strategic policy is developed. It is reflected in how the Soviets postulate they could win a nuclear war and their assessment of the possibility of limiting a nuclear conflict.

The Decision-Making Context The view that the Soviet Union is surrounded by hostile nations eager to pounce on their homeland (combined pragmatically with the need for military forces to quell potential internal differences in one of the world's most ethnically diverse countries) provides the Soviet military, under strict political supervision, with a very strong societal influence. Richard G. Head points out that this position is influential in strategic policy making and content: "Soviet strategic thought has largely emanated from professional military men who have never made . . . the sharp distinction between deterrence and defense."[32] The perspective that the Soviet professional military brings to the prospect of nuclear conflict is reflected in basic conceptual differences: "The principal differences between American and Soviet strategies are traceable to different conceptions of the role of conflict and its inevitable concomitant, violence, in human relations; and secondly, to different functions which the military establishment performs in the two societies."[33] This role is reinforced by the way in which Soviet society views military science: "In the Soviet Union an article of faith and ideology is that war is a science. To Soviet officers, military science is a unified system of knowledge about preparation for, and the waging of, war in the defense of the Soviet Union and other socialist countries against imperialist aggression."[34]

Russian historical experience, partly Soviet and partly pre-Soviet, serves to reinforce the importance of military preparation and the ability to absorb massive losses. Much of Russian history has been punctuated by the need to repel foreign invaders from Kublai Khan to Napoleon to the Germans twice in the twentieth century, adding a distinctly Russian experiential basis to the notion of encirclement. In the process, and somewhat abetted by the communist experience, the Russians have learned to absorb enormous losses. That this accumulated experience affects the way the Soviets would view devastation associated with nuclear war is suggested by Pipes: "But clearly a country that since 1914 has lost, as a result of two world wars, a civil war, famine, and various 'purges,' perhaps up to 60 million citizens, must define 'unacceptable damage' differently from the United States."[35]

Fighting a Nuclear War Soviet objectives in a nuclear war should deterrence fail are straightforward: "to fight it, win it, survive it, and recover."[36] The belief that it is plausible to accomplish these objectives arises from a Soviet view of the consequences of nuclear war quite different from conventional American views: "Soviet doctrine, by contrast, emphatically asserts that while an all-out nuclear war would indeed prove extremely destructive to both parties, its outcome would not be mutual suicide: the country better prepared for it and in possession of a superior strategy could win and emerge a viable society."[37] The basis on which winning a nuclear war could be accomplished is composed of five interrelated parts: "(1) preemption (first strike), (2) quantitative superiority in arms, (3) counterforce targeting, (4) combined-arms operations, and (5) defense."[38]

American deterrence theorizing has occasionally flirted with the idea of preemptive attack (for example, during the massive retaliation era) and has generally dismissed the strategy as destabilizing. The Soviets apparently view the situation differently in their planning process. As Goure et al. point out, "Since the early 1960's Soviet strategy has emphasized the critical importance of the first nuclear strike for the course and outcome of a war and the need for the Soviet Union to anticipate an enemy attack by launching a preemptive attack" on the basis that "major and possibly decisive advantages can be gained from a nuclear surprise attack."[39] Although they seem to believe that "advantage" can accrue from preemption, the Soviets apparently realize that such an attack would not be disarming: "Soviet spokesmen indicate that they do not expect to succeed in destroying the entire U.S. strategic force in a pre-emptive attack, and in particular appear to recognize the difficulty of dealing with the U.S. submarine-launched ballistic missile (SLBM) capability."[40] This preemptive emphasis in Soviet thinking has been a particular concern to American strategists on two grounds. First, belief in the profitability of first-striking weakens disincentives built into secure second-strike systems, thus affecting adversely the theoretical basis of stable deterrence. Second, the obvious target of such a preemptive strike would be the American ICBM force, reinforcing concerns about ICBM vulnerability. On the other hand, it is not entirely certain that privately the Soviets believe this bellicose position. It may be a case of "nuclear posturing," to borrow Tsipis's term,[41] designed to assure the United States of the seriousness of Soviet intentions and thus to strengthen their deterrent posture. Possession of quantitative superiority of arms is, of course, necessary to carry out an effective first-strike strategy (as discussed in Chapter 2), but may as well reflect the Soviet penchant for large stockpiles of hardware (at the beginning of World War II, for instance, the Soviets had more tanks than the rest of the combatants combined).

Preemptive possibilities, combined with an interest in having something worthwhile to have "won" once the war was concluded, help explain the Soviet penchant for counterforce rather than countervalue targeting. As Snyder explains, "The preponderance of Soviet thought . . . has shown a preference for the unilateral approach to damage limitation by means of unrestrained counterforce strikes."[42] Rather than aiming at cities and industries, the Soviets' "primary aim is to destroy not civilians but soldiers and their leaders, and to undermine not so much the will to resist as the capability to do so."[43] Reducing collateral damage is congruent with Soviet war-winning emphases, because "the Soviet plan in the event of war is to defeat NATO forces, and to disarm and occupy Western Europe. Thus it is to the Soviet advantage to limit damage to that which is necessary to achieve these objectives."[44] Reflecting the Soviet belief that nuclear weapons represent a quantitative rather than a qualitative change in military calculations, these purposes are traditional military goals associated with the general mobilizations of World Wars I and II. From this perspective, the integration of nuclear and conventional forces flows. As Snyder observes: "In general, Soviet military writings reflect the long-standing notion that nuclear war (whether theater or intercontinental) will be waged with simultaneous strikes against the opponent's military forces, political-military command infrastructure, and economic-administrative centers."[45] It should also be obvious from Soviet goals in a nuclear war that they assume fighting would be protracted rather than a spasmodic exchange that was over quickly.

Finally, the Soviet goals of survival and recovery, in deference to the American arsenal's destructive capability, require a much greater emphasis on passive and active defenses than is present in American thought. Containing "counterforce-capable ICBM's, heavy active defenses, and urban evacuation"[46] as elements, the goal of these efforts is to ensure the viability of the Soviet state after a general nuclear exchange. As pointed out in Chapter 4, the possibility of effective defense against nuclear attack has not been regarded highly in the United States, and the effectiveness of the extensive Soviet system is hotly debated in this country. Whether they seriously believe nuclear defense will work or not, the Soviet assertion that a nuclear war can be won requires attention to ensuring, to the greatest degree possible, that a maximum part of Soviet society remains intact at that war's conclusion.

Soviet Views on Limiting Nuclear War Because the possibility of a graduated nuclear conflict applying limited nuclear options doctrine as a means to allow restoration of peace short of general exchange has been prominent in the recent American debate, Soviet views on the subject are relevant. Generally speaking, the Soviets disdain the possibility that such a war could be limited. As Pipes observes, "Limited nuclear war, flexible response, escalation, damage limiting, and all the other numerous refinements of U.S. strategic doctrine find no place in its Soviet counterpart."[47] Rather, "Soviet analysts appear not to allow a realistic possibility of avoiding an escalation involving the U.S. and the USSR to fullblown nuclear proportions."[48]

This position, of course, is consistent with the general Russian position on fighting a nuclear war: if one is fighting with the intention of winning and feels one can win, then it only makes sense to use as much force as is necessary to accomplish the purpose. Moreover, such a war would, in all likelihood, begin in the European theater, where early nuclear employment is necessary for NATO to avoid conventional defeat. Knowing this and basing their war-winning capacity on preemptive counterforce attack, the Russians would likely cross the firebreak early in a European war. Thus it is hardly surprising, that, as Jeffrey Record observes: "Soviet doctrine strongly suggests that any major [Warsaw] invasion of Europe would be accompanied if not preceded by a 'mass employment of nuclear weapons' designed to destroy or paralyze NATO's own tactical nuclear capabilities and deployed conventional forces."[49]

Soviet views that such a war could not be restrained doubtless reflect the opinions of the professional military, notions shared by many Western military personnel. Arthur S. Collins, Jr., for instance, says, "Ask any NATO commander what he would do if his units were hit with . . . nuclear weapons. The response would be a request for all the nuclear firepower he could get."[50] On the other hand, this position may be another example of posturing, as Record suggests: "Many observers believe that the Russians do privately recognize the difference between strategic and tactical nuclear war and that public statements of their inseparability are designed simply to enhance overall deterrence of military actions by the West."[51] Moreover, the conceptual and perceptual landscape is so uncertain that one must view Soviet statements with some caution. As Schelling admonishes: "Soviet declarations that no nuclear war could be limited do not mean that Soviet leaders believe it or, if they do, that they would not change their minds if a few nuclear bombs went off."[52]

As I have sought to demonstrate, the Soviet Union, at least in public utterances, has conceptions on the dynamics of deterrence that are significantly at odds with American formulations. Russian stated ideas are not only alien to American "wisdom" on the subject. By stressing war-winning capability as the basis for maintaining the deterrence system, they have a bellicose, threatening tone that reinforces many American images about the Soviets and lends itself to alarmist interpretations, particularly when viewed alongside the growth in the Soviet arsenal that is the subject of the next section.

Some conditioning of these positions needs to occur before making a rush to judgment about what the Russians are up to. Although all the possibilities are not exhausted, two points can be made. First, the Soviet conception is, in some ways, a reflection on early American strategic doctrine. At heart, after all, the credibility of massive retaliation (to the extent one wished to argue that the doctrine was believable at all) was based very much on U.S. ability to fight and win a nuclear war: if we were to threaten massive devastation of Soviet society, it could best be believed in the context that we could deliver such a blow without receiving a corresponding attack in return. As was pointed out in

Chapter 3, this was indeed the case until the Soviets achieved ICBM capability, and the Russians knew it. Whether the knowledge of overwhelming nuclear inferiority deterred them from any action they might otherwise have undertaken is problematical. It does seem entirely likely that Soviet military leaders were impressed with the advantages nuclear superiority produced, whether the United States planned to use it or not. Although Soviet doctrine has gone well beyond massive retaliation in attention to planning detail, their belief in the efficacy of nuclear weapons for attaining other foreign policy goals is an indirect, if unstated, tribute to the doctrine of the Eisenhower years. It can be countered from an American perspective that our posture was essentially defensive, whereas theirs is offensive, but such a position is no more convincing from their viewpoint than their position that they are the defensive power is to Americans.

The second point relates to whether the Soviets truly believe they can win a nuclear war. There is, of course, great debate in the United States about whether assured destruction deters the Soviet Union, which often refers to Soviet doctrine as a source of refutation. The Soviets, however, recognize that more than ten thousand strategic warheads are aimed at their homeland. The peaceful coexistence principle is testimony to their belief in deterrence as the major goal and, in context, should remove some of the shrillness from the debate about Soviet intentions arising from their stated doctrine. And though the Soviets do not "go public" with their policy debates, it is not unreasonable to question whether such a debate is not in fact going on. It undoubtedly occurs to the Russian leadership, as it does to ours, that any nuclear victory might well be Pyrrhic, leading Secretary Brown to muse, "It is extremely difficult to believe that the Soviets would ever seriously consider using their forces, and it is even more difficult to believe that they would contemplate any nuclear employment except in the gravest of crises."[53]

Soviet Nuclear Forces The major stimulus for the debate about Soviet nuclear intentions, resulting in the intensive examination of Soviet doctrine, has been the meteoric growth of Soviet strategic forces. As Donald G. Brennan observes and as has been noted in earlier chapters, "Beginning in about 1966, the Soviets undertook a major expansion of their ICBM forces that has continued at least until the recent past—in some senses it may still continue."[54] The result of this buildup, according to Lee, is that "the Soviets have achieved or they will soon achieve numerical parity or superiority in almost all important types of weapons systems."[55] Numerical counts of strategic systems are, of course, a simplistic method of comparing forces, but Goure et al. believe there is evidence that Soviet activities to date are part of a long-range strategy: "Soviet public discussions indicate that the leadership has not thought of resting content with the present situation, but that it intends to press for further and even more decisive shifts in the balance of world forces against the U.S. and in favor of the USSR."[56] As evidence of this trend, Head contends that "for reasons that we do not fully understand, the Soviets are pushing hard on scientific research in antiballistic missile systems, high energy lasers, charged-particle beams."[57]

Why the Soviets are engaged in these activities and the meaning that should be attached to the buildup have been the source of considerable controversy and discussion. Writing in the FY 1975 Defense Department *Annual Report*, then Secretary Schlesinger posed the basis of discussion: "Primarily at issue are the answers to two major questions. To what extent have the Soviets simply responded to and tried to counter U.S. initiatives? And to what extent have they sought (and do they continue to seek) something more ambitious than a capability for second-strike massive retaliation against the United States?"[58] These questions are closely related to the issues framing the internal debate about the adequacy of American doctrine and forces and provide an excellent example of external environmental factors influencing the internal environment and strategic doctrine. Obviously, both questions cannot be answered in the same way: if the Soviet arms buildup is essentially an action-reaction phenomenon, that is explanation in and of itself; similarly, if it is not an ARP, it follows that the Soviets have other things in mind. In turn, the answers one has for these questions affect the way one views the overall strategic balance and support of or opposition to ongoing arms control efforts.

Although it risks some oversimplification, there is a close correspondence between those who do not view Soviet force capabilities with enormous alarm and those who maintain that strategic parity poses no additional threat and thus need for doctrinal change. This position essentially argues that force comparisons, within reasonable bounds of symmetry, are largely meaningless (and that the United States is ahead on the important measures anyhow), that the only salient measure is U.S. retention of assured destructive capability, and that the conclusion of arms control agreements will not jeopardize U.S. security. At the other extreme, those who contend that the Soviets have something in mind other than assured destruction generally conclude that they are intent on gaining a meaningful war-winning capability, that current Soviet comparative force advantages may in the future become decisive through technological advance, and that current arms control proposals potentially place the United States in a position of decisive strategic inferiority.

Those who argue that Soviet buildups are reactive to American force initiatives contend essentially that the Soviets have, over the last decade, been engaged in a "catch-up" game. Representative of this position is a statement by Brodie: "Since we are looking so hard for the reasons for the Soviet build-up, one possibility that ought to be considered is that it was simply triggered by ours, and that it continues to be stimulated by a desire to catch up."⁵⁹ As evidence of this assertion, he states: "Actually, since 1970 the United States has built and deployed more ICBM's than the Soviets (550 versus 330), the difference being that our newer ones replaced older types which were retired, while the Soviet Union has kept active nearly all the ICBM's it has ever built—which incidentally tells us something about the difference in quality between the two forces."⁶⁰ Further support comes from the fact that the U.S. MIRV program has produced a substantial warhead gap between the two countries and a large "lethality gap" arising from greater targeting accuracy. Arms control advocates maintain that MIRV, as discussed earlier, may have been the major stimulus to current Soviet activities.

Those critical of the American comparative stance, most vocally represented by the Committee on the Present Danger, reject the reactive explanation of Soviet activity. Gray, for instance, asserts, "The Soviet Union seems to be locked into a weapons development and procurement rhythm that is largely independent of American arms race moves." Rather, he contends: "The Soviet Union is seeking to purchase military options that should provide it with no unreasonable prospects for (a) deterring crises, (b) winning crises, and (c) *winning wars (at any level)*. It is not suggested either that Soviet leaders will seek confrontation—in order to road-test their military capability—or still less, that they would welcome nuclear war" (emphasis added).⁶¹ This interpretation, of course, derives almost directly from the reading of Soviet doctrinal statements: deterring and winning arise from Soviet beliefs in the role of nuclear weapons in attaining other foreign policy goals; and war-winning capability, without necessarily aggressive intent, is the basis of Soviet deterrent capacity. This position argues a unity between doctrine and forces which is not always clearly present in American conceptions. Head, however, maintains that such a linkage does exist in the Soviet Union: "Soviet military doctrine is authoritative and comprehensive, providing statements of how and with what weapons military forces are expected to fight. Military doctrine is used to define requirements for new weapons, develop operational plans and tactics, influence resource allocation for military investment, indoctrinate military members in their tasks, and mobilize popular, professional, and political support."⁶²

It is within this atmosphere of American doctrinal-related disagreement that Soviet forces must be viewed. Soviet strategic forces are not symmetrical to their American counterparts, although we will use the same categories to analyze them: ICBMs, SLBMs, and air-breathing components. The Soviets have never, however, developed a true triadic configuration or balance to their forces: the Strategic Rocket Forces (ICBMs) have always been paramount in numbers and attention, whereas the submarine forces have lagged somewhat in emphasis, and a strategic bomber component has largely been missing. At the same time, active and passive defense systems, following from strategic doctrine, have played a much more important role than in the United States.

Strategic Rocket Forces The Soviet ICBM program, as represented by the Strategic Rocket Forces, has been the cornerstone of Soviet nuclear capability since the successful development of ballistic missilry. The current force is at about 1,400 ICBM launchers (compared to 1,054 American ICBMs), down from a peak figure of over 1,600 "due to the dismantling of SS-7 and SS-8 launchers in accordance with the Interim Agreement [reductions dictated by SALT I]."[63] The Soviets have produced a large number of missiles that generally represent incremental improvements in earlier modes, reflecting "a distinctive national style for weapons development and acquisition that produces large numbers of weapons of increasing quality to meet the requirements of Soviet military doctrine."[64] As has been mentioned, the Soviets prefer much larger missiles (with larger payloads) than does the United States, but these have generally been considerably less accurate. Most of the Soviet missiles have used liquid propellants, and the Soviets lag behind the United States in MIRVing of their force. Using Walker's figures cited in the last chapter, the Soviets have a much larger amount of their force capability invested in ICBMs than does the United States (over 80 percent of raw megatonnage, as opposed to about 50 percent for the United States).

Although the problem of ICBM vulnerability has largely stunted American interest in additional fixed-site land-based ICBMs, the Soviets have continued extensive work in this area. As Polmar states: "The size of the Soviet ICBM program, in terms of both deployed missiles and research and development efforts, and Soviet superiority in missile payload demonstrate that the Soviet Union is embarked on an intensive effort to improve the quality and quantity of its ICBM program."[65] This improvement program (which was allowed under the Interim Agreement on Offensive Arms of SALT I) has involved qualitative improvements through development and deployment of new ICBM systems.

Secretary Brown, in the FY 1979 *Annual Report*, provides an overview of the current Soviet program: "It is now clear that all three of the 'fourth generation' ICBM's the Soviets are now deploying—the SS-17, the SS-18, and the SS-19—have the potential, with feasible accuracy improvements, to attain high single-shot kill probabilities against U.S. silos."[66] Of most concern to strategic planners is the advanced SS-19. Head provides a description of this system's capabilities: "The SS-19 represents a 15 percent increase in length and diameter, which produces a 32 percent increase in hard-target kill capability, or lethality. A series of technological improvements provide the SS-19 with multiple independently targetable warhead capability, increased payload efficiency,

Soviet Doctrine and Forces

and increased accuracy."[67] Secretary Brown also relates that the Russians are continuing development efforts: "The Soviets have a fifth generation of ICBM's in development, estimated to consist of four missiles. Flight testing of one or two of these could begin at any time, with the others following by the early 1980's."[68]

Given the issue of ICBM vulnerability, Soviet knowledge of ongoing U.S. warhead accuracy programs, and primary Soviet reliance on liquid-fuel rockets, this emphasis seems strange on the face of it. The enigma is made greater because Soviet reliance on ICBMs makes their vulnerability more of a problem for them than it does for the United States: if hard-target kill is achieved, the United States could destroy a much larger proportion of Soviet retaliatory forces in a preemptive strike than the converse. There are three possible answers to this problem.

The first and most obvious answer is that the Soviets do not plan to use their ICBMs in a retaliatory manner. As was pointed out earlier in this chapter, Soviet war-fighting doctrine includes the use of preemptive counterforce attacks, and it is this possibility that weighs most heavily on those concerned with American ICBM vulnerability. It is, however, hard to imagine the entire Soviet ICBM force being used to attack U.S. silos, because the Russians would exhaust so much of their capability in the process. At the same time, American hard-target capability would seem to make holding back a residual force questionable ("damage limitation" notions have consistently included remaining Soviet ICBMs as a target).

A second possible explanation is that land-basing of strategic systems is dictated by Russian geography. The Soviet Union possesses the largest land mass in the world, and much of it is virtually uninhabited, thus providing a great number of places to deploy ICBMs and forcing the United States to expend a large number of missiles targeting them all. Also, the other basic medium for strategic forces, the ocean, is not a particularly attractive alternative because of port problems and the absence of forward bases.

The third possibility, related to the second, is that the Soviets may elect to go mobile with future systems. Sidney D. Drell points out that such an option is geographically attractive: "All analyses show that mobile missiles not deceptively based are of advantage to the Soviet Union because of their much larger land area available for deploying such systems."[69] The Russians already have an intermediate range mobile missile, the SS-20, which is deployed on large trucks for use in Europe. The Defense Department, in the FY 1979 *Annual Report*, raises the possibility of converting the SS-20 into an ICBM by adding a third booster stage: "The Soviets have essentially completed development of a fourth ICBM—the SS-16—which we believe to be intended as a land-mobile system; although it can be placed in silos. It is a solid-fuel, three-stage missile with a post-boost vehicle (PBV). However, it currently carries a single warhead."[70]

The same observations as were made about MX apply to any Soviet mobile system: its virtue lies in restoring ICBM invulnerability, which is strategically stabilizing from a second-strike perspective, but it has negative arms control implications. The fact that the SS-16 currently has a single warhead also points to Soviet inferiority in targeting accuracy: the problems of programming a warhead to target are more difficult when you do not know in advance exactly from where you will fire, and having to program multiple warheads from a single mobile launcher compounds the problem. Projected MX capabilities in a highly accurate, MIRVed mode thus only highlight Soviet technological deficiencies and may discourage the move to mobility.

The SLBM Force According to the provisions of the 1972 Interim Agreement on Offensive Arms, the Soviet Union was allowed a larger SLBM force than the United States: 62 missile-carrying submarines to 41 and 950 launchers (tubes) to 740. The Soviets have moved rapidly toward their quota, as General Brown reports: "The Soviet ballistic missile submarine force continues to grow in size and capability, a reflection of its high priority. As of January 1, 1978, the Soviets had almost 900 SLBM launchers on SSBN's [missile submarines] in operational status."[71] Although the Soviets have tried aggressively to improve their SLBM capabilities, it is generally conceded that they lag significantly behind the United States: the submarines are slower, noisier (a major factor in detection for ASW purposes), and require more frequent maintenance; their missiles have larger CEPs than American SLBMs; and they have yet to wed MIRV capability to their sea-based deterrent.

There are at least two reasons the Soviets trail the United States in the SLBM field. First, "the USSR has severe geographic disadvantages that reduce the on-station rates and increase the vulnerability of their submarines."[72] The Soviets' geographic problem is that they have fewer sheltered ports other than on the Black Sea (and especially ports that are not icebound part of the year) than the United States in which to harbor and maintain their fleet. Second, the technological problems of targeting SLBMs are similar to those for mobile missiles (having to be prepared to fire from a multiplicity of points, not all of which can be predetermined). With their more limited technological base, the Soviets have thus emphasized the conceptually simpler ICBM problem of targeting from one fixed place to another, to the detriment of SLBM improvements. Newer Soviet models are improving their capability, but Trident deployment will probably actually widen the current technological gap.

The Soviet Union was allowed a higher launcher ceiling than the United States because of its on-station disadvantage arising both from the geographic problem (because of the location of its ports, Soviet submarines must spend more sea time reaching locations from which they can fire their missiles) and because their vessels require longer maintenance ("down") periods. There has been some criticism of this provision (actually it is a series of provisions relating to various classes of submarines and containing requirements for retirement of ICBMs as the upper limits are approached[73]) on the grounds that, as Soviet SLBM missile range increases, so will on-station time and that improved design will reduce down times. As evidence of this contention, critics point to the SS-N-8, which has a purported range of 5,800 nautical miles, which, according to the FY 1979 *Annual Report*, "will permit the Soviets to cover targets in the United States from patrol areas as distant as the Barents Sea and the waters of the North Pacific."[74]

As in ICBMs, the Soviets have a welter of submarine problems reflecting their incremental development style. Numerically, the largest part of their force is in the so-called "Yankee-class": a total of 34 boats and 540 missile tubes. Emerging as the backbone of the force, however, are the newer "Delta-class" vessels: "The Soviets now have a total of 27 DELTA submarines. The DELTA I's and II's carry the SS-N-8, a single warhead with a range of at least 7,800 kilometers. A new submarine, the DELTA III, is now undergoing sea trials. The Soviets are also testing the SS-NX-18—a very long-range liquid-fuel missile with a post-boost vehicle and up to three MIRV's."[75] Although the development of these new systems will greatly increase Soviet SLBM, and thus strategic, capability, they are generally not viewed in the United States with the same alarm as are ICBM improvements, because of American preference for sea-based systems based on notions of second-strike capability (we view SLBMS as basically stabilizing).

Air-Breathing Forces The most visible asymmetry in American and Soviet strategic capabilities has been the absence in the Soviet arsenal of an impressive bomber, or air-breathing, component. Well symbolized by the designation of Soviet forces as part of Long-Range Aviation as opposed to the American Strategic Air Command, the Russians have not, until recently, made a serious effort to close the quantitative or qualitative gap in this area. Nor do the Soviets have any equivalent to the American cruise missile program, although the Department of Defense has estimated that they could develop a cruise missile with a 600-kilometer range (which, of course, would require penetration of U.S. air space by its carrier to attack most targets) "within the next five-to-ten years."[76]

The deployed strategic air component of Soviet forces is small and has not increased since 1962, although a new aircraft, the Backfire, has entered the force. General Brown describes what has been the heart of Russian bomber capacity: "For a number of years the Soviet Strategic Long-Range Aviation (LRA) bombers have included the four-engine turboprop BEAR and the four-engine jet BISON. The current strike force has about 140 BEAR and BISON."⁷⁷ Only forty of those planes are jet-propelled Bison, but a new jet bomber, the Backfire, is entering the force and has been the source of some controversy.

The Backfire bomber has a reported operating range of 3,000–3,500 miles, according to most accounts (making it the rough equivalent of the projected American FB-111 H). The Soviets have maintained consistently that Backfire has a theater role in Europe and possibly against China and that therefore it should not be considered a strategic system within the SALT process (a position parallel to the American stand on fighter-bombers stationed in England and West Germany). The aircraft is, however, capable of one-way missions against American targets (and then reaching airfields in Cuba) or, with in-flight refueling (a technique at which the Soviets are not conspicuously proficient) of two-way flights. Some would like to count Backfire against SALT launcher totals because of the ability to hit American targets and reach "friendly" Cuban bases (although it would be, under the circumstances, remarkably and unexplainably beneficent on the part of the United States if we had not vaporized those bases before the Backfires could reach them). At the same time, Backfire strategic capability prompts some observers to believe the United States should shore up its miniscule active defense system, which on January 1, 1978, numbered 57 surveillance radar and 324 interceptor aircraft.

The question, in light of a strong American belief in the efficacy of strategic air-breathing forces, is why the Soviets have never seriously gotten into the business. As has been mentioned, one obvious answer is that they did not have a large bomber force at the time they developed nuclear capability. Polmar presents a more comprehensive explanation: "The Soviet Union did not improve its long-range bomber force, apparently because of (1) the geographic limitations to direct USSR-to-US bomber strikes, (2) the lack of overseas bases for Soviet bombers and tankers, (3) the lack of a politically strong air force organization to compete for long-range bomber development, and (4) the logic in an approach to weapons development that differed from that of the United States."⁷⁸ Polmar's point regarding the Soviet air force (LRA) arises from the fact that the Strategic Rocket Forces (SRF) is an independent arm of the Soviet military, rather than subsuming the ICBM program within the air force, as the United States has done. Kahan adds another reason for not developing bombers that has doubtless been used to effect by the SRF: bombers are "expensive to build and operate."⁷⁹

Soviet Defense Systems The most recent "lightning rod" in the debate about Soviet strategic doctrine and capabilities has been the extensive Russian active and passive defense system. Those concerned about the Soviets gaining a credible war-winning capability cite Soviet efforts to protect their society as dangerously eroding of American retaliatory capacity and thus destabilizing to deterrence. Critics rejoin that the systems are ineffective (and that the Russians know it) and that any effects they have can easily be overcome by U.S. offensive forces, meaning the program has little if any effect on the strategic balance. Because this debate has become prominent (and occasionally acrimonious), the arguments warrant examination.

Soviet Active and Passive Defense Programs Soviet active defense systems (those intended to destroy incoming American strategic forces) consist of air defenses against American air-breathing forces, ABM defenses, and an emerging "hunter-killer" antisatellite program. In aggregate, they represent an extensive array that has no equivalent American counterpart.

The Soviet air defense program consists of 6,500 surveillance radars, 2,500 jet interceptor aircraft, and 10,000 surface-to-air missiles (SAMs) designed to interdict an incoming B-52 attack (as pointed out above, the United States has 57 radars, 324 interceptors, and no SAMs).[80] These defenses are countered by American offensive force plans that include blasting "corridors" through the defenses with ballistic missiles and low-flying penetrating aircraft and cruise missiles (described in Chapter 4). The Galosh ABM site around Moscow is a "light" installation erected in 1964 consisting of "64 interceptor missiles, with one-or-two-megaton warheads and an estimated range of 200 miles."[81] The system is not designed to protect against a U.S. missile attack nor would it be effective against such an attack (one Poseidon submarine, for example, carries over two and a half the number of warheads as Galosh has interceptors). The system, rather, is justified on other grounds: "Political realities make us heed other threats to security, including those posed by quarters other than the United States."[82] In plain language, Galosh is a defense against a nuclear attack by the People's Republic of China.

The emergence of "hunter-killer" satellites is a new phenomenon. Secretary Brown says, "We credit the Soviet Union with having an operational anti-satellite interceptor that could be intended for use against some of our critical satellite systems."[83] As an active defense element, satellite destruction would serve two purposes: disruption of American monitoring of target destruction during a nuclear war (and thus the ability to retarget warheads on the basis of that knowledge); and negation of emerging U.S. accuracy programs involving RV and warhead interface with central command-and-control computers using a satellite link. The obvious countermeasures are orbiting redundant, including decoy, satellites (so that all critical satellites could not be destroyed) or developing offensive weapons to destroy the antisatellite interceptors (prohibited by the treaty barring the use of outer space for offensive military use).

Soviet passive defenses (those intended to minimize the effects of American offensive forces penetrating active defenses) consist of dispersal of Soviet industry and population, urban evacuation in the event of crisis, and blast shelters to protect key personnel (such as political cadres, industrial managers, and high-ranking military officers). These plans are augmented by an extensive educational program to teach the population what to do in the event of nuclear attack.[84] The potential that particularly passive defense could weaken the hostage effect has been the most controversial aspect of the debate about Soviet defense.

"Soviet Defenses Pose a Threat to Deterrence" As Pipes points out, "Nothing illustrates better the fundamental differences between the two strategic doctrines than their attitudes to defense against a nuclear attack."[85] American doctrine defining deterrent stability as the mutual hostage effect is doctrinally hostile to civil defense and views Soviet civil defense suspiciously, whereas survival and recovery elements of the war-winning capability on which the Soviets posit deterrent stability dictate attention to civil defense. The basic problem that Soviet defense measures present is summarized by Secretary Brown: "If the Soviets believed that they could protect most of their population, and simultaneously cause major damage to the United States, they might calculate, on this basis, that they could gain a meaningful military advantage."[86]

A number of analysts believe that the Soviets are indeed attempting to gain such an advantage, at the expense of the American deterrent. Lehman and Hughes state emphatically: "Soviet civil defense programs, when coupled with the extensive air defenses already deployed and a strategy that emphasizes survival and a war-winning capability, *are dangerously eroding the U.S. deterrence posture*" (emphasis added).[87] Nitze agrees, directing his concern specifically to the impact of Soviet programs on the hostage effect: "As the Soviet civil defense program becomes effective it tends to destabilize the deterrent relationship . . . the United States can no longer hold as significant a proportion of the Soviet population as a hostage to deter a Soviet attack."[88] Gray, in turn, believes that American force deployment and doctrinal planning help stimulate Soviet civil defense efforts and thus contribute to their potential effectiveness: "Given the current American enthusiasm for (relatively) low-yield warheads that should minimize collateral damage—and for limited strategic options—the Soviet Union has every incentive to devote considerable resources . . . to civil defense and the protection of key industrial machinery (rather than buildings)."[89] Arthur A. Broyles and Eugene P. Wigner maintain that specific aspects of the program could prove destabilizing in a crisis: "If evacuation is undertaken during a crisis, it will greatly aggravate the situation. It can . . . serve as an aggressive move."[90]

The most frequent criticism of the destabilizing impact of Soviet civil defense is that it is so obviously unworkable that the Soviets cannot possibly take the prospects of success seriously. Goure, among others, disagrees diametrically with this interpretation: "Pronouncements and directives by Soviet political and military leaders leave no doubt they regard civil defense as a factor of great strategic importance. In their view, it is a logical, essential, and integral part of a rational defense posture and a war-winning capability."[91] Such conclusions about the seriousness with which the Soviets take the program and the effectiveness they attribute to it are based on civil defense manuals distributed as part of the educational program. These manuals estimate that full implementation of the program could limit civilian casualties to 5 to 8 percent of the urban population (thus 3 to 4 percent of total population). Viewing the totality of Soviet civil defense efforts, Admiral Elmo Zumwalt maintains that "the Soviets have concluded that if they cannot avoid a nuclear war, . . . then they must be able to (*and can*) survive and recover from one" (emphasis added).[92]

"Soviet Civil Defense Poses No Meaningful Threat" The various studies and pronouncements regarding the effectiveness of Soviet civil defense have been subject to extensive and sharp criticism. These critiques can be grouped into three basic objections: that effective civil defense against nuclear attack is impossible; that the studies on which allegations of effectiveness are based contain erroneous information and inferences; and that, at any rate, it would be relatively easy to overcome these efforts with offensive weaponry.

The Soviet, Georgi Arbatov, states the case of the ineffectiveness of civil defense succinctly: "Hardly any serious military expert today believes that civil defense can help win a nuclear war or even tangibly affect its possible outcome."[93] Viewing specifically the Soviet program, Fred M. Kaplan agrees and goes a step further: "The Soviet civil defense program, even on its own terms, is unimpressive and . . . even Soviet leaders [like Arbatov] seem to doubt its effectiveness; . . . even if it were all its celebrators claim it to be, the United States could counter such a program with forces far below even the present level."[94] John C. Culver cautions that "Soviet civil defense efforts must be carefully monitored. But there are fundamental uncertainties regarding their effectiveness."[95] Kincade, on the other hand, turns the tables on those who contend the program's effectiveness, saying such statements are the most dangerous aspect of Soviet efforts: "The real concern of American military planners should be to avoid giving Soviet non-military passive defenses more public credit than they deserve, lest the Soviets place an unwarranted faith in them and err in their own strategic calculus."[96]

The information on which those who fear Soviet defenses make their warnings, as well as the extrapolations from that data, are also subject to criticism. Kaplan issues a blanket indictment: "All studies expressing concern over Soviet civil defense suffer from unrealistic assumptions, leaps of faith, violations of logic, and a superficial understanding of the dynamics of a national economy."[97] The figures given in Soviet defense manuals on which some estimates have been based are the target of particular fire; these figures, passed out to civilians as part of training, have the obviously propagandistic, exhortatory purpose of convincing the populace that the exercise is worthwhile. Specific parts of the Soviet program are criticized as well. Of the blast shelter program, Congressman Les Aspin says: "Even those most worried about Soviet civil defense have argued that 80 million people would die in the Soviet Union if urban shelters are relied on."[98] Similarly, the practicality of urban evacuation is questioned because "evacuation plans decrease in effectiveness as urban agglomerations continue to grow larger.... Their utility is also a function of the warning time available to execute the mass movement of city populations, of the availability of life-supporting resources in the host areas, of the season and weather conditions, of the availability of transportation, and of the training and obedience of the evacuees."[99] In other words, a large number of imponderables can complicate evacuation, and these are particularly difficult for the Russians: "The population within large Soviet cities is much more concentrated than in the United States. Indeed, there are more than twice as many people per square mile in the 150 largest Soviet cities compared to the 50 largest U.S. cities."[100] Finally, any estimates of effectiveness are compromised by the fact that "the Soviets have never staged an evacuation exercise in any major city nor, even in small towns, has an entire community been evacuated."[101]

Finally, the projections are criticized for being largely beside the point. On the one hand, Kincade alleges that the civil defense effort can be overcome by retargeting the American arsenal:

> All of the defensive measures in the Soviet Union can be frustrated by: 1) avoiding protected facilities in favor of striking unprotected or unprotectable but equally critical economic targets, 2) allocating enough warheads to penetrate protected facilities, 3) holding adequate retaliatory forces in reserve to target urban areas or expediently-protected industrial sites, if either are reoccupied after an initial exchange, 4) attacking before defensive preparations can be completed, or 5) any combination of these measures.[102]

On the other hand, and related, large-scale population killing is not a necessary part of effective assured destruction. Improvement in guidance technology "means that vast civil damage would not be an inevitable by-product of military operations but would require a deliberate choice."[103] Population destruction is not necessary to accomplish basic goals and "is considered an incidental effect . . . of our strategic nuclear deployments. In fact, our deterrent is based on our ability to destroy what the Soviet leadership values most—the Soviet state as a functioning entity, the economic base which is the pride of the Soviet regime, and the nation's ability to recover from a nuclear war."[104]

Assessing the Soviet Threat The remarkable growth in Soviet strategic capabilities and the realization that the Russians do not look upon the dynamics of the nuclear balance in a manner entirely symmetrical with American views has, more than anything else, been the major causal variable in the ongoing internal doctrinal debate. The tone of that debate, particularly as it has been extended to measuring Soviet capabilities and imputing their strategic nuclear intentions, has been intense, sometimes acrimonious, and upon occasion curious.

The curiosity of the discussion is nowhere better focused than in the interpretation of Soviet doctrine and the import of their passive defense program. Normally hard-line, conservative analysts whose natural proclivity is to look with jaundice on the truth value of what Soviet leaders say generally find themselves accepting almost as litany Soviet public literature and pronouncements about strategic concerns. More "liberal" analysts, fighting to maintain mutual assured destruction as the grim conceptual base of American deterrent posture, are placed in the reactive position of counterinterpretation.

What the Russians are up to thus becomes the central feature in the American debate, coloring one's position both on the adequacy of American doctrine and forces and the collateral question of arms control agreements. Because the way one assesses the Soviet nuclear capability is pivotal to one's overall strategic position, the nature of the Russian challenge as it has been presented should be interpreted summarily to presage the impact of that assessment on the ongoing arms control process that is the subject of the next chapter.

The Nature of the Threat That American and Soviet views about the nuclear balance differ significantly is by now obvious. Soviet perceptions are not mirror images of their American counterparts, but instead are the result of Soviet ideology and national experience that lead them to look at security, and thus nuclear, relationships in their own unique manner. Understanding the dynamics of Soviet nuclear perceptions is thus prefatory to interpreting Soviet intentions and predicting likely Soviet strategic behavior. At the risk of some oversimplification, three concepts emerge as overriding in Soviet doctrinal formulation: peaceful coexistence; capitalist encirclement; and war-winning capability as the basis of stable deterrence.

The most basic concept is peaceful coexistence: the consistent position by the Soviet leadership since the 1950s that superpower nuclear war is an unacceptable form of conflict between systems. From this basic position, the paramountcy of deterrence as the basic value in nuclear relations arises, and there is little in the literature that indicates the Soviets have deviated from this tenet. Although there may be significant differences about how to make the strategic balance between the two countries maximally stable, there is little disagreement about whether it must be stable. Peaceful coexistence is, however, a limited concept: it describes Soviet views on nuclear relationships, not overall interactions between the United States and the U.S.S.R. By ideological definition, the systems are competitive and will remain so. Peaceful coexistence rules out certain means of fighting the competitive struggle in much the same way that kidney punches are outlawed by the rules of boxing.

The notion of capitalist encirclement, arising from both communist ideology and Russian national experience, defines the Soviet Union's hostile view of the external environment in which it exists and thus its assessment of security needs. The Soviet leadership has a traditional Russian view of the deterrent purposes of military forces generally: potential invaders (who, by definition of the encirclement concept, do exist) are deterred from aggressive military action by the prospect of military defeat. As was pointed out in Chapter 1, this is the historic, potential war-fighting deterrent role of military forces and remains the deterrent thrust of American (and allied) conventional forces.

Soviet and American conceptualizations differ significantly in the integration of nuclear capabilities into this traditional war-fighting deterrent role. Despite doctrinal challenges like limited nuclear options, the notion that nuclear forces are useful for anything but nuclear deterrence has generally been theoretically disreputable in American strategic thinking. The Soviets, however, seem to rely more on the nuclear threat as part of the "overall correlation of forces" in a war-fighting calculus, and American discussion about nuclear war fighting is in some respects a form of mirror imaging Soviet concepts.

The notion that stable deterrence rests on Soviet ability to win a nuclear war flows from these definitions: a hostile environment creates the need to be ready to repel an aggressor, and that aggressor is best deterred if he knows he will lose. In a sense, this notion should not be overly troubling, if it serves Soviet psychological nuclear security needs. The posture is, after all, defensive in nature, and, as was argued earlier, deterrence operates at the psychological level. For example, if they think their capability deters an attack we are not planning, and if we believe our capability deters an attack they are not planning, both can be satisfied simultaneously. Much of the ongoing debate has centered on denying the Soviets' war-winning capability (which is why there is such great concern about Soviet forces). Although understandable, these discussions are potentially destabilizing to the extent that they deny the Soviets' belief in a stable deterrent base.

The dilatory effect of Soviet confidence in their nuclear capability may be more indirect and arise from their apparent belief in the utility of nuclear weapons to accomplish other foreign policy goals. It is at best problematical that the Soviets believe they have achieved the strategic capacity meaningfully to win a nuclear war, but it is less uncertain that they believe their nuclear position has removed important constraints on other foreign policy goal achievements. Nuclear parity and a more aggressive Soviet foreign policy have emerged at about the same time, and those intent on denying the Russians a superior nuclear capability would be better served by making that linkage than in becoming obsessed with Soviet intentions to prepare for nuclear war.

The obvious reason for the overall concern has been the growth in the Soviet arsenal and why they have built up their forces to a level comparable with the American arsenal. This growth is a legitimate concern for defense planners and analysts, but there are paradoxical elements in the discussion.

The simplest and most compelling explanation of the Soviet buildup, arising from the Cuban experience, is that they determined that marked nuclear inferiority was no longer tolerable. Such a conclusion is hardly radical for defense planners, and, although the condition of decisive nuclear superiority was certainly more comforting for Americans, it strains credulity to argue that it was not legitimate and understandable that the Russians would seek to rectify a situation that, if reversed, we would find most intolerable. In fact, the debate about the Soviets gaining decisive nuclear superiority as a result of their present programs is, in all probability, a virtual replay of Soviet discussions in the early 1960s that eventuated in their decision to catch up in nuclear armaments.

It must also be remembered that the period during which the Soviets began to narrow the gap was not exactly a time of weapons deployment dormancy in the United States. The present configuration of ICBMs and SLBMs was a product of the early 1960s, and the decision to increase dramatically the lethality of the American arsenal through MIRVing a large part of the force occurred in the latter 1960s. That these factors have some causal relationship to ongoing Soviet activities is well summarized by Moulton: "In all probability it was the premature American decision to deploy large numbers of strategic weapons systems during the 1960's that eventually led to the major Soviet challenge that today our leadership describes as a Soviet bid to gain strategic nuclear superiority over the United States."[105]

The debate about where we stand in the overall weapons balance is basically reducible to the question of quantity versus quality. Central to the description of Soviet strategic systems is the fact that they have a penchant for building a large number of big weapons, but those weapons are qualitatively inferior to American strategic arms, as Lee describes: "Despite impressive progress, however, the Soviets have not done as well in the qualitative competition. They still lag in most weapons technology, and probably will continue to do so for sometime."[106] There are two basic reasons why the Soviets compete unfavorably at the qualitative level: their more limited scientific and technological base and a bureaucratic style that does not promote dramatic technological breakthroughs.

At the scientific level, the Soviets are constrained by a more limited base, particularly in the critical area of computer technology that fuels modern scientific and technological endeavors. In the vernacular, the Russians are several "generations" behind the United States in computers, a problem that is self-perpetuating: newer generations are able to process and analyze information, including data on improving computers themselves, much more rapidly than older models and can process increasingly complex demands. The differences in computer technology, including miniaturization, are largely responsible for the commanding U.S. advantage in targeting accuracy. This gap is likely to continue, paradoxically, because of Soviet emphasis on military programs: a much larger portion of the Soviet scientific effort is directly weapons-related, thereby diverting human and other resources from basic scientific and computer development efforts that undergird the application of technology to military ends.

The Soviet bureaucratic system also impedes dramatic technological development, as Robert Perry explains: "The central difficulty of Soviet strategic weapons development would appear to be incapacity to match the doctrinal and procedural flexibility—and the sometimes perplexing variability—that is a normal concomitant of American military research and development."[107] Rather than producing highly innovative techniques, this bureaucratic influence results in "incremental growth or cumulative product improvement as opposed to the U.S. tendency to favor whole new weapons systems."[108] This greater breadth and depth of scientific inquiry combined with looser direction of the scientific process helps explain why the United States has been the first to develop every major weapons systems since the ICBM itself and why, unless attention erodes markedly, it is likely to continue.

The bottom line of force comparisons thus reduces to saying that Soviet forces are bigger, but American forces are better. Many analysts are concerned that, as the Soviets perfect technologies like MIRV, and qualitative gap will narrow and, as a result, their quantitative advantages will loom far more significant. This is a legitimate concern, of course, but it rests implicitly on the assumption of a static American technological position. If the history of the arms competition and American strategic weapons technology is any indication, however, such an assessment is open to question.

Linkages The threat level that the Soviet challenge presents and thus the relative American position in that competition cannot be viewed in isolation from the overall Soviet-American relationship. The emergence of the Soviet Union as a nuclear coequal has apparently had an impact on more adventuristic Soviet postures in places like Africa and focuses attention on the outcomes of arms control processes.

In the American debate, the relationship of Soviet nuclear capability to the broader issue of Soviet-American relations reflects the conceptual differences between detente and peaceful coexistence and has focused recently on the concept of "linkages." The linkages idea is at heart an application of the detente principle of a generalizable relaxation of tensions between the two countries, saying, in effect, that progress in strategic arms limitation is dependent upon (or linked to) improvement in relations in other areas.

The linkage notion is controversial. From the Soviet viewpoint, there are no linkages between efforts to lower the likelihood of nuclear war and competition elsewhere that flow directly from the doctrine of peaceful coexistence. In the United States, some observers and policy makers (who generally oppose arms control) maintain that allowing the Soviets to achieve success in SALT while engaging in provocative behavior elsewhere (such as supporting brushfire wars in Africa) is letting them have it both ways and that if they expect American cooperation in SALT, they must act responsibly in other areas as well. Others, like Jeremy Stone, however, maintain that arms control measures should stand on their own merits: "SALT is and ought to be, an overriding shared imperative of both major powers. It *should* not be used to try to influence other matters. It *cannot* be used effectively to influence other matters" (emphasis in original).[109]

Whether strategic arms questions are independent of more general foreign policy concerns (peaceful coexistence) or whether such matters are all part of a larger picture that must be considered as a whole (detente), linkages have become part of the debate. President Carter, speaking at the June 1978 commencement exercise at Annapolis, recognized this fact as he expressed his own position on the linkage question: "We have no desire to link the negotiations for a SALT agreement with other competitive relationships. . . . In a democratic society, however, where public opinion is an integral factor in the shaping and implementing of foreign policy, we do recognize that tensions, sharp disputes or threats to peace will complicate the quest for a successful agreement."[110] The president, of course, was warning the Russians that the prospects for ratifying SALT II are compromised in some senatorial eyes by the linkage problem. Although the merits of linkage are debatable, one impact of the debate is fairly clear from past experience: once an idea enters the strategic debate, it tends to become durable (the bargaining chips notion is a good example). As a result, the impact of linkages is likely to become a consideration in the arms limitation debate well beyond SALT II.

Chapter 6:
The Arms Control Process

The movement to eliminate or limit strategic nuclear armaments is almost as old as nuclear weapons themselves. In 1946, the Baruch Plan proposed unilateral nuclear disarmament by the United States in return for renunciation of weapons development by the other nations of the world. Although nuclear disarmament proposals have foundered in a suspicious and distrusting world, the very failure of early attempts at arms limitation helped provide the impetus for an arms control process that is an important and enduring feature of the strategic balance and hence of doctrinal formulation.

The influence of early arms control failure on later success was convoluted: the absence of agreements resulted in the development of the frighteningly powerful arsenals we know today. When the Cuban crisis brought man uncomfortably close to the nuclear precipice, it became apparent that something had to be done to lower the likelihood of nuclear conflict. The result was the wedding of national defense and arms control, as Paul C. Warnke, former director of the U.S. Arms Control and Disarmament Agency, explains: "The primary motivation for arms control is *to preserve the necessary military balance* and to lessen the chances of war which today could escalate uncontrollably" (emphasis added).[1] Joseph Kruzel concurs in the tying together of security and arms limits: "National survival in the nuclear age depends *not only on maintaining a strong deterrent posture but also on developing arms control agreements* which reduce strategic uncertainties and the risk of nuclear war" (emphasis added).[2] Emphasis has been added to these quotations for a reason: before the idea of arms control could gain the serious attention of the national defense and security community, it was necessary intellectually to take it from the disarmament movement into the realm of mechanisms supporting American defense. Nuclear disarmament has always been looked upon as disingenuous and utopian by members of the defense community, and as long as arms control was intellectually connected to disarmament, it was not to be taken seriously by "realistic" men. The intellectual wedding of arms control and national security has not been without tensions and incompatibilities, but it has made an enduring and influential consideration of the arms control idea that would in all likelihood have been impossible otherwise.

SALT, of course, is the most visible outcome of the arms control movement. Entering its second decade of existence, SALT has become a process more than a discrete set of negotiations, and the achievement of SALT II will serve as the prelude for SALT III and beyond. The influence of strategic arms control has extended into collateral "spin-offs" like the Comprehensive Test Ban Treaty and has been linked to success in preventing so-called "horizontal" proliferation of nuclear weapons to countries in the developing world. The movement has also had domestic impacts, such as the requirement that proposals for all new weapons systems contain an arms control impact statement for congressional consideration.

Arms control has thus become a vital component of the environment in which strategic doctrine is formulated. To understand this phenomenon, this discussion begins by examining the conceptual dynamics of arms control and the various forms it takes, followed by description and analysis of the history of arms control proposals, from early unsuccessful attempts through the ongoing SALT II process and including matters not directly related to SALT. This examination in turn will allow distillation of some of the primary lessons that have been learned from experience with arms control.

The Dynamics of Arms Control Successful arms control agreements have tended to rest on two primary requirements: assurance that such agreements minimally do not detract from physical security and maximally enhance that security; and confidence by the parties that the other(s) will not violate provisions of the agreement. These two requisites are, of course, interrelated: the incentives to violate agreement (normally treaty) provisions are lessened if one believes the agreement preserves a satisfactory security level; and cheating on an agreement can threaten the security that the accord seeks to establish.

Successful arms limitation, thus, is the obviation of the arms race, either by freezing some agreeable arms level or by reducing an arms level which the parties agree is excessive to legitimate security needs. As a result, the key to establishing limits on armaments is overcoming the prisoner's dilemma that is the principal dynamic of the arms race. At the heart of overcoming the prisoner's dilemma is the ability either to trust an adversary or to establish a basis other than trust that fosters confidence that cheating is not occurring. In turn, the question of violation relates to a cost-gain analysis of the advantages that can accrue from such violations and relates closely to the armaments levels established in any agreement, and as such is a matter of the incremental benefits from cheating as opposed to the risks of being caught.

Overcoming the Prisoner's Dilemma The arms race which arms control processes seek to harness arises from two basic and interrelated security fears: that a given armaments level provides inadequate security because of the exigencies of the environment in which one operates; and that the failure to continue arming will place one at a strategic disadvantage. This, of course, is an obverse way of stating the dual requisites of successful arms control agreements mentioned above, and the two concerns are sequential. Before one can entertain seriously the possibility of placing limits on the arms one has, one must first have confidence that the present arsenal provides adequate security, as Warnke points out regarding the durability of agreements: "Unless an arms control agreement provides each side with the assurances it needs for its own security, it is certain such an arms control agreement won't last."[3]

The second fear is the basic dynamic of the prisoner's dilemma.[4] At heart, the prisoner's dilemma attempts to capture the situation where, in the absence of trust in the actions of the other party in a two-party situation, each is compelled to act in a manner injurious to both. The game is weighted toward mutually injurious behavior, because if one person cooperates and the other does not, the noncooperator receives maximum gain while the cooperator incurs loss far in excess of the losses involved in noncooperation. Both, however, can gain moderately if they cooperate.

The basic situation that gives the prisoner's dilemma its name is as follows. Three prisoners share a cell, and one morning one of them is found murdered. The warden separates the remaining prisoners and brings them individually to his office, offering each a deal. The deal is that if one prisoner turns the other one in as the murderer, the prisoner who "rats" on his cellmate will receive a reduction in sentence; the other prisoner will be convicted of murder and receive a life sentence. If both prisoners accept the deal and turn the other in, both will be convicted, but of a lesser crime, and will receive an incremental increase in sentence. If, on the other hand, both refuse the warden's deal, he will be unable to prove that either one committed the crime and thus neither will receive an additional sentence. The situation is depicted in game-theoretical matrix form in Figure 6.1.

Figure 6.1

		Prisoner 2		Worst Outcome
		Don't Rat	Rat	
Prisoner 1	Don't Rat	0,0	−100,+100	−100
	Rat	+10,−100	−5,−5	−5
Worst Outcome		−100	−5	

In this depiction, the possible outcomes by following each option for Prisoner 1 occur before the comma in each cell, and the outcomes for Prisoner 2 are after the comma.

The margins contain the calculation of worst possible outcomes for each prisoner (Prisoner 1 on the horizontal rows, Prisoner 2 on the vertical columns), because the end result of the situation depends upon what option each independently chooses. Thus, if Prisoner 1 chooses not to turn in his companion (Don't rat), he will get no additional sentence (0) if the other prisoner does likewise or will get a life sentence (−100) if his companion turns him in. On the other hand, if he chooses to accuse the other prisoner (Rat), he will receive a reduction in sentence (+10) if the other remains silent or will be convicted of a lesser crime (−5) if both inform. The outcomes are symmetrical for Prisoner 2 (as can be seen by reading the outcomes following the comma down the columns).

By adding the minimax principle to the situation (the idea that, in a single play of a game, it is rational to minimize your maximum possible loss, as described in Chapter 1), one can determne the prisoner's dilemma. From the marginal calculation of worst outcomes, it is rational for each player to turn in the other: the worst that can happen following that strategy is an incremental increase in sentence (and the possibility of getting a reduction), whereas keeping quiet runs the risk of a life sentence. As a result, it is "rational" for each player to act in a manner injurious to both.

The prisoner's dilemma, as it applies directly to the strategic arms race, is shown in Figure 6.2:

Figure 6.2

		Don't Arm	Arm	Worst Outcome
USA	Don't Arm	+2, +2	−100, +100	−100
	Arm	+100, −100	−5, −5	−5
Worst Outcome		−100	−5	

The values are somewhat altered in this depiction to reflect the situation: if both the United States and the U.S.S.R. cease additional armaments, the resources (+2 for each) invested in armaments (−5) can be diverted to other, presumably more productive, uses (the smaller positive figure reflects the fact that this diversion in all likelihood would not be complete). Continued armament is depicted as a negative value for each because additional deployments by each side cancel one another out and thus reflect unproductive spending. The heart of the dilemma, of course, is that if one ceases armament while the other does not, the noncooperator has the possibility of gaining great strategic advantage (+100) over the cooperator (−100). In strategic nuclear terms, this advantage potentially translates into first-strike capability with its accompanying destabilizing impact. As in the original prisoner's dilemma, rational behavior dictates continued armament (at least in a single play) and thus mutually unproductive behavior.

The key to overcoming the prisoner's dilemma, regardless of the context in which it is stated, is developing the ability to cooperate. As the situation is structured, this in turn requires the capacity of the players to trust one another so that each can engage in the cooperative behavior that is mutually beneficial. Trust, of course, is not a characteristic of the operational milieu that produced the U.S.–U.S.S.R. strategic arms race, and thus a means to gain confidence in an agreement's terms being honored that does not require trust was necessary. The obvious alternative to trusting the adversary is being able to see what he is doing. Thus, monitoring agreements, or what is known in the strategic lexicon as verification, have become the key element in breaking out of the arms race prisoner's dilemma.

Verification It is not unfair to characterize the history of the postwar arms control movement as having been divided into two phases: the preverification period and the verification period. Before nonintrusive means of monitoring military activity (methods that do not require on-site inspection or close aerial reconnaissance that violates national air space) became available, arms control efforts foundered, largely because the Soviet Union refused to allow the intrusive monitoring that was at the time the only technology available to ensure compliance. Because Soviet-American tensions were high and their relations were marked by distrust and animosity, arms control proposals were generally brushed aside as utopian and naive.

The signal event that broke the verification barrier was the introduction of the aerial reconnaissance satellite, the so-called "spy in the sky." Jules Moch describes the development of this capability:

> The SAMOS (Satellite and Missile Observation System) has developed since 1960. It was later divided into two types. First there are the Area Surveillance Satellites which . . . provide coverage of the Soviet Union and of China, and which, since 1968, have been able to provide immediate transmission of data through Communication Relay Satellites. Secondly, there are the Close-Look Satellites which are heavier than the Area Surveillance Satellites, and which carry a camera with longer focal length and wider apertures.[5]

These small satellites, orbiting the earth's atmosphere at an altitude of several hundred miles, have remarkable capacity. According to Herbert Scoville, Jr., "The U.S., and probably also the U.S.S.R., have a capability to resolve objects with a dimension of a few feet or less."[6]

The orbiting satellite allows monitoring of deployed weapons systems, particularly anything as large as an ICBM silo or a missile submarine construction facility, with great precision, and it accomplishes this purpose without the need for the intrusive monitoring measures to which the Soviets have always refused to accede (and which are opposed by many members of the American defense community as well). This form of surveillance is also more effective for certain purposes than on-site inspection (the inspector knows what he is seeing, but not what he is not shown) and, as Scoville points out, aircraft overflight surveillance as well: "Satellite reconnaissance has a number of major advantages over that carried out by aircraft. A satellite in an orbit of 100–300 miles can survey very large areas in a very short time period. If a satellite were launched in a north-south polar trajectory, then the entire earth could be covered, once in daylight and once at night, every twenty-four hours."[7]

Arms Control 167

It is difficult to overstate the importance of satellite detection in making serious arms control negotiations possible. The development and increasing sophistication of this technology closely paralleled in time the Cuban crisis that activated American and Soviet leaders to the dangers of nuclear war and the growth of the Soviet arsenal to the point where it felt sufficient confidence in the security of its nuclear deterrent to entertain arms limiting ideas without sacrificing national defense needs. Satellite reconnaissance was enshrined as a basic "right" in the arms control arena through prohibitions on impeding these so-called "national technical means of verification" in SALT I, and the verification issue is at the heart of ongoing SALT discussions. Robert Perry states this importance flatly: "Assured Detection is likely to be as fundamental to American concepts of strategic deterrence in the 1980s as was Assured Destruction in the 1960s."[8]

Although satellite reconnaissance and parallel progress in seismographic techniques and remote sensors located near the U.S.S.R. to monitor warhead testing have allowed precise verification of deployed weapons systems and reasonably precise monitoring of weapons systems testing, they do not allow detection of all the characteristics of the nuclear weapons cycle. There are two major areas important to effective arms control where national technical means of verification are of no assistance: qualitative improvements in existing weapons systems and research and development.

The problem of upgrading weapons arises primarily in the area of improving warheads and particularly adding MIRV capability to existing ICBMs and SLBMs. In fact, the casings on MIRVed warheads are similar enough to the coverings on single-warhead weapons that even on-site inspection not involving removal of the casings would not reveal violations of agreed limits. Thus, it is not surprising that satellite photographs, no matter how precise their detail, are not helpful.

Satellite pictures are also ineffective in monitoring the R&D phases of the development process described in Chapter 4. Research and development efforts represent a kind of sub-prisoner's dilemma in the armaments process, one that in many ways is more vexing than the problem of developing assurances about deployed systems. This dilemma arises because "up to eight years of preflight development is usually required to bring a new missile to the flight test stage. During the early development period, there is usually little or no evidence of an identifiable new program."[9] As Klaus Knorr explains, the ability to obscure R&D efforts creates powerful incentives to cheat on any limitations that might be negotiated: "Unfortunately, the suspension of the qualitative arms race, of the pursuit of military research and development, is practically impossible to verify. Given the profusion of laboratories in each country, the evasion of an agreement, whether formal or tacit, is relatively easy;

and *if only for protection against the other's evasion, the inducement to evade is considerable*" (emphasis added).[10]

The emphasized words define the prisoner's dilemma in the R,D,T,&E area. Observation is impossible until the weapon undergoing development has reached the stage of active testing. Testing occurs late in the overall process, and there is considerable lead time before a new weapon reaches that point. Thus, a state making a real breakthrough (for example, a hard-target capability, an effective, possibly laser-guided, ABM) that had not been countered or duplicated by ongoing R&D in the other could emerge with a crucial advantage for which the other would need literally years to compensate. Both sides would gain by being able to expend R&D efforts in nonweapons areas, but a conservative, minimax-oriented analysis would have to conclude that the costs incurred in a continuing weapons development program are far less than the potential disadvantages of not continuing the effort.

The dual limits on monitoring qualitative improvements and R&D place an inherent limit on verification as it now exists: such techniques are far more effective for monitoring quantitative arms limits than they are in monitoring qualitative aspects of the arms race. Current capabilities allow the monitoring with high confidence of numbers of deployed weapons systems (at least launcher numbers) and have the potential, through testing bans, of limiting the ability to deploy systems that have not yet undergone testing (discussed later in this chapter). These same techniques are ineffective in monitoring the deployment of systems improvements that have already undergone testing or of monitoring the critical basic scientific and engineering activity that allows new weapons to come into existence. This limitation is the most difficult problem facing arms control advocates and represents a prisoner's dilemma the solution for which has not been found.

Different Forms of Arms Control Proposals to limit strategic arms fall into three categories: arms freeze, arms reduction, and disarmament. As was pointed out in the last section, attempts to monitor compliance to agreements are not perfect, and the ability to cheat on agreed limits is inevitably present in any arms control situation. This fact creates the single greatest tension between those who view one or another form of arms control as desirable prima facie and those who are basically suspicious of arms control but will accept agreements if they can be convinced that particular measures are not detrimental to national security.

This dynamic tension relates directly to the incentives or disincentives that attach to cheating on any given agreement. In turn, the incentives issue has two basic elements: the level of armaments enshrined in the agreement and, consequently, the amount of confidence the security community has in the security afforded; and the amount of advantage that can be gained by cheating on an agreement. These factors, as we shall see, are related and interact with the plausibility of avoiding detection of violations in the calculation of cheating incentives or disincentives. As Perry points out, "If the rewards for successful tampering seem great, the probability of being caught slight, and the prospective penalty small, evasion becomes increasingly attractive."[11] Conversely, if there is little gain to be accrued and the possibility of detection is considerable, disincentives are maximized.

Taking advantage of opportunities to evade arms control provisions thus represents a cost-gain analysis and one where those charged with policy making must reach judgments based on their assessments about the relative strengths of incentives and disincentives. In areas where verification is uncertain, the question of whether cheating is beneficial enough to warrant the effort, if answered negatively, becomes an alternative way to escape the prisoner's dilemma. The dynamics of this incentives-disincentives calculus, however, vary depending on the kind of arms control proposal put forward.

Arms Freeze The form of arms control most generally acceptable to members of the security community is the simple arms freeze: agreement to put ceilings on the number of strategic arms at a level consistent with actual deployments (already in place or in production) at a given point in time. This form of agreement, almost axiomatically, is the least objectionable to defense and security planners: since arms levels are not reduced, there is a minimal threat to basic security, which is the prerequisite condition for entertaining arms control notions (presumably, existing armaments levels were developed because it was calculated that they were necessary for deterrent stability). Because both sides to an agreement must be satisfied with the preservation of their security, "it is inevitable that some kind of rough balance will have been reached between the major powers in numbers and capabilities of delivery systems"[12] before agreement is possible.

SALT I and the Vladivostok accords are examples of applications of the arms freeze model. They are the easiest form of agreement (which is not to say that reaching any arms limitation accord is easy), because they represent formalization of a status quo with which both parties presumably are satisfied, although satisfaction is a relative and subjective matter. Such agreements do not force the parties to give up anything they already have (which makes them more palatable to the military) and can be argued as producing benefits even by those not automatically receptive to arms control as a general movement. Thus, Donald G. Brennan says, "In so far as the United States and the Soviet Union have any common interest, it would be in limiting the scale of potential or actual damage in a major nuclear war. Hence, both superpowers have a common interest of major proportions in limiting the size of existing strategic offensive forces."[13] Many would argue that, given the size of nuclear arsenals, arms freezing is inadequate for meaningful damage limitation. These forms of agreement, when limited to quantitative ceilings on items such as launchers and types of launchers, are amenable to reasonable verification, although projected systems like concealed mobile and cruise missiles potentially present formidable monitoring difficulties.

Of all arms control forms, the arms freeze presents the least incentives to cheat, for at least two reasons. First, an arms freeze presumably institutionalizes a status quo which the participants at least implicitly have accepted as providing deterrent stability. From the American assured destruction viewpoint, for instance, stability is defined as secure second-strike capability, which our current arsenal provides (leaving aside the future problem of ICBM vulnerability). An arms freeze simply reinforces this capability, whereas any reduction in arms (or for that matter, asymmetrical vertical proliferation) potentially threatens the defined basis of stability. This dynamic leads Fred C. Ikle to conclude that "this order is so constructed that it cannot move toward abolition of nuclear weapons. It demands, as the necessary condition for avoiding nuclear war, the very preservation of these arms, always ready to destroy entire nations."[14] If an agreement preserves a stable and acceptable situation, then incentives to violate it are minimal; and disincentives for being caught cheating, which would include international approbation and possible treaty abrogation resulting in the possibility of a new and potentially destabilizing arms race, are maximized.

Second, cheating incentives are minimized by the bulk of arms permitted in a freeze agreement, which also relates to the dangers of being detected. With launcher ceilings in excess of two thousand apiece and with deliverable warheads hovering near or in excess of ten thousand, small violations are unlikely to produce substantial advantage. Conversely, to gain meaningful strategic advantage (something seriously threatening second-strike effectiveness or, in other words, approaching first-strike capability) would require such massive violations of agreed limits that it would be impossible to conceal them, resulting in the negative consequences already listed. Such behavior would be nonsensical unless one's original intention in entering an agreement was duplistic (for example, part of a plan to gain strategic advantage) and could probably be accomplished with less international disapproval without a freeze.

Arms Reductions Those who advocate arms reductions disagree that it is necessary to maintain current armaments levels, because the destructive capabilities of both sides are so enormous that assured destruction could be guaranteed with far fewer weapons than are currently in nuclear

Arms Control

arsenals. Thus, arguments (such as that of Brennan quoted above) are specious when applied to any notion of limiting the devastation a nuclear war would produce, and only a reduction in the number of arms can possibly avoid an exchange being disastrous.

Arms reduction proposals call for either "shallow" or "deep" cuts in the number of weapons in arsenals. Although there is no precise breaking point between these two ideas, "shallow" cuts generally refer to fairly modest decreases in strategic vehicles (the SALT II agreement, calling for roughly a 10 percent reduction in launchers by 1981, is a primary example), whereas "deep" cutting involves substantial arsenal reductions in the direction of nuclear disarmament. Following the logic of arms reduction and the definition of current capabilities it puts forward, the deeper the cuts, the more desirable they are. Shallow cuts such as those proposed in SALT II do not seriously affect the deadly calculus of nuclear destruction (among other things, they involve phasing out the oldest and least capable weapons that both sides would just as soon get rid of anyway) and are primarily symbolic in their value. The possibility of meaningfully reducing the destructive potential of nuclear war therefore requires deep cuts in arsenals that would factually reduce damage capabilities.

The logic of this argument is intuitively and individually appealing (one cannot avoid being drawn to the possibility of personally surviving a nuclear war), but, paradoxically, the results of arms reduction could make nuclear war more likely, and the potential for destabilization increases the more deeply arsenals are cut. Colin S. Gray presents the basic antireduction argument:

> Heretical though this may sound, major force-level reductions through SALT are not merely of little interest to negotiators but could be positively dangerous. The basis for the SALT negotiations is a healthy redundancy in the means for conducting strategic retaliation. The lower the strategic force levels, the more rigorous must be the rules for verification, the more important any major technical surprise, and the more obvious any differences in the structures and capabilities of the strategic forces of the superpowers—not to mention the encouragement that lower strategic force levels would give to another (nuclear-armed) power desirous of diminishing the strategic differential between the Big Two and itself.[15]

Deep-cut arms reduction thus strikes at such cherished values as the planned redundance that is the basic justification of TRIAD, as well as raising specters of an unpredictable future situation when both sides are relatively content with present realities. This position can be subdivided into two related points in opposition to arms reduction.

The first point arises from, and is partially supportive of, the arms reduction contention that only deep cuts would make any real difference in the destructive balance. As Rathjens points out: "Reductions on the order of 50 percent in the numbers of weapons delivered against urban-industrial targets are likely to mean reductions of, at most, 10 to 15 percent in damage levels. Thus, while reductions of, say, 30 to 50 percent in force levels may be politically impressive, and depending upon how taken . . . , they can hardly be characterized as radical or dramatic if the measure is damage-inflicting capability."[16] This analysis arises from the fact that the number of large (and thus "prime") urban targets in any country is limited and that, as a result, additional warheads are aimed at smaller, more peripheral targets or redundant warheads are aimed at prime targets as a hedge against malfunction or interdiction. The less important targets would be the first to be removed from targeting lists under reduction limits, meaning that any sizable reduction in damage infliction potential would require even greater cuts.

This fact leads to the second and related point: the severe cuts in warheads (and consequently launchers) that meaningful damage reduction requires make each remaining launcher more vital. As Kruzel states in the specific context of ICBM cutbacks: "The phased reduction of ICBMs is an appealing but paradoxical proposition, and it must be approached with great care. The paradox is that if such reductions were ever agreed upon, the effect would be to place greater importance on each remaining ICBM. This no doubt would spur each side to maximize qualitative improvements."[17] If truly deep cuts (beyond 50 percent) were agreed upon, there would, in all likelihood and depending on how the reductions were defined (for example, launchers, throw-weight, warheads), have to be reductions in all TRIAD legs, so that the ICBM analogy can be extended to all systems. In turn, the effects of reductions relate to verification and cheating incentives and thus to strategic stability.

As argued earlier, the principal virtue of current force deployments is their redundancy and the consequent inability of a potential adversary to calculate survival after a preemptive strike. To the degree one agrees that present force configurations provide secure second-strike capability, an arms freeze does not remove the considerable disincentives to try to overcome that strategic reality. As one begins to reduce armaments levels, however, the cost-gain analysis of the benefits to be derived from cheating begins to change, and the deeper the cuts are, the greater the incentives become.

The shallow cuts proposed under SALT II do not, in any real way, affect the destructive balance: since only obsolete weapons will be involved (and not very many of those), qualitative improvements of current systems will easily compensate. As a result, incremental, small-scale violations that one might be able to get away with will have little effect on the balance, and only large-scale, detectable cheating could make a difference. Thus, incentives to evade the agreements are virtually as absent as in a freeze.

Truly deep cutting that would have measurable impact on destruction levels would require cutbacks in excess of half of existing arsenals, and, at that point, the calculus shifts, because the incremental benefit of each violation increases The impact, for example, of hiding 100 mobile launchers with hard-kill capability for use against 1,054 American ICBMs is not particularly threatening, but hiding that same number against a force of 500 is more tempting (it threatens 20 percent of capacity). The deeper the cuts, in other words, the more precious is each remaining weapon, and thus the greater is the incentive to create covert means to destroy that weapon. The situation is exacerbated by multiple warhead missiles, because destroying one launcher has a multiple effect on retaliatory destructive threats. Put another way, the calculation of decisive advantage through cheating (gaining first-strike capability) increases as the number of targets one has to preempt against decreases.

Disarmament The ultimate form of arms reduction is complete and total nuclear disarmament. A world of no nuclear weapons minimizes absolutely the amount of destruction such weapons could produce and therefore is appealing. Unfortunately, a condition of genuine disarmament would maximize the incentives to cheat in areas where detection would be extremely difficult if not impossible. Since the arguments supporting this contention are extrapolations from the effects of deep cuts in armaments, they can be treated more summarily.

The incentives to evade disarmament provisions arise from the ease of maintaining covert weapons stockpiles (with the accompanying knowledge that the other side can do so as well), the potential strategic advantages of keeping a covert stockpile, and the ease of constructing a new stockpile should the need arise. The first two points relate to the basic arms race prisoner's dilemma; the third is a product of the R&D and technological subdilemma.

Although national technical means of verification would make it relatively easy to monitor the dismantling of strategic delivery systems in a disarmament agreement, warhead destruction would be much more difficult to verify. Some form of on-site inspection (Soviet attitudes toward which have already been discussed) would undoubtedly be required, but that process would be imperfect: the inspectors could attest to the defusing of the warheads they saw destroyed, but they would have no means to know if other warheads were hidden somewhere else that were not being destroyed.

There would be powerful incentives for each side to cheat on a disarmament agreement that would make such tactics likely. First, the incremental advantages would be considerable: one hundred hidden nuclear devices might not make much strategic difference in a world where both sides have several thousand, but they could be terribly decisive if the other side has none. Second, if this calculus occurs to one side (which it obviously would), it would also occur to the other, which would have to take into consideration the probability that the other side was violating the agreement in its planning. The result of this analysis is a new prisoner's dilemma, and in a sense, one that is more difficult to overcome than the dilemma attached to less severe arms control measures: verification and the absence of cheating disincentives are no longer available as alternatives to trust. To exacerbate the situation, small delivery systems like the cruise missile also make delivery vehicle concealment easy, further weakening the basis of a disarmament agreement. Consider, for example, a hypothetical situation that seems far-fetched but is technically feasible.

The Boeing wide-bodied 747 aircraft contains, on each wing, an extra power pack (or pod) beneath the wing to act as an emergency power source. With slight modification that would not be externally visible, an ALCM could be fitted into each of these pods with very little danger of detection, making, metaphorically speaking, the "friendly skies of United" a potential strategic delivery system. This example may appear outlandish and open to ridicule, but the idea that parallel schemes would not be devised in a disarmed world is naive.

Even if all warheads could be destroyed, the problem remains that the knowledge of how to build nuclear weapons would not disappear with the last warhead's demise. Even if nations were sufficiently committed to disarmament not to "fudge" on agreements with residual stockpiles, simple prudence would require keeping R&D at a sufficient level to allow quick rearmament in the event of evidence that the other side might be cheating.

The disarmament dilemma (which also applies to deep-cut arms reduction) is that in making the prospects of nuclear devastation less horrendous, it might simultaneously make nuclear war more likely. This conclusion, of course, derives from conventional notions about deterrence based upon second-strike assured destruction (or nuclear war-winning capability) as the basis of stable deterrence. If one finds those arguments incredible or repugnant, one will disagree with those conclusions. Regardless of agreement or disagreement, however, this line of argumentation underlies much of the ongoing arms control debate and discussions, as the quotation from Gray in the section on arms reduction indicated.

The Evolution of Arms Control SALT has occupied the arms control spotlight for a decade, but it is only the most visible aspect of the arms control movement. The desire to limit strategic nuclear armaments has been a feature of the international strategic system since 1946 and the announcement of the Baruch Plan for nuclear disarmament. Early attempts at arms control foundered, partly because limitations on monitoring technology did not allow escape from the prisoner's dilemma and partly because the world had not been sufficiently sobered by the prospects of nuclear war to perceive enough incentives for serious arms control to begin the process in earnest.

The Cuban crisis, closely following the development of satellite surveillance, provided the incentive and the means to get arms control moving. Scoville describes the directions that arms control processes took in the wake of Cuba: "Following the successful negotiation of the Limited Test Ban Treaty and the Treaty on Outer Space, attention in arms control negotiation shifted first to the multilateral Treaty on the Non-Proliferation of Nuclear Weapons and, secondly, to bilateral negotiations between the U.S. and the U.S.S.R. to limit strategic delivery systems."[18] It is not the purpose here to treat in detail the multifaceted forms that arms control efforts have taken, but rather to look at those aspects of arms control that have had an important impact on the strategic balance. Scoville's description is helpful in creating the necessary basic distinctions and indicating the directions that movement has taken. One analytical category has been in trying to reach agreements that are not directly SALT-related. These agreements share the characteristic of being multilateral and thus beyond the scope of SALT (although having an impact on SALT outcomes). These efforts have focused on the Non-Proliferation Treaty (NPT) and more recently on various forms of limitation on nuclear testing. The other analytical category encompasses the SALT process and will require examination of the various stages of SALT: SALT I, the Vladivostok Accords, SALT II, and beyond.

The Early Period and Cuba As described in Chapter 3, the period between the end of World War II and the middle to latter 1960s was characterized by enormous growth in destructive capabilities and dramatic environmental changes that affected the development and evolution of nuclear doctrine. Particularly in the period before the Cuban missile crisis, weapons arsenals were evolving too rapidly for policy makers to have the requisite confidence in their arsenals and doctrinal bases to feel adequately secure enough for arms control measures to succeed. Samuel F. Wells, Jr., for example, explains how the major effort of the Eisenhower years, the so-called "open skies" proposal, was subverted by the environmental pressure of technological dynamism: "In the spring of 1955, American officials headed by Harold Stassen began a series of comprehensive arms control negotiations with the Soviet Union covering conventional and nuclear arms, limits on testing, nuclear free zones, and restricted aerial inspection. Prospects for agreement faded when

Soviet successes in rocket development threatened to alter dramatically the strategic balance."[19]

The impact of the ICBM on notions of viability need not be repeated here (see discussion in Chapter 2), but adjustment to the new condition dampened serious arms control until the wake of the Cuban missile crisis created the perception of adequate incentives to reactivate arms control. The result of the post-Cuban stimulation were the Hot-Line Agreement (providing for quick teletype communication directly between Washington and Moscow) and the Limited Test Ban Treaty. This stimulus proved to be temporary, however, because of the evolving dynamics of the nuclear balance. As Moulton explains: "Apart, therefore, from the achievement of the partial test ban in 1963, all other efforts through 1965 to agree with the Soviets on major arms control arrangements . . . failed to make any progress whatever. This was due in large measure to the continually widening margin of strategic superiority being achieved by the United States during this period."[20] In other words, the United States was more committed to its programmed deployments of nuclear weapons than it was to arms control. Kahan states bluntly: "During the early 1960s the United States was simply not ready to take serious steps toward controlling strategic systems."[21] At the same time, the Soviet Union showed a parallel reluctance until it achieved some measure of strategic parity in the latter part of that decade.

Thus, fruition of the movement toward strategic arms control came in the latter 1960s. The technological means to guarantee adequate assurance that agreements would be honored, within the limits already described, were available. Soviet achievement of a secure and, to them, adequate strategic capability provided the prerequisite confidence in armaments to allow consideration of limitations, just as American deployments ending in 1967 had met our defined needs for an assured destruction deterrent force. Within that context, we can examine the two arms control "tracks" with obvious strategic nuclear implications.

Nuclear Proliferation and Testing Limits The Limited Test Ban Treaty, signed in Moscow by the United States, the Soviet Union, and the United Kingdom on August 5, 1963, and ratified by the United States Senate on October 7, 1963, was the first successfully concluded agreement to place limitations on nuclear testing. Stimulated by negative public opinion to atmospheric radiation levels caused by testing above ground, the treaty prohibited all nuclear explosions in the earth's atmosphere and was followed by a series of additional agreements extending testing prohibitions that are described by Wayland Young (Lord Kennet): "There followed a spate of agreements not to do things that nobody would want to do anyhow; not to plant nuclear weapons on the seabed, not to use nuclear weapons in an accidental or unauthorized way, not to militarize the moon, or to put nuclear weapons into orbit round the earth, not to introduce nuclear weapons into Latin America."[22] The effect of these agreements among the nuclear possessors was to leave underground explosions the only permissible form.

There have been two emphases in extending arms control provisions beyond these early efforts. One emphasis has been to avoid so-called horizontal proliferation of nuclear weapons to nations that do not already possess them, as represented by the Non-Proliferation Treaty of 1970 (NPT) and given added impetus and urgency by demands for nuclear power generators in the Third World stimulated by the OPEC oil boycott of 1973 and the Indian explosion of a "peaceful" nuclear device in 1974. The second emphasis has been on reducing the size or eliminating altogether nuclear testing, a movement made possible because "seismic detection between earthquakes and explosions has improved markedly since the Limited Test Ban Treaty of 1963."[23] This second emphasis is represented by the Threshold Test Ban Treaty (TTBT) of 1974 and its companion, the Peaceful Nuclear Explosions Treaty (PNET) of 1976 and proposals for a Comprehensive Test Ban Treaty (CTBT).

These two thrusts are related to one another and are complementary to the SALT negotiations. This relationship can be seen in William Sweet's description of the Indian position on accepting NPT and the subsequent safeguards on nuclear usage arising from it: "India ... will accept full-scale safeguards only if the nuclear weapons states negotiate a comprehensive test ban treaty, agree to terminate production of weapons-grade material, and take significant steps toward nuclear disarmament."[24] The relationship to SALT arises from the disarmament issue: Third World states have been particularly critical of the "vertical proliferation" of superpower nuclear arsenals. Thus, "an effective SALT agreement ... , will meet the commitment which was made by the nuclear-weapons states that signed the Non-Proliferation Treaty of 1970."[25]

NPT The NPT has been lavishly described as "in many ways the most significant" of the arms control agreements, because "it circumscribes the freedom of both nuclear powers to pursue policies that are practicable and even, from many points of view, desirable."[26] It has also been one of the most unsuccessful arms control measures. To understand the dynamics and difficulties of NPT requires first looking briefly at the problem the treaty seeks to solve, then at what NPT tries to do, and, finally, why it has been resisted by many of the nations whose nuclear behavior it seeks to restrict.

The $N+1$ Problem: Controlling the spread of nuclear weapons to nonweapons states, and particularly politically unstable states whose responsibility with nuclear devices is questionable, is a serious problem for the international strategic system. As Paul Doty, Albert Carnesale, and Michael Nacht capsulize the problem, "For of all the elements that can contribute to a less stable world, that of horizontal nuclear proliferation is perhaps the most threatening."[27] The chilling prospect of numerous nations possessing nuclear capability raises grim possibilities ranging from regional nuclear conflicts into which the superpowers might catalytically be drawn to the threat or use of these explosives by terrorist groups, leading Lewis A. Dunn to observe, "A world of many more nuclear states is likely to be an increasingly competitive and nasty place, one in which the United States and other nations would find themselves directly threatened in a variety of ways."[28] Michael A. Guhin summarizes what that variety of destabilizing ways could be: "The emergence of several small nuclear powers may present no direct threat to the nuclear superpowers. But it would surely introduce (1) complications, as in achieving international or regional restraints on nuclear weapons, (2) some unknowns, such as the possible transfer of nuclear capability to other countries or groups, and (3) substantial risks, including an increase in the possibilities of technical accident, unauthorized use, preemptive surgical military actions by others, and third-country provocative attack."[29]

The problem gains focus if one examines the list of what Alan Dowty would call "near-nuclear" states (a group he maintains has not been given adequate attention in nuclear thinking[30]). The present nuclear weapons "club," including the year each first exploded a nuclear device, is comprised of the United States (1945), the Soviet Union (1949), the United Kingdom (1952), France (1960), and the People's Republic of China (1964). To this group, many would add India by virtue of its 1974 self-proclaimed peaceful nuclear explosion (PNE) and Israel, which is widely suspected of having stockpiled ten to twenty weapons from materials extracted from their unsafeguarded Dimona nuclear reactor. A minimal list of near-nuclear states (countries that have, or will soon have, the ability to exercise the nuclear option) includes Argentina, Brazil, the Federal Republic of Germany, Iran, Japan, Libya, Pakistan, the Republic of China (Taiwan), South Africa, and South Korea. A number of West European states have had the capacity but not the desire to produce nuclear weapons for years (for example, Sweden and Denmark), such other states as the Arab Republic of Egypt, Mexico, Nigeria, and Syria are not far from joining the near-nuclear club, and, by the end of the century, the list will be greatly expanded.

The need to give serious consideration to the problems of proliferation was highlighted by two events in 1974, as Joseph S. Nye explains: "One was the Indian explosion of a 'peaceful' nuclear device using plutonium derived from a Canadian-supplied research reactor. . . . The other was the oil embargo and fourfold increase in oil prices, which

Arms Control

created widespread insecurity in energy supply."[31] The economic effects of the OPEC-induced oil problem have been felt especially strongly in the developing world and have resulted in the search for alternate, including nuclear, energy sources. In turn, this energy need has translated into a considerable international market for nuclear power-generating reactors that, in most cases, produce weapons-grade material as a by-product of their cycles.

Access to weapons-grade plutonium is the critical element in gaining the nuclear option, as can readily be seen by examining the requirements for producing a weapon supplied by Ted Greenwood, George Rathjens, and Jack Ruina:

> The requirements for any organization to make a simple, practical, militarily useful fission bomb are:
> (1) an understanding of the nuclear theory involved in fission;
> (2) data on the physical and chemical properties of the basic materials in a nuclear weapon;
> (3) technical facilities to fabricate a weapon and to test implosion or gun devices and other components;
> (4) availability of fissile materials;
> (5) the will to allocate the necessary resources to develop a weapon.[32]

Combined with the political desire to build weapons, access to weapons-grade material completes the cycle, because "the first three requirements are met by virtually every country with a significant industrial capability, for the open literature, much of it stemming from nuclear reactor technology, contains a wealth of material which was very hard to come by twenty years ago."[33]

The problems created in the nuclear balance when the number of nuclear states possessing weapons ("N") increases ("+1") arise in this manner. Hidden within this concern is an additional aspect of the N+1 problem: each new weapons state tends not to view itself as the problem, but to look at other states as "+1." As Albert Wohlstetter explains: "So far as long-run world stability is concerned, the nth country tends to think of the problem as beginning with N plus 1. The original irony intended by the label 'nth-power problem' was seated in the fact that the United States and the Soviet Union thought of the trouble as the third-country problem, Great Britain thought of it as the fourth-country problem, France as the fifth-country problem, and so on."[34] The degree to which one considers proliferation a problem is thus a matter of perspective: one tends to view it as a matter of concern after one joins the nuclear club. As a result, the dialogue between those who possess nuclear weapons and those who aspire to that status operates at different levels, with neither side fully appreciating the other's position.

The Substance of NPT: The basic dynamic of the NPT is simple: in return for non-nuclear states signing the treaty and thus renouncing acquisition of nuclear weapons, the nuclear states would refuse to aid proliferation and seek to reduce existing nuclear arsenals. As Guhin explains:

> Nuclear-weapons states party to the treaty agreed not to transfer nuclear explosive devices to any other country (nuclear or non-nuclear) and not to facilitate, in any way, the acquisition of such devices by any non-nuclear-weapon state. Non-nuclear parties to the treaty agreed not to construct or otherwise acquire nuclear explosives for military or nonmilitary purposes and to allow IAEA [International Atomic Energy Agency] safeguards on all their nuclear activities, whether indigenous or based on materials and equipment imported from other countries. All parties assumed an obligation to require the application of IAEA safeguards on any exports of fissionable materials and related processing or production equipment and materials.[35]

The effect of NPT is thus to create a kind of nuclear weapons caste system consisting of two categories of states, as Bertrand Goldschmidt points out: "It aims at dividing the world between the five nuclear states, which undertake not to assist any country to manufacture a nuclear explosive device, and the other states, which undertake not to proceed with such a fabrication."[36]

The importance of NPT as a means of nuclear stabilization has been extensively heralded, but with its obviously discriminatory aspects noted. As George Quester states, "A halt to proliferation will probably now indeed require the general acceptance and ratification of the NPT. But NPT in several ways enshrines the existing five nuclear powers, subjecting all other states to IAEA inspection, acknowledging the birthright to these weapons in five cases and eliminating it in all others."[37] This means that the nonweapons states are required to make sacrifices within the treaty that are not required of the weapons states (for example, provision for IAEA inspection is required *only* of the nonweapons category). Young states this distinction dramatically: "The nuclear weapons countries solemnly swore not to part with their privileges, and most non-nuclear weapons countries solemnly swore not to tempt them to break that oath."[38]

Verification of conformity to the treaty comes through the IAEA safeguards, including the right of IAEA to station inspectors in countries party to NPT. The adequacy of these safeguards to prevent a state from clandestinely diverting materials to weapons use has, however, been criticized on two basic counts: not enough inspectors are available adequately to monitor all nuclear activities in a country; and the inspectors lack intrusive rights to investigate suspected violations at reactor sites. Because they can be excluded from the sources of potential cheating, the inspectors must rely on more indirect methods, such as analyzing electricity flow rates from nuclear generators (to extract fuel from a generator, either its output must be sharply curtailed or the facility must be shut down entirely). These limitations lead Guhin to ponder, "Whether IAEA safeguards, based as they are primarily on national materials accounting records, can effectively meet the limited objective of detecting national diversion, and thereby deterring it, remains subject to question."[39]

Resistance to NPT: Four groups of states have been involved with NPT: the five nuclear weapons states that have sought to limit proliferation; states that either have not the capability or interest in the foreseeable future to develop nuclear weapons; states whose behavior can be influenced by one of the major weapons states; and states that have or will soon have the ability to exercise the nuclear option. This categorization is useful for distinguishing among states that are parties or nonparties to the agreement, as Sweet explains: "States that are very far from being able to build nuclear weapons have ratified the treaty. And countries that are allied with nuclear powers . . . have ratified the treaty. *But no completely non-aligned country capable of building nuclear weapons has joined the treaty*" (emphasis in original).[40] Renunciation of adherence to NPT is a mechanically easy task accomplished by simply stating the intention to do so (although no state that has ratified NPT has done so), indicating reluctance to forego the possibility of going nuclear. Two related arguments are generally made against NPT acceptance.

First, it is argued that the treaty imposes unequal burdens on the nuclear and non-nuclear states and that the nuclear weapons states have been reluctant to live up to their minimal vertical proliferation obligations under NPT. John Maddox summarizes this position: "Non-nuclear parties to the treaty, even if they regard as desirable the prevention of what is awkwardly called horizontal proliferation, are also concerned that the nuclear powers should implement their explicit undertaking on nuclear arms control, an argument also advanced (not usually in isolation) by non-signatories."[41] Sweet summarizes the position of India, possibly the most prominent NPT nonsignatory, succinctly: "The non-Proliferation Treaty seeks to disarm the unarmed while allowing the armed to keep arming."[42] Second, the obligations are asymmetrical, requiring more sacrifice from the non-nuclear than the nuclear parties: "Non-nuclear states have argued that the general requirements imposed on the nuclear powers to work toward measures of nuclear arms control are an insufficient quid pro quo for the obligations non-nuclear states are required to make."[43] This argument is increasingly accompanied by demands for greater inducements to sign, such as guaranteed fuel supplies in exchange for accepting IAEA inspection.

It is significant that acceptance or nonacceptance of NPT, in the final analysis, has amounted to whether a state must make substantial sacrifices under its provisions: the nuclear states, committed to the SALT process anyway, give up essentially nothing, whereas the nonsignatories are asked to renounce a potentially geopolitically important option. As is pointed out later in the chapter, this fact is one of the lessons of arms control generally. With the failure of NPT to act as a satisfactory means to dampen the prospects of nuclear proliferation in the face of increasing dangers created by nuclear generator proliferation, a second approach, at least providing some symmetry in sacrifices and requirements, has emerged as the means to prevent the spread of nuclear weapons: limits or bans on nuclear testing.

TTBT and PNET The U.S.–Soviet signing of TTBT on July 3, 1974, and PNET on May 28, 1976, were of enormous symbolic importance to arms control advocates, as Robert Helm and Donald Westervelt explain:

> Perhaps the most significant aspect of the new test ban treaties bearing upon further, more restrictive test bans is the fact that the TTBT and the PNET together represent the first move away from the status quo in thirteen years. Since the signing of the Limited Test Ban Treaty (LTBT) in 1963, efforts in the Conference of the Committee on Disarmament (CCD) and elsewhere to achieve a follow-up agreement that would prohibit all underground nuclear tests have been stalled on the issue of verification.[44]

The two treaties form a tandem for the dual reason that they establish a common limit on the admissible yield of a nuclear explosion and together they close the loophole of weapons testing under the guise that explosions are made for nonweapons purposes (a factor largely occasioned by India's self-proclaimed PNE in 1974).

TTBT establishes a yield limit for underground tests of 150 kilotons (KT) or less, a level where "verification . . . is more-or-less assured . . . with the exchange of the test site geological and calibration data required"[45] to allow adequate seismographic detection. The limit established is not arbitrary, but reflects the state of seismology and thus the confidence of seismologists in their capacity to differentiate nuclear tests from natural geological phenomena like earthquakes when such tests take place at established locations about which they have adequate geological data (which is required in the treaty).

The 150-KT level is a relative measure of reasonable assurance of verifiability rather than an absolute measure (the ability to discriminate a blast from an earthquake at that level is not absolute), and it has been suggested that a nation could cheat on the provision by conducting tests in conjunction with natural geological phenomena. Because the treaty does not prohibit all testing, George Rathjens and Jack Ruina suggest that it does not solve the problem of demands for a comprehensive ban and could even exacerbate the problem:

> A 150 kiloton threshold treaty will fuel cynicism and distrust of the genuineness of United States and Soviet interest on curbs to the nuclear competition, since it will be widely understood that, given the current state

Arms Control

of seismology, the 150 kiloton limit has not been determined by verification capabilities but is rather a consequence of mutual interests in continuing tests at a level high enough to have minimum impact on nuclear programs.[46]

PNET extended the TTBT constraints to all nuclear explosions by limiting "to 150 KT the yield of any individual peaceful nuclear explosion (PNE) and the 1,500 KT the combined yield of any group of PNEs detonated in rapid succession."[47] Although tests below the threshold may be conducted without restriction, Helm and Westervelt detail the mechanism for monitoring larger explosions: "Under the cooperative PNET inspection arrangement, the party conducting a PNE is obliged to invite 'designated personnel' of the other party to conduct on-site observations of specific PNE events that fall under this requirement of the treaty. This provision applies only to multiple-device PNEs whose aggregate yield exceeds 150 KT and whose time and location are completely under the control of the host country."[48] The combined effect of the two treaties thus closes the incentive loophole to maintain that weapons tests are PNEs, as Donald G. Brennan explains: "Since the parties are allowed unrestricted underground weapon testing up to 150 kilotons, they naturally would have no incentive to conduct a weapon test of lower yield in the guise of a PNE shot."[49]

CTBT The Threshold and PNE treaties represent compromise solutions to a total ban on nuclear weapons testing. Proposals for a comprehensive ban have been made for some time, but have always been hamstrung by the verification problem and particularly by intrusive monitoring below the level of confident seismographic determination. As Rathjens explains: "A CTB could have been negotiated at virtually any time during the '60s but for American insistence on, and Soviet resistance to, on-site inspection as a means of verifying that remotely observed, suspicious signals had their origins in earthquakes rather than in clandestine underground weapons tests."[50] Amid speculation during the summer of 1978 that the Carter administration would unilaterally declare a moratorium on testing,[51] progress toward CTBT remains stuck on this crucial point.

A comprehensive ban on nuclear weapons testing would have much more dramatic effects than the TTBT-PNET combination. As Richard Falk notes: "A comprehensive test ban (CTB) would represent a more momentous step down the denuclearization path because it would substantially inhibit further nuclear weapons innovations, provided it is imposed before accuracy, reliability, and throw-weight goals have been attained."[52] Warnke points out that a CTBT would parallel and reinforce the SALT process as well: "A comprehensive test ban treaty would complement our efforts at SALT by rendering it more difficult to develop new types of nuclear weapons."[53] By stunting warhead technology at the testing stage of R,D,T,&E, a CTBT could also provide a precedent for limiting or banning delivery system testing within SALT (discussed later in the chapter), as well as meeting one of the critical demands of the potential weapons states.

The CTBT idea has not, however, met with unanimous acceptance. Two objections have been raised by Brennan: the danger of cheating and the potentially deleterious effects on the American nuclear arsenal. Regarding the verification problem, he cites two incentives for the Soviets to violate CTBT provisions: "The importance of covert Soviet tests would be greatest for two possibilities: a) Soviet confidence tests of one or more weapons in their deployed stockpile, as an immediate prelude to an attack on NATO and the United States, and b) Soviet development of nuclear weapons based on wholly new technology and economics, leading to wholly new military applications of nuclear weapons."[54] The second problem he sees is for the American nuclear program and is, in fact, a mirror image of the problem he envisages as a cheating incentive for the Soviets: the effect on the U.S. arsenal. Maintaining that it is necessary occasionally to detonate a deployed warhead to be sure that it has not degraded and that testing is necessary to keep the American development effort vital, he cautions that "the problem of weapon-stockpile reliability constitutes an important potential cost for the United States if we undertake a CTB."[55]

Steinbruner and Garwin disagree that the warhead degradation problem is sufficient to warrant violations, because "there is no practical test program a potential attacker can undertake which will provide much confidence that his forces will perform in the real event according to the required specifications."[56] Since a CTBT could not be precisely monitored given current verification methods, the possibility for violations remains. To deal with this difficulty, Brennan proposes an "escape clause concept" whereby the treaty would be abrogated if there was strong suspicion that any nation had performed a clandestine test. He explains the mechanics: "It is important to note that the failure involved need not be confined to signatory states, and may happen entirely among non-signatory states. Once the determination of a failure is made, it would nullify the entire treaty, for all signatories."[57]

SALT The SALT process formally began in 1969 and has continued, with occasional breaks, ever since. The series of agreements signed on May 26, 1972, by President Nixon and First Secretary Brezhnev that are collectively known as SALT I has been the most visible achievement of those negotiations, but other notable accomplishments include the 1973 Prevention of Nuclear War Agreement and the 1974 Vladivostok Accords that set the basic parameters for the SALT II agreement.

The American impetus for discussions on limiting strategic nuclear armaments is a product of the Johnson administration. At the time, then

Secretary of Defense McNamara explained the motivations behind this decision: "Since we each now possess a deterrent in excess of our individual needs, both of our nations would benefit from a properly safeguarded agreement first to limit and later to reduce both our offensive and defensive strategic nuclear forces."[58] In other words, strategic parity gave both sides sufficient confidence to consider arms control, as Tammen asserts: "The stage for the successful SALT I negotiations was set primarily by the concerted Soviet buildup in land- and sea-based systems and the thorough modification of existing U.S. forces by MIRV."[59] There was even support for the process in the most generally anti-arms control sector of government, the defense establishment, because "many senior Pentagon planners and others looked to the SALT Talks [sic] as the best means available for limiting the further growth of Soviet strategic strength."[60]

SALT I The negotiations leading to the series of agreements cumulatively referred to as SALT I began in 1969 and culminated in the dramatic, last-minute signature of the agreements on May 26, 1972, by Brezhnev and Nixon in Moscow. Likened by its most thorough chronicler to the Congress of Vienna,[61] SALT I was a multitiered series of discussions involving both highly technical negotiations and strictly political considerations. Within the American decision process, SALT I provided a microcosm of the strategic doctrinal debate orchestrated by National Security Advisor Kissinger, contrasted with the tight control by top political and military officials of the Soviet delegation.

This process resulted in a lengthy series of agreements, joint declarations by the United States and the U.S.S.R. explaining what various provisions meant, and unilateral interpretive declarations where differences remained.[62] Four agreements were reached in SALT I, two of which were of major importance (the other two involved improvement of the "hot line" through satellite transmission and a convention on procedures in the case of accidental war). The two major accords were the Interim Agreement on Offensive Arms (IOA) and the ABM Treaty, followed in 1973 by the Prevention of Nuclear War Agreement.

Interim Agreement on Offensive Arms: The major thrust of the IOA was to place a five-year freeze on permissible number of strategic launchers (as opposed to warheads) for the two superpowers, including limited rights of substitution and rights to modernize offensive strategic systems. The agreement sanctified "national technical means of verification" as the way to monitor compliance, while leaving open a number of concerns, such as forward-based systems (FBS) and mobile missile deployment.

Under the agreement's terms the Soviets were allowed a larger number of ICBMs and SLBMs than the United States. Kruzel summarizes the general effect: "The United States granted the Soviet Union a three-to-two superiority in ICBMs, and an advantage of well over 100 sea-based missiles."⁶³ The mechanism to enforce these limits, which were partially compensation for the American "air-breathing" advantage and the inability to resolve the FBS controversy, was "a 'freeze' on new construction of ICBM launchers with numerical limits on ballistic-missile submarines and SLBMs."⁶⁴

John B. Rhinelander summarizes the two articles that set ICBM limits within the IOA:

> Article I leaves the Soviets with a maximum of 1,618 ICBM launchers and missiles with larger throw weights than the U.S. The U.S., according to present plans, will deploy MIRVs on 550 of its Minuteman missiles, for a total of 2,100 independently targetable warheads on its 1,000 Minuteman missiles. . . .
>
> The effect of Article II is to limit the U.S.S.R. to 313 launchers for "modern heavy" missiles. The conversion of ICBM launchers for older heavy ICBMs (the U.S. Titan, the Soviet SS-7 and SS-8) or smaller ICBM launchers (the U.S. Minuteman, the Soviet SS-11 and SS-13) to launchers for "modern heavy" ICBMs (such as the Soviet SS-9) is explicitly prohibited.⁶⁵

This conversion provision is generally viewed as a partial compensation for higher Soviet overall ICBM launcher numbers because it places a limit on the deployment of the 25 MT SS-9 that is not reciprocally limiting on the United States, which had no plans to deploy "modern heavy" ICBMs (not specifically defined, but generally thought of as large multiple-megaton launchers). In the process of allowing force modernization, it was agreed that silo volume could be increased by 15 percent to allow some increase in missile size. This provision, in turn, was something of a concession to Soviet technological inferiority: newer Soviet systems must be larger to gain additional capability, whereas newer U.S. missiles like MX can be fitted into existing Minuteman silos.

IOA does not place limits on ICBMs. Rather, as Kruzel points out, "the Interim Agreement is incorrectly thought to have established a freeze on ICBM's. Actually, ICBM's themselves are unlimited; the Agreement put a freeze only on silos for fixed landbased ICBM's."⁶⁶ This apparent loophole was largely a matter of verification: it is much easier to count silos than missiles, which can be hidden. As a result, "we should not be shocked (but probably would be) if some day it is revealed that the Soviets have a thousand or so additional ICBMs in inventory; if so, it would not be a violation of the SALT agreement."⁶⁷ This provision (or lack thereof) thus permits reloading of ICBM silos for subsequent firings and was a primary stimulus to the "silo-busting" counterforce provisions attached to the notion of damage limitation in the so-called Schlesinger doctrine (see Chapter 4).

Rhinelander also details the effects of the various provisions related to the SLBM balance:

1. The U.S. may not have more than 656 and the Soviets more than 740 "modern SLBM launchers" on "modern" ballistic-missile submarines unless other ICBM or SLBM launchers are dismantled or destroyed.
2. After exercising full rights of substitution, the U.S. cannot have more than 710 and the Soviets more than 950 "modern SLBM launchers" on "nuclear-powered" submarines.
3. The U.S. cannot have more than 44 and the Soviets more than 62 "modern" ballistic-missile submarines after exercising full rights of substitution.
4. SLBM launchers on additional "modern" ballistic-missile submarines may be substituted for "older heavy" ICBM launchers or for launchers for "modern" SLBM missiles on nuclear-powered submarines.
5. The launchers being replaced must be dismantled or destroyed, under agreed procedures, when the replacement submarine begins sea trials.[68]

These replacement procedures allow the United States to implement the Trident program by first substituting the new submarine for Titan ICBMs and, as the force is deployed, early Polaris submarines. Reciprocally, as Polmar explains, "the Soviet Union must halt SLBM production with 744 missiles unless it is willing to phase out older, land-based SS-7 and SS-8 missiles. . . . This replacement option would permit the replacement of some 200 older ICBMs with 200 SLBMs."[69] As was pointed out in Chapter 5, the Soviets have exercised these options, reflecting some acceptance of the U.S. "one-way freedom to mix" position.[70]

The IOA specifically legitimizes "national technical means of verification" and prohibits attempts to evade these nonintrusive monitoring means. As Kruzel explains in reference to national technical means, "This arcane phrase refers primarily to photographic satellites, but also includes communications and electronic means of surveillance operating outside the country being inspected,"[71] thereby setting the stage for seismographic surveillance within TTBT and PNET and electronic surveillance in countries like Turkey. Satellite photography is an adequate means "to determine the number of ICBM launchers the Soviets had which were operational or under construction and to identify those which are for test and training purposes."[72] SLBM deployment can be monitored in this manner, because "missile submarines are about 400 feet in length and, in the later stages, construction is normally carried out in open pens so that observation from above is quite easy."[73]

Two issues, FBS and a ban on mobile ICBMs, were left for future discussions. The FBS question (which, it will be recalled from Chapter 4, Kruzel maintains was the basic reason for an interim rather than a permanent treaty) reflects a Soviet position that Newhouse summarizes: "Moscow's aversion to a cluster of American air bases within striking distance of the motherland is genuine, long-standing, and quite unaffected by Washington's argument that these planes are tactical, not strategic, and essential to NATO war plans, notably to counter the 650 or more Soviet missiles targeted on Western European cities."[74] Although this problem has been overcome within SALT II by consigning FBS to the agenda for future (SALT III) discussions, it vexed SALT I. The United States position that mobile missiles should be banned on the basis of verification difficulties also could not be resolved, because "[T]he Soviets argued that such missiles should be exempt from the Agreement, and that specific limitations on strategic offensive forces should be discussed in SALT II."[75]

The ABM Treaty: The most permanent outcome of SALT I was the ABM Treaty, described by Rhinelander as "clearly the most important and most detailed agreement reached at SALT."[76] The treaty, which contains no time limitations other than periodic reviews, limits each side to two "light" ABM sites each: one around the national capitol (National Command Authority or NCA) and one around a designated ICBM field at least 1,500 kilometers from NCA. In addition, "Article VI of the treaty provides that not only shall air defense missile launchers not be given the capabilities to counter strategic ballistic missiles, but that they shall not be tested in an ABM mode. It is this prohibition on testing that provides additional confidence that the treaty is not being violated by deploying ABMs disguised as anti-aircraft systems."[77]

Although the motivations, details of negotiation, and changes in position by each side on the ABM issue are too complex to describe here, the ABM Treaty was successfully negotiated in the form it eventually took as a result of two essentially political bargains. The first of these was insistence by the United States on tying the treaty to IOA, as Brennan explains: "The Soviets were interested in limiting Safeguard, coupled with some slight willingness to limit offensive forces. The United States, on the other hand, was interested in limiting offensive forces, coupled with a not-quite-so-slight interest in limiting ABM. This resulted roughly in a deal."[78] American leverage in establishing this "linkage" arose because of Soviet insistence on ABM limitation, a position seemingly anomalous to their heavy interest in nuclear defense. Goure and his associates, however, posit what became conventional wisdom about Soviet motivations: "The persistence of the Soviet search for a war-winning and war-survival capability suggests that Soviet adherence to the ABM Agreement was the result of technical difficulties in Soviet ABM development and concern that the US might achieve a technological breakthrough in its ABM program."[79] Given generally divergent technological levels, this position seems reasonable.

The second bargain, more strictly political than strategic, was the agreement on actual ABM deployments, reflecting political realities in the two countries. That there was provision to protect national capitols resulted from the fact the Soviets already had the Galosh system around Moscow, an option that had been discarded for Washington for political reasons. At the same time, the American debate was still ongoing for an

ABM system around the Grand Forks ICBM field, an option in which the Soviets had shown little interest. The result was strategically anomalous, as Jack Ruina points out:

> The Soviet ABM system . . . is essentially of no military consequence whatsoever, as far as the United States is concerned. Yet there were indications that the Russians were unwilling to take it down. Perhaps they have an internal political problem about taking an ABM system down; if one invests so much money and if one has a constituency that wanted this thing, it is very hard to say no now. In the circumstances, then, one would have thought that the United States would have been big enough to let the Russians have their ABM and agree to have nothing themselves—but there was concern about symmetry. Of course from the military's point of view, the whole thing is nonsense.[80]

From the American vantage point, the justification for maintaining an ABM option arose both from asymmetry and senatorial reaction had the United States not gotten "something back" in return for the Soviet Galosh system and from the defense community, which needed an ABM site for further R&D.

The result of the ABM Treaty was to place a moratorium on active missile defenses. As Schlesinger argued in the FY 1975 Defense Department Annual Report, "The salient point, however, is that the ABM Treaty has effectively removed the concept of defensive damage limitation (at least as it was defined in the 1960s) from contention as a major strategic option."[81] Mason Willrich agrees: "The ABM Treaty reflects a political conclusion derived from a technical judgment. The technical judgment . . . is that ABM defenses of territory and population against a sophisticated nuclear attack are unworkable . . . The ABM Treaty implies recognition . . . that nuclear deterrence by means of an assured destruction capability is vital to the security of both sides."[82]

Prevention of Nuclear War Agreement: Although not a formal part of the SALT I agreements, an important spin-off of the agreement was the Prevention of Nuclear War Agreement signed in June 1973, in which the superpowers agreed to the inadmissibility of thermonuclear conflict. Raymond Garthoff explains why this accord has generally been overlooked in the American assessment of SALT: "In SALT I and II there was one major political quid which we could have used as a bargaining chip in getting some substantial quo, but it was never so recognized by Washington. . . . Regarded as highly important by Moscow, and as 'harmless' but not important by most in Washington, it was used simply to lend momentum at a time of stalemate in SALT II. One reason is that only a handful of people in the Administration knew about it before it was publicly unveiled after signature."[83] Whether declarations that nuclear war is intolerable would make any difference in a crisis is, of course, problematical. The formal declaration, however, represents American acceptance of the peaceful coexistence principle and thus has symbolic importance for the Soviets. Reflecting this position, Milstein and Semeiko say, "The outstanding importance for the preservation of peace on our planet of the affirmation of the inadmissibility of nuclear war as a principle of the foreign policy of states, and particularly of the two major nuclear powers—the USA and the USSR—is obvious."[84]

Assessing the Impact of SALT I: The major outcomes of SALT I have been both criticized and praised. Of the two most significant agreements, the ABM Treaty has received the lesser negative commentary, partially because it settled an internal debate that most wanted to see foreclosed anyway and in part because limiting ABMs was seen as a reinforcement of ongoing American doctrine, as Newhouse explains: "Security's handmaiden is MAD. . . . SALT, then, is supposed to persuade each side to abandon steps that might weaken or jeopardize the other's assured destruction capability. That is what SALT is all about in the eyes of in-house scholastics."[85] Although ABMs can contribute marginally to second-strike posture by placing them in defense of retaliatory forces, removing ABM as a threat to the mutual hostage relationship supports MAD doctrine.

IOA has been more controversial, and because offensive weapons limits are the primary object of SALT II, disagreements have continued. There have been two basic criticisms. On the one hand, it is argued that the agreement elevated and sanctified asymmetries in forces with dangerous consequences as the Soviets approach American technological levels (to extend the argument in the conclusion of Chapter 5, their forces could become bigger *and* better). This position is countered by the assertion that the agreement inhibits further Soviet force growth without affecting the American configuration. As Newhouse concludes, "As for offense, a ceiling on strategic missiles would constrain Soviet programs, but not American. Although embarked on MIRV, the United States was not building new ICBM's or Polaris submarines or bombers."[86] This counterargument, which has become a staple part of ongoing SALT defenses, views the SALT process as a means to keep asymmetry within reasonable bounds that would be exceeded in the absence of SALT.

IOA has also been condemned as stimulating the arms race. On the one hand, SALT (through the Nixon administration's advocacy) created the bargaining chip, and, in the process, "the armed services have gained an argument in the annual budget battles that helps convince skeptics to fund new nuclear weapons."[87] On the other hand, nuclear arsenals have actually increased since SALT I was signed, both in numbers of launchers (the Soviets increasing their forces rapidly to reach agreed limits) and in warheads (largely the result of the MIRV program). This latter criticism is really an admission of the limitations on verification of qualitative aspects of the arms competition, reflecting the fact that it is possible to count gross considerations like launchers but not more subtle arms sophistications like MIRV. As we shall see, this is a recurring problem in the ongoing arms limitation process.

The Vladivostok Accords Because IOA was of limited duration and would expire in 1977, it was necessary to begin negotiations almost immediately on a follow-up agreement. The principle issue to be dealt with was achieving acceptable levels of weapons deployments, an issue that had not been dealt with comprehensively within the IOA freeze on existing systems. The first step in this process was the Vladivostok accords of November 1974 between President Gerald Ford and Leonid Brezhnev.

The Vladivostok accords serve the dual purposes of acting as a kind of "heads of agreement" between the two leaders on an eventual SALT II agreement and, in the process, of providing parameters within which negotiators could reach a final treaty. Specifically, in the accord, "both the United States and the Soviet Union agreed to limit until 1985: 1) the number of their strategic bombers and missile launchers, either sea- or land-based, to 2,400 for each country; and 2) the number of missile launchers with multiple independently targeted reentry vehicles (MIRV), either sea- or land-based, to 1,320 for each country."[88] In addition to including bombers in overall force configurations and delimiting overall ceilings and a MIRV sublimit, the "heavy modern" ICBM issue was included: "It carries over from the Interim Agreement a restriction on new fixed ICBM launchers, and limits the number of what are called modern large missile (MLBM) launchers to those operational or under construction in May 1972—308 to 320 on the Soviet side and none on the US side."[89] Beyond setting deployment limits and moving SALT II along toward completion, Secretary of State Kissinger defended the agreement in the same way SALT I was upheld: "Both sides would have equal ceiling on missiles, heavy bombers, and on multi-warhead missiles, and *this would require the Soviets to dismantle many weapons while our planned forces would not be affected*" (emphasis added).[90]

The Vladivostok agreement has been championed and condemned on much the same grounds as IOA. On the one hand, although equal launchers limits are established within the accord, some observers, such as Lehman and Hughes, feel the terms were asymmetrically favorable to the Soviets: "In applying the broader concept of the current SALT negotiations, there are a number of important asymmetries (either existing or developing) that are not addressed in the Vladivostok guidelines. The issues of missile throw-weight, cruise missiles, Backfire, and civil defense are all matters that may directly affect the overall strategic balance—and they are all crucial to an overall essential equivalence in

U.S. and Soviet strategic capabilities."[91] The throw-weight asymmetry "seems destined to energize a long simmering debate over the vulnerability of the land-based missile component of the United States strategic forces,"[92] because, although the number of missiles that can be MIRVed on each side is set, the yield of warheads in each MIRV is not covered. As a result, "the payloads of Soviet ICBMs would permit them to deploy several times the number of reentry vehicles available to the U.S. missile forces."[93]

On the other hand, some observers believe that the MIRV launcher limits and the absence of limitations on qualitative improvements stimulate the arms race. As Leader and Schneider argue:

> SALT agreements like the November 1974 Vladivostok agreement, allow greater numbers of strategic weapons than exist in current inventories. These high limits on MIRVs and strategic delivery vehicles may serve as "lures" rather than as limits to future arms production.
> ... A ... major flaw in the Vladivostok accord is that no limits were placed upon the qualitative arms race, that is on improvements in strategic nuclear weapons. No restrictions were placed on spending for improvements in accuracy and none on the creation of mobile ICBMs, or antisubmarine warfare technology, or on MARV.[94]

These problems, once again, reflect the verification limits on the qualitative arms race and the difficulty of escaping the prisoner's dilemma in these situations. Honoring the MIRV sublimit, for instance, will have to be a matter of mutual trust and faith, since it cannot be verified. Monitoring R&D is likewise highly imperfect, although Scoville offers a partial solution congruent with the earlier analysis of verification: "A key goal in SALT II...will, therefore, be to place restraints on the replacement of old weapons by new models, to restrict the development of new missile technology and, particularly, to place some controls on MIRVs. Limitations on missile testing seem the most likely approach to halting missile improvements."[95] The effectiveness of testing limitations is, of course, somewhat attenuated by the temporal lateness of testing in the R,D,T,&E cycle. It has, however, been a recurring theme, along with the other problems, in the movement to SALT II.

SALT II The negotiations for a successor to SALT I have taken over five years, the agreed limit of SALT I. While discussions dragged through 1978 and the ratification debate extended through 1979 after the agreement was signed in June, both the United States and the Soviet Union implicitly agreed to honor the limits set at Vladivostok. In September 1977, the first successful review of the ABM Treaty was completed, in which "the U.S. and the U.S.S.R. jointly stated a belief that the treaty serves the security interests of both countries and reaffirmed their commitment to the Treaty."[96]

President Carter and First Secretary Brezhnev signed SALT II in Vienna on June 18, 1979. The long period to reach agreement and the lengthy period of Senate ratification debate are partially attributable to environmental changes since SALT I was completed, not the least of which have been two changes in American administrations. The agreement that finally emerged represents an extension and elaboration of the IOA, and a number of issues related to qualitative aspects of the strategic balance are included in the treaty.

The SALT II Environment: The SALT II negotiations have been conducted in an atmosphere of intensified doctrinal debate in which the consequences of both SALT I and II have been a primary issue and onto which the Carter administration has yet to make its distinctive, clarifying doctrinal contribution (see Chapter 3). As new weapons systems possibilities emerge on each side, the overall debate has crystallized around the likely outcomes of SALT, and the likelihood of Senate ratification has been clouded both by apocalyptical predictions about the relative strategic position of the United States arising from the treaty (or its absence) and the question of linkages to Soviet global foreign policy behavior.

The growth in Soviet strategic power has been the major source of contention surrounding SALT II, just as it was in SALT I and Vladivostok. The basic question, of course, has been whether a follow-on accord serves to stabilize the weapons balance at an acceptable level or whether it will consign the United States to a strategic inferiority making no agreement a better outcome. Because the Soviets have a more extensive R&D program, particularly in ICBMs, than the United States, Rathjens concludes that the Russians are in an advantageous position: "The blunt fact is that the Soviet Union now appears to be in a better bargaining position than the U.S. for SALT II, and with time, its position is likely to improve further. To get a satisfactory offense forces treaty, the U.S. will have to make significant concessions in other areas."[97]

Many U.S. observers have taken the position that the United States has made too many concessions to the Soviets in recent years generally, a posture seemingly adopted by the Carter administration in summer 1978 in the form of warnings of a "new Cold War" and declarations and demonstrations of American military strength. Fearing the consequences of aggressive Soviet force modernization, these observers, most vocally represented by the Committee on the Present Danger, have cautioned against a treaty that does not ensure minimally long-term strategic parity. Opposing these fears is the official administration position that the likely consequences of failing to conclude an agreement will lead to an unrestrained arms race: "It is estimated that, without a SALT II agreement, the Soviets could have over 3,000 strategic delivery vehicles by 1985."[98] Extrapolating from these projections what it would cost the United States to counter Soviet growth in this circumstance,

Congressman Les Aspin estimates that "if we reject SALT II we can have the privilege of spending an additional $20 billion to end up in exactly the same position we would be in if we ratified SALT II."[99]

Within this occasionally raucous debate, SALT II has become a much more politically charged issue than SALT I, a situation which one analyst, Colin S. Gray, finds regrettable: "It is really only in military strength that the Soviet Union can be considered a serious competitor with the United States. Hence, the attention focused upon, and the importance attached to, SALT risks the encouragement of public misperception of the overall competitive strength of the Soviet Union."[100] Whether they merit the level of concern they have spawned, both the provisions included in SALT II and matters not covered by the agreement are likely to remain contentious regardless of the outcome of the treaty itself.

SALT II Provisions: As reported by Richard Burt in July 1978 and confirmed in the treaty text, the SALT II agreement consists of three elements: "a *treaty* lasting until 1985, a *protocol* lasting for three years and a *statement of principles* to guide negotiators during the next phase of negotiations" (emphasis in original).[101] The treaty, consisting of thirteen components, requires a shallow-cut reduction in launchers and places some restrictions on ALCMs (they can be deployed only on "heavy" bombers such as the B-52) and an exclusion of the Backfire bomber from agreed limits, with restriction on production rates and possible restrictions on refueling and basing. The three-year protocols place restrictions and bans on new system testing and deployment, and the statement of principles "is intended to provide a framework for future negotiations, meant to begin shortly after the conclusion of SALT II."[102] Of the three documents, the treaty and protocols are of the most immediate concern.

Aspin sets out the basic numerical balance the agreement encompasses for both sides: total launchers not to exceed 2,250 by the end of 1981; a maximum total of 1,320 MIRVed missiles and ALCM-carrying bombers; further limits on MIRVs, to a total of 1,200 and no more than 820 MIRVed ICBMs; and a limit of 308 modern large ballistic missiles.[103] These limits represent a reduction in total launchers of about 10 percent from Vladivostok and roughly an equal reduction of MIRVs, although the inclusion of provision for 70–120 cruise missile carriers leaves the overall multiple-weapons launcher total at the Vladivostok MIRV limit. The 308 MLBM limit reflects the number of SS-9s and SS-18s currently in the Soviet arsenal.

The protocols break new ground within the SALT process by placing temporary limits on qualitative aspects of the arms competition through a three-year testing and deployment ban. Burt lists the four elements in the protocols:

1. A ban would be placed on the testing and deployment of "new types" of missiles, with certain exceptions. . . . There are indications that, as the two sides narrow their differences on what new systems would be exempted, they may move this provision from the protocol into the longer-term treaty.
2. The significant improvement of existing systems will be restricted, with substantial upgrading of boosters, post-boost vehicles and guidance systems ruled out. . . . Electronic improvements—to existing guidance units . . . would not be restricted because they cannot be verified.
3. The testing and deployment of mobile ICBMs would be banned. . . .
4. The testing and deployment of armed (nuclear or conventional) cruise missiles with ranges exceeding 2,500 km. would be barred.[104]

The continuing contentious issue of FBS has been consigned to the statement of principles, and each side is constrained to deploying only one "new" weapons system.

These limits represent a compromise for the arms-reduction-conscious Carter administration, which early in 1977 presented the Soviets with a "comprehensive" package containing deeper arsenal cuts. Perry summarizes the original Carter proposal:

The American proposals reportedly extended to: (1) reducing the number of strategic delivery vehicles from the 2,400 accepted at Vladivostok to perhaps 1,800; (2) reducing the number of permitted "heavy" Soviet ICBMs from 300 to about half that total; (3) reducing the ceiling on MIRVed launchers from 1,320 to a reported 1,000; (4) imposing limitations on the development and deployment of improved strategic weapons, continuing the freeze on new silo construction, limiting flight tests (and presumably operational training launches) to a small annual set for each type of ICBM, and banning the test and deployment of new ICBMs; and (5) banning mobile ICBMs, limiting cruise missiles to a range of 2,500 kilometers, and obtaining from the Soviets "a list of measures to assure that the Backfire would not be used as a strategic bomber."[105]

The administration was unsuccessful in obtaining the lower numerical limits, which would have forced the Soviets to dismantle a more significant number of weapons than the United States and which, congruent with the earlier analysis of the effects of arms reduction, would not have affected the destructive balance particularly. At the same time, limited bans on testing and deployment emerged in the negotiated document.

Anticipating the effects of SALT II on American strategic plans, Secretary Brown assessed the impact in his 1979 *Annual Report:*

If the eventual SALT II agreement meets our expectations, it will:
—mean somewhat lower levels of strategic delivery vehicles and MIRVs than was envisaged at Vladivostok or in later talks—and lower than we estimate we would face if there were no agreement;

—introduce an important new sublimit . . . on the total number of MIRVed ICBMs;
—permit us to deploy an air-launched cruise missile (ALCM) force . . . ;
—Constrain to some degree the pace of technological change, but preserve U.S. flexibility to continue R&D . . . ;
—meet specific allied concerns by omitting forward-based systems . . . ;
—place some limits on BACKFIRE.

From this assessment, he concludes that the agreement "will help to preserve perceptions of essential equivalence and will contribute to military equivalence and stability."[106] Aspin reports that these goals are achieved with only minor effects on U.S. deployment plans in MIRVs and ALCM-armed bombers: "Current plans call for 1,483, but SALT II would limit us to 1,320, a 10 percent reduction."[107]

Omissions in SALT II: The new agreement must be judged both for what it does and does not do in limiting strategic armaments. SALT II does produce quantitative limitations on armaments that include modest arsenal reductions and stabilize basic armaments levels. The gross limitations are verifiable, and the MIRV and cruise missiles are constrained by limiting how MIRVs are counted (any system tested as a MIRV is considered a MIRV), permissible RVs on a MIRV (essentially the number that have already been tested on a system), and the number of ALCMs that can be put in any carrier (20).

SALT II also takes a first and modest step into the morass of the qualitative arms race through temporary restraints on new missile testing and deployments. Restraints in this area are, of course, the most difficult to develop and to monitor, and the measures in SALT II are tentative. Because the agreement does not harness long-term development efforts (although future extensions of testing and deployment bans or inclusion of additional systems into the treaty could), it is in this area that the agreement will probably be most criticized. As a result, slowing or stopping the destructive spiral will remain an ongoing problem in future SALT discussions.

Pressures for new weapons systems arise from a number of sources: the fear that the absence of development programs will result in advantage to the other side (the R&D prisoner's dilemma); the bargaining chip approach to future arms agreements[108] because, once conceived, new systems are perceived to have utilities that create deployment pressures (such as damage limitation and hard-kill capability); and to counter actual or anticipated developments by the other side (such as the "need" to counter Soviet conversion of the mobile SS-20 to intercontinental range with MX).[109]

These problems are most severe regarding new systems developments that can be deployed deceptively or threaten retaliatory capabilities. Deceptively deployed (or deployable) systems present verification difficulties that are more or less critical depending on launcher levels (see discussion earlier in this chapter). Mobile missiles and cruise missiles are particularly vexing in this regard, as Metzger points out with regard to cruise: "Force level assessment, a critical preliminary to successful negotiations, and verification of any agreement reached, become murky propositions at best, due to the very military advantages that cruise missiles possess: mobility, dispersal and concealment of launchers, and flexibility in range, warhead choice and target selection."[110] The pressures for deceptive deployment are, of course, related to the degree of threat one perceives to retaliatory forces. Hard-kill targeting accuracy is a major incentive to mobility and concealment (this problem is discussed at length in Chapter 7), and a breakthrough in antisubmarine warfare (ASW) would be particularly devastating to retaliatory notions conceptually underpinning deterrence. The ability of future arms limitations discussions to come to grips with these and related issues[111] will be a major test of the long-range success of the entire SALT process.

Lessons from Arms Control Efforts Beginning with the Limited Test Ban Treaty of 1963, the arms control process leading to substantive agreements is over a decade and a half old. Although the process has been imperfect and the agreements reached not as comprehensive or as foolproof as advocates would prefer, there has been gradual success, and discernible patterns have emerged that are instructive for prospects of future activity.

Viewing the sweep of the arms control process, four interrelated observations arise as lessons from past experience and as guides to assessing future progress. First, accord is most likely where the basis of agreement describes a reality with which the parties are content. Second, the ability to provide adequate verification is a prerequisite for successful agreements. Third, bans or sharp limitations on testing new systems are the most effective means to impede qualitative aspects of arms races that arms control agreements seek to stop. Fourth, arms control has achieved a momentum that makes it a part of all strategic considerations, but the process is, under some circumstances, reversible. The first two observations arise almost axiomatically from the interface between arms limitation and security, whereas the third and fourth are the result of actual experience with the process.

The Descriptive Nature of Agreements Generally speaking, the arms control proposals that have met with the greatest success have been descriptive of a status quo that supports the security needs of the parties and that is viewed by the parties as worth preserving. As Doty and his colleagues observe, "A coincidence of some perceived interests is a prerequisite for any meaningful agreement."[112] The perceived interests are generally dictated by security needs assessments heavily influenced by military analysts concerned that an agreement not jeopardize strategic security, argued earlier as the basic prerequisite for willingness to consider arms control proposals. At a more pragmatic level, such agreements do not force either side to give up anything significant, thereby avoiding entrenched interests in particular weapons systems and not unduly arousing suspicions that an agreement will undermine basic security.

This dynamic can be seen operating in both arms control tracks. Within testing limitations, the Limited Test Ban Treaty was possible partly because of public clamor against atmospheric testing, but also because both the United States and the U.S.S.R. had conducted enough testing in the atmosphere so that they did not need to conduct any more for basic research purposes (underground testing data were adequate). As we saw earlier in this chapter, the NPT required absolutely no sacrifice for the superpowers while fulfilling a mutual interest in dampening proliferation of weapons, and the potential weapons states who would have to sacrifice the nuclear option have refused to sign the treaty. Similarly, TTBT was negotiable because neither side needs to engage in large warhead testing any longer, and a major factor inhibiting CTBT has been that it impedes new warhead technology (which is its main purpose) and thus confronts established interests in continued R&D.

The factor is also apparent in the SALT process. Two of the points that Mason Willrich regards as crucial to success in SALT I bear directly on the point: "the mutual interests of the United States and the Soviet Union in preventing nuclear war" and "the relationship of rough equality or parity between their strategic nuclear forces."[113] SALT I was, of course, a simple freeze, thus not threatening to existing systems, and some of the difficulties that have plagued SALT II are attributable to attempting to create a new status quo. As Garthoff explained in 1977, when progress was stalled: "It may be observed that SALT I succeeded—and SALT II and MFBR to date have failed—because of the fact that SALT I reflected, codified, and stabilized an existing balance in force levels, while our SALT II proposals at least up Vladivostok . . . have avowedly attempted to alter that balance and to create a new balance."[114] The Soviet rejection of the Carter "comprehensive" proposal, which involved reasonably deep cuts, exemplifies this point and leads Gray to the wry comment that "President Carter is but the latest senior American to learn that Soviet leaders cannot be persuaded to agree to arms control plans that are not in the best Soviet competitive interest."[115] The most modest cuts in the SALT II agreement, reflecting mostly phaseouts of obsolete weapons, were clearly easier to accept.

Closely related to this basic point is the complexity of the proposal and its consequent likelihood of successful resolution. Comprehensive and complex issues inevitably are more difficult to deal with than simpler proposals, if only because more interests will be affected in the political tradeoffs and compromises attendant to broad proposals. Discrete issues like TTBT, PNET, or the Prevention of Nuclear War agreement are conceptually easier to deal with and do not activate as wide an array of potential opponents as does a broad area of concern such as that covered by SALT. The separation of the IOA and ABM Treaty within SALT I speaks to this point, and the division of the various measures included in SALT II into three separate categories is further indication of the difficulty of dealing with comprehensive proposals in a single format. Complex problems raise particular problems in light of the "essential equivalence" criterion, and General Taylor comments on this difficulty: "The item count is plagued by the same difficulty which has been encountered in our negotiations with the Soviet Union over the limitations of strategic weapons—the asymmetries between force structures and the difficulties of equating dissimilar elements."[116] Even fairly specific and symmetrical sublimits on launcher categories in SALT II have not overcome this problem entirely.

Adequate Verification A means of verifying compliance has been a prominent feature of all arms control agreements, a testimony to the enduring need for an alternative to trust as the basis for escaping the arms race prisoner's dilemma. Malcolm Hoag states the need simply: "A mutual ban upon weapons systems whose observance is not verified cannot be trusted."[117] And, as Perry points out, verification has been a recurrent "dilemma" in the arms control process: "While there remains any substantial doubt of U.S. capabilities to detect Soviet deceptions, the U.S. will balk at accepting unverifiable Soviet assurances. So long as the Soviet Union views all non-client states as potentially malevolent, the Soviets will resist verification that enlarges U.S. knowledge of their strategic capabilities. That is the dilemma of arms control."[118]

The central importance of verification provisions, implicit or explicit, is obvious from quickly viewing arms control evolution. As has been pointed out, successful conclusion of agreements required the development of nonintrusive monitoring techniques. On the test ban side, LTBT could be monitored by reading radiation levels in the atmosphere, TTBT and PNET levels were largely the result of seismographic capabilities, and a major source of opposition to CTBT has been inability totally to monitor its provisions. Similarly, the SALT agreements have been limited to those items—launcher numbers in SALT I and launchers and testing in SALT II—amenable to national technical means of verification.

The limits on verification also pose a problem to future developments in arms control. R&D efforts cannot be monitored nonintrusively (it is doubtful that they could be totally monitored under any circumstances), resulting in exclusion of R&D in agreements in conformance with the dictates of the R&D sub-prisoner's dilemma identified earlier. As Gelber points out, this problem also extends to deploying qualitative improvements: "One of the major difficulties in the attempt to extend these limitations in SALT II is the formulation of qualitative restraints"[119] that can be verified.

The importance of assurance that agreed limits are being assiduously honored varies, of course, with the type of arms control arrangement under consideration and thus the incentives or disincentives of the parties to cheat on provisions. Because the limits established in the shallow cuts demanded by SALT II leave arsenals essentially intact, Paul Nitze concludes that precise monitoring is unnecessary: "I personally take the verification issue less seriously than most because the limits are so high that what could be gained by cheating against them would not appear to be strategically significant."[120] In an era of "Assured Detection," however, verification provisions will remain a central and necessary part of any acceptable agreement. Rathjens, however, warns that "excessive emphasis" on monitoring could have undesirable effects on the whole arms control process: "There is a danger that the present concern about the adequacy of verification capabilities may divert us from more serious problems with respect to future agreements, and it may lead to our not concluding agreements that would be advantageous."[121]

Test Bans With SALT II placing effective limits on the quantitative arms race (at least in strategic delivery vehicles), the major remaining problem for arms control is placing a cap on the qualitative spiral. As has been argued, imposition of a qualitative ban requires interruption of the development process at a point prior to when a nation gains sufficient confidence in a new weapon to make the decision to deploy. The ongoing verification requirement that provides sufficient confidence that the balance will not be upset by new weapons programs dictates interruption of the R,D,T&E process at some point where observation of compliance is possible. Since testing is the first visible phase in the cycle, it is the most logical point at which to impose limits or bans.

The concept of bans on weapons is precedented and has developed support in the strategic community. The ban on ABM testing in the SALT I ABM Treaty serves as precedent, as do the time-limited bans on new weapons testing in SALT II. The entire warhead test ban track provides history and experience to the effectiveness of this idea, as indicated in a backhanded way by Edward Teller in testimony opposing the LTBT as a deterrent to ABM warhead development: "To try to build up our missile defense forces without proper and complete experimentation that can be performed faithfully and in a relevant way only in the atmosphere, *to do without this experimentation is most hazardous*" (emphasis added).[122] If one seeks to limit the deployment of new weapons systems, the more "hazardous" the decision is, of course, the better.

Considerable support has developed within the arms control community for either banning or severely limiting weapons systems testing. Rathjens, who describes test limits as "the most interesting prospect on the SALT horizon,"[123] has been a leading advocate, and he explains how such restrictions would work: "One can conceive of writing an agreement proscribing such tests in which one can have some confidence as regards compliance. I believe this is possible, particularly if there could be included, as part of the agreement, a requirement that tests be conducted only in certain specified areas where observations by the adversary power could be made."[124] Observation would, of course, be conducted by nonintrusive electronic surveillance (such as satellites and offshore electronically equipped vessels). Gelber agrees that the test phase is the proper place to impose limits: "If verified controls are to be added to changes of incentive, they are likely to be most effective if imposed in the area of advanced development and prototype testing."[125]

Barry Carter believes that testing limits could be an effective means to slow the movement toward hard-target kill capability and thus defer the issue of ICBM vulnerability: "An indirect way to limit or impede accuracy improvements through SALT would be by placing a strict limit on the number of missile tests. This would make it more difficult to develop advanced guidance techniques and to test them often enough so that the military would have confidence in them."[126] Since hard-target capability leaves little margin for error, nurturing lack of confidence would seem to be an effective deterrent. Congressman Downey agrees that such limits could be effective, but wonders if the defense establishment is not already too committed to the program to impose limits: "While tight qualitative controls on ICBMs, including a total test ban, might deny the Soviets a high-confidence countersilo capability if implemented in the near future, this is probably not a realistic political possibility within the time available."[127] Carter also warns that the data obtained from observing testing are indirect and thus not conclusive, because such data "tell one about the ballistic coefficient (or pointedness) of the warhead, its reentry speed and similar information. . . . An outside observer, however, can never be sure what the actual target is."[128]

The discussion of limiting hard-kill attainment illustrates the possibilities of testing bans, but also suggests two major limitations. First, if tests are limited but not banned altogether, observation is not entirely conclusive, a problem complicated by the fact that not all new weapons require complete testing. Indirectly referring to converting the Soviet mobile

SS-20 to the SS-16, Perry notes that full-range testing "is not essential to individual tests of key subsystems or to their successful integration; a two-stage MRBM can rather easily become a three-stage ICBM."[129] James L. Garwin adds that this is a particular problem with cruise missiles, because "1. Testing of the cruise missile is unobtrusive"; and "2. Cruise missiles need not be tested at full range to have high confidence that they work at full range."[130]

Second, the ability to negotiate test bans or limits is a matter of timeliness and symmetry, a lesson learned from the MIRV experience in SALT I. If bans are to be imposed, they must work an equal hardship on each side. Soviet refusal to accede to American suggestions of a MIRV ban reflects this need for symmetry: neither side is going to give up a capability (or the possibility of developing one) that the other side possesses (the United States had completed MIRV testing when SALT I opened), reflecting the legitimate fear of negotiating oneself into a position of strategic inferiority. Asymmetrical sacrifice is unlikely to be acceptable, suggesting the need for timeliness: test bans are likely to be acceptable only on systems neither side has tested. From this realization, it is extremely unlikely that long-term meaningful limitation on Soviet cruise missile development will be attainable. At the same, the Carter administration's 1978 decision to delay testing of MX and Trident II undoubtedly allowed the limits on these systems in SALT II.

The Arms Control Momentum With the successful completion of an arms control agreement fifteen years old and a decade of the SALT process in hand, the arms control process has achieved a level of permanence that makes it an important environmental consideration in ongoing strategic nuclear concerns, affecting both the internal and external environments and thus doctrinal choices. The fact that new weapons proposals must now be accompanied by an Arms Control Impact Statement is but a single symbol of the degree to which arms control concerns permeate the strategic calculus. At the same time, however, the failure of arms control processes to gain control of the qualitative arms race could mean the reversal of this influence.

The SALT process is the most visible aspect of arms control momentum. As Warnke states, "SALT should be viewed as a continuing process. The SALT II agreement won't be the last word on strategic arms control."[131] Summarizing the arguments of SALT supporters (including himself), Downey agrees, calling SALT II the "son of SALT I" and the "father of SALT III."[132] The ongoing nature of the SALT process has been institutionalized in the ABM Treaty through the Standing Consultative Commission, which Rhinelander says "could be one of the significant results of SALT I." As he explains, "The Commission will have two primary functions—the first relating to the implementation of the treaty and the Interim Offensive Agreement, *and the second relating to further negotiations after the conclusion of SALT II*" (emphasis added).[133]

Although the actual SALT negotiating sessions have been conducted privately and with little fanfare, the overall process has attracted considerable attention. Ruina believes this has had a salutary effect: "I think an important bonus from the total SALT negotiation is that in the process . . . there has developed now a much more widespread knowledge about nuclear issues in both countries. It is no longer limited to a very small group."[134] Testimony to this visibility is the degree to which strategic concerns have been associated with SALT outcomes.

Although arms control has achieved a prominence that even its most committed advocates would have found unlikely as little as twenty years ago, the process is not irreversible. In a sense, the agreements that have been reached to date, dealing primarily with quantitative aspects of the arms race, have succeeded in the comparatively "easy" aspects of the problem. Putting a lid on the qualitative spiral represents a much more difficult set of dilemmas, but is vital to the long-term vitality of arms control. Emerging technologies like hard-kill capability have potentially destabilizing and negative arms control consequences that could be devastating to effective arms control (as argued in Chapter 7). If the movement cannot overcome these obstacles, much of its positive impact on the arms race could be unraveled.

Chapter 7:
The Uncertain Future:
American Policy in the 1980s

The major purpose of the preceding pages has been to try to present, in a reasonably complete and coherent manner, the evolution of the strategic environment in which American doctrine is developed and to which it must be responsive. The major thesis has been that the environment surrounding strategic doctrine is extremely dynamic, presenting new challenges and problems to decision makers and policy analysts. The problems likely to emerge in the 1980s must be viewed in light of changes since the last comprehensive statement of American strategic preference, the doctrine of mutual assured destruction, was enunciated, both in terms of the need for doctrinal adaptation and deterrent stability.

In the midst of discrete changes and dynamic trends that have an impact on strategic policy formulation, one ongoing element remains constant: the purpose of the strategic nuclear system remains the prevention of nuclear war. This purpose in turn acts as a standard against which to measure changes and potential changes in both the environment and doctrinal choice: to the extent that these dynamics contribute to making nuclear war less likely, they are positive; to the extent that they lower the firebreak, they are destabilizing and thus undesirable.

Assessing the trends in the strategic nuclear system is the purpose of this final chapter, and as the title suggests, it is indeterminate and uncertain, subject to vicissitudes and events that are not entirely predictable. With that caveat in mind, I will begin by identifying and analyzing, from the mass of concerns discussed in earlier pages, three problem areas that I believe are likely to be particularly crucial in the 1980s. These issues are interrelated, and outcomes and decisions in one will have direct bearing and salience for the others, providing the rationale for the second major section in the chapter. One of the major theoretical weaknesses of scholarship on nuclear deterrence, as suggested in Chapter 1, has been a tendency to treat issues and problems in a fairly narrow and discrete manner rather than in terms of their overall impact on the strategic system. The framework identified in the first chapter was intended to help alleviate this difficulty by presenting a comprehensive set of categories and possible interactions between them as a way to facilitate assessing overall impacts. Thus, the discussion will conclude by placing the identified set of problems into the framework and showing how these issues relate to and affect one another.

Major Problems in the 1980s Of all the various challenges facing the strategic system in the wake of SALT II, three stand out with particular clarity. The first, and most difficult, is stemming or coping with the qualitative arms race. This issue focuses on emerging qualitative improvements in weapons system and guidance capabilities that can result in the attainment of true counterforce capability through "hard-target kill" against fixed land-based offensive systems and has implications for the viability of fixed ICBMs, disincentives for preemptive firing, and arms control as it has evolved. Second, and related, is the continuing nature of the evolving Soviet threat. This issue involves the degree to which qualitative improvements in Soviet forces in light of quantitative limitations affect the strategic balance, prudent but effective means by which the United States can counter negative consequences of Soviet developments, an assessment of Soviet intentions in their programs, and arms control impacts of this action-reaction phenomenon. Third, the danger of horizontal proliferation of nuclear weapons will be accelerated or attenuated partially by the other issues and particularly by arms control consequences of strategic determinations. This issue arises against the theoretical backdrop of an understanding of nuclear balance developed and understood primarily in the bilateral Soviet-American context and includes problems such as how nth power nuclear possession affects the superpower balance, how interjecting nuclear weapons could affect local or regional conflicts, and what special problems proliferation creates for the likelihood that nuclear weapons could be used by national governments or extremist groups for terrorist purposes.

Technology, Counterforce Capability, and the Strategic Balance As has been discussed primarily in Chapters 1 and 4, there is a dynamic tension between strategic doctrine attempting to create a stable nuclear balance minimizing the probabilities of nuclear holocaust and the fruits of technological endeavor. Ideally, technological ingenuity contributes to overall strategic purposes, but at the same time "technology can destabilize the strategic balance and provide less, rather than more, security."[1] Although this statement was made in direct reference to the MIRV impact on the balance, the targeting accuracy movement toward hard-kill capability threatens to have at least as dramatic an impact.

The hard-target problem (as expressed primarily in terms of ICBM vulnerability) is a complicated issue with complex ramifications. To unravel it, I will begin by reviewing how the capability is attained, including the contribution of increased targeting accuracy and warhead yield. I will then examine the impact of attaining this capacity on incentives and disincentives to move from the second-strike firing posture that has underlaid deterrence notions to a first-strike firing doctrine and what can be done to attenuate the incentives to make the doctrinal shift. Such decisions, in turn, have strong implications for effective arms control, and doctrinal justifications of counterforce capabilities must be assessed in light of the problems created. Finally, I will address the question of how, if at all, the movement toward hard kill can be deferred or stopped.

The Basis of Hard-Kill Capability This concept is, of course, defined as the capacity, with some high level of assurance, to destroy an unfired weapon that is in a hardened container. What constitutes a "high level" of belief in the ability to kill an offensive weapon is a contentious matter, because if one fails to do so, that weapon and its destructive force can be fired in retaliation following the initial attack. The fratricidal effect (successive incoming missiles aimed at a target being detonated by the blast effects of previously exploded warheads) in turn dictates that calculation of hard-kill attainment be measured in terms of the likelihood that a single incoming warhead can destroy the hardened target and gives rise to the "single-shot kill probability," or SSKP, as the basic measure (expressed as a percentage probability, with hard-kill capability improving as SSKP approaches unity).

The hard-kill concept can be applied to any weapons system, although it is most frequently attached to ICBMs. As has been pointed out, the air-breathing leg of TRIAD does not use hardening (for examples, reinforced or buried hangars) as a means to overcome vulnerability because of structural difficulties in shelter reinforcement and the impossibility of hardening runways to any appreciable extent. Invulnerability is achieved through dispersal and ground alert, and the principal means to achieve hard kill against the bombers is perfection of depressed-trajectory SLBM launch that could deliver warheads to target before alert procedures could have the force airborne.[2] An effective threat against the submarine leg requires effective ASW, primarily in the areas of detection and development of effective "hunter-killer" atomic submarines. Detection is the most difficult ASW problem and the area in which the most aggressive R&D will continue. Although breakthroughs in

ASW are not likely in the near future, their impact would be considerable, as Tsipis points out: "Counter-SLBM ASW threatens the vectors of deterrence of an opponent, provoking a destabilizing escalation."[3] Because SLBMs are generally agreed to constitute the most stable second-strike arsenal element, developments in ASW are a particular concern.

The greatest concern, of course, has arisen around the vulnerability (or survivability) of the ICBM force. The issue involves the relationship between the degree to which a missile silo is hardened against the effects of atomic blast and the combination of accuracy and yield necessary to overcome that protection: the harder the site, the greater the blast effect necessary to overcome that hardening. With the most recent improvements in U.S. silo construction, approximately half the Minuteman force is housed in silos hardened to withstand a blast overpressure of 1,000 psi (see discussion in Chapter 4), providing a finite protection as explained by General Brown: "These improvements, while adequate to protect MINUTEMAN silo survivability in the near term, may be insufficient in the long term as forecast improvements in Soviet counterforce capabilities are fielded. Unless these improvements can be restrained . . . , silo survivability could be reduced in the mid- to late-1980s."[4] This reduction arises from the relationship between weight and yield to hardening capability: although it is theoretically possible to reduce CEPs to zero while increasing yield, there are structural limitations on the reinforcement of concrete, which is the basis of silo hardening (the practical limit is at about 3,000 psi). Alternatives to conventional hardening such as deep cave siloing (which the Chinese employ for some of their intermediate-range missiles) are enormously expensive and can be overcome eventually by offensive methods.

As Tammen explains, there are three basic means of achieving hard-kill capability, which vary in effectiveness: "Hard-target capability rests on three factors: accuracy, megatonnage [i.e. yield], and degree of defensive hardening. Of these three, accuracy is the most important because the radius of damage does not increase proportionately with an increase in yield."[5] The sensitivity of accuracy reflects the lethality factor (see Chapter 4). Leader and Schneider add warheads as a factor as well (which, of course, MIRVing accomplishes): "Hard target capability is more sensitive to the number of warheads in a country's arsenal and to their accuracy than to their size or yield."[6]

The present American arsenal contains more warheads with lower CEPs than their larger Soviet counterparts, giving the United States a current advantage. The Soviets, of course, are working diligently to reduce CEPs, and their larger payload missiles offer the possibility of carrying larger numbers of MIRVs. The *Strategic Survey, 1977,* explains how the United States is conducting its program to improve accuracy through guidance technology: "These guidance systems use for homing either those characteristics of a target which distinguish it from its surroundings . . . or highly accurate navigation to strike fixed targets with known locations . . . by such means as terrain contour matching (TERCOM), advanced inertial guidance systems, or navigation satellites."[7] Reporting as SSKP in the 45–60 percent range for the MK 12 warhead currently deployed in Minuteman III, the *Survey* shows the dramatic increase in MN III's successor: "The MK 12a warhead . . . with its 350 KT warhead and a CEP of 500 feet, is expected to have an SSKP . . . of up to 90 percent."[8] They further report that advances programmed for the SLBM force (currently the least accurate in the American arsenal) will have a similar effect: "The introduction of the Trident II (D5) missile, planned for the late 1980s, would give the American SLBM a true counterforce capability."[9] Testing of Trident II, of course, is prohibited under the protocols of SALT II, but remains a future counterbalance to Soviet progress (as we shall see, slightly less precise accuracies are projected for the Russians).

The Effect on Deterrence As stated at the outset of this discussion, the most basic consideration about any weapons system or capability is the degree of its contribution to overall deterrent stability. Since that stability has been defined as the mutual possession of a secure second-strike capability, anything that threatens the dual criteria of penetrability and survivability of deterrent forces erodes the system's basis of stability. "By the mid-1980s deployment of the new technology currently being developed will erode the seond-strike capability of land-based missiles,"[10] and, as a result, "the question of the decade, for both superpowers with respect to the structure of their strategic forces, should be what to do with their land-based ICBMs."[11]

At the theoretical level, ICBM nonsurvivability presents a dual problem, particularly when hard-target capability is mutually possessed: there is the incentive to use weapons first (or before they have absorbed an attack in a launch-on-warning mode) to guarantee they are fired at all; and knowing one has the capacity to knock out an opponent's forces gives a measure of the possibility of surviving a nuclear war through a preemptive attack. This ability to calculate profitability from first-striking is absolutely anathema to developed deterrent notions rooted in the realization that any preemption is suicidal. Although those ideas were created in a technological environment where counterforce targeting was not feasible, they have produced an environment in which nuclear war has been successfully avoided. Whether an environment in which first-striking might, under some circumstances, be appealing would produce the same result is problematical. Warnke doubts it, saying, "The fine tuning of . . . nuclear weapons and delivery systems could create fear of counterforce attacks on the other side and hence be destabilizing."[12]

The possibility that one side could attain hard kill when the other does not raises the specter of true superiority: the situation where one side has first-strike capability and the other has a lesser capability. First-striking, in that case, is attractive for the dominant state, because "real second-strike or assured destruction is not a possessed capability by the attacked nation,"[13] and it is necessary for the weaker state to fire first if at all. The actual system, of course, appears to be moving toward mutual silo-busting capabilities (a limited form of the situation where both sides have first-strike capability), but that situation only exacerbates first-strike incentives on each side, since both can calculate gain (damage limitation) by doing so. As Conover points out, "The mutual possession of counterforce capabilities large enough to threaten very large portions of each side's land-based missiles would lead to crisis instability and preemption incentives that are undesirable."[14] Downey spells out the incentives: "If the proposed new strategies and programs are to have any effect at all, they will provide the Soviet Union with positive incentive to launch a first strike against the United States, will remove its greatest disincentive against such an attack, will destroy crisis stability, and will increase the probability that a 'limited' nuclear exchange will quickly escalate to total spasm war." He adds that mutual possession of the capability does not change these incentives: "To respond to an opponent's counterforce capability by creating one's own . . . is an exercise in irrelevance since it does nothing to reduce the impact of his counterforce upon one's own countervalue."[15]

The degree to which hard-kill capability against ICBMs presents a problem to overall deterrence stability is lessened somewhat by the fact that there is not a coordinate threat to other TRIAD systems. Rathjens, for instance, says: "I contend that even if the Soviet Union had multiple warheads in large numbers, each of which could destroy an American ICBM with a probability of near unity, it would still not likely lead to a pre-emptive attack on the United States, or vice-versa. I say this, having in mind the fact that for an attack to be a 'success,' missile-launching submarines and bombers would also have to be destroyed with virtually 100 per cent effectiveness."[16] The rejoinder limits somewhat the overall problem from a deterrence stability standpoint (although adherents of the Schlesinger Soviet surgical strike against the ICBM fields scenario would contest how much). At a minimum, the problem becomes instead the continuing deterrent contribution of the ICBM leg to the overall TRIAD configuration, and the defenders or opponents of the need to keep all three legs viable differ significantly on what to do about the problem.

Solutions and Their Arms Control Implications The basic reason that the ICBM force is the first strategic weapons system to become vulnerable, of course, is that, in a fixed mode, it is the easiest to target. Given that targeting accuracy will soon allow adequate lethality to destroy these weapons, the question becomes what to do about them. There are two basic positions within this debate: either phase them out altogether and rely on a survivable "dyad" of air-breathing and SLBM components (a position the detractors of which disdainfully label "minimum deterrence") or find an alternative land-based system. As Kahan phrases the latter option, "Vulnerable strategic systems should either be protected to the extent feasible or phased out and replaced by survivable systems."[17]

Since protection through hardening is basically a temporary measure that can be overcome with increased accuracy, adherents of the TRIAD have had to look for other methods to ensure ICBM survivability, and the most common response has been the MX mobile system. This potent weapon is discussed in Chapter 4. Gray describes in detail how it could be deployed and how it would counter Soviet hard-kill capability:

> The MX concept envisages the following: a missile weighing 192,000 pounds with slightly more than 8,000 pounds payload, to be deployed in numbers ranging between 150 and 300, in a multiple aim point (MAP) mode. The multiple aim points may consist of buried trenches (13–20 miles in length), dispersed and hardened horizontal or vertical shelters or pools of water. . . . The intention is to provide . . . 5,000 individual aim points—i.e. points that Soviet strategic forces would have to cover.[18]

The obvious virtue of the system is that it requires the Soviets to target a large percentage of their warheads to destroy all the locations at which all the missiles might be at any time. The equally obvious reason that all 5,000 MAP sites would have to be covered is that the missiles would be concealed, showing the Damoclean effect on arms control: "Allowing mobiles would complicate verification, but it would permit the option of constructing a more secure and stable strategic force."[19]

The question of contribution to deterrent stability thus becomes the central consideration, either within or outside the arms control context, and there are merits on both sides. Defenders of mobility point out that it is the only effective hedge against targeting accuracy and that failure to deploy the system weakens deterrent posture and is therefore destabilizing. At the same time, increasing accuracy is not amenable to arms control: "Accuracy improvements are generally accepted as being among the most difficult weapons characteristics to limit in an arms-control agreement, because of problems of both definition and verification. Drafting a workable, direct limit on accuracy seems impossible."[20] Nonverifiable arms control arrangements, where the likelihood of detection of violations is low and the potential gains are substantial, are unstable themselves, as was pointed out in Chapter 6.

Simultaneously, the concealment that overcomes vulnerability threatens the ability to conclude arms control arrangements that are meaningful (the same cheating incentives apply). As Newhouse points out, the move to mobility could be destabilizing to arms control efforts: "A mobile missile force with no fixed launching points would be more secure than a heavily defended one; but a mobile force would be destabilizing if the missiles could not be counted by the other side. Stability relies on each knowing what the other has."[21] Drell states the problem created by mobile missiles more bluntly: "The verification problems raised by this option could mean the end of arms control by negotiated agreement."[22] Conover agrees and extrapolates the effect, stating that "mobile deployments would touch off a destabilizing arms race."[23]

The same problem would arise should cruise missile carriers become vulnerable to preemptive attack. Their small size and the variety of platforms from which they can be launched make prelaunch concealment relatively simple, and, if their known carriers (presently projected as B-52s) could be knocked out on the ground, the option would become attractive. McLucas offers a specific example: "My principal concern is the effect of the emphasis on cruise missiles on our ability to keep track of offensive weapons. The cruise missile is small and easily stored in warehouses. Thousands of them can be stored in a small building, such as an aircraft hangar."[24] Moreover, ALCMs and SLCMs can be converted easily, making verification of the range limitations prescribed in SALT II problematical, as Lehman and Hughes point out regarding Russian versions of the weapons systems: "The changes could more than double the range of the Shaddock without making any external configuration change."[25] Undoubtedly reflecting official Soviet policy on cruise, Arbatov summarizes the negative arms control effects of the cruise missile: "The arms race could take a course . . . which would make new agreements on limiting and reducing armaments far more difficult, if not altogether impossible, due to the insurmountable obstacles for their verification. This is, for instance, one of the negative consequences of the development of the strategic cruise missile."[26]

The central dilemma is that both sides of the argument in this debate have merit. If the movement toward hard-kill capability and thus ICBM vulnerability (and, by extension, depressed trajectory SLBM launch and cruise missile carrier vulnerability) is inexorable, then mobility and concealment are the obvious antidotes, and the alternative is a diminished deterrent posture that could make preemption more attractive and therefore be destabilizing. It is equally true, however, that the mobility-concealment response leads to systems that are not amenable to verification, thereby creating undeniably negative arms control consequences. One simply cannot have it both ways regarding the impact of hard kill and the resulting consequences for strategic stability. Before reaching a judgment, however, it is useful to look at how the targeting accuracy improvement movement has been justified and to analyze its overall effects on the strategic system. This need arises because, as

Conover points out: "It does not appear that accuracy really buys any militarily significant additional capability nor does it expand flexibility beyond what is already available in current forces; hence, the burden of proof still lies with advocates of improved counterforce to demonstrate why the risks of moving in that direction are worth taking."[27]

Justifying Improved Accuracy Although most of the discussion (reflecting emphasis in the literature) has focused on the difficulties Soviet hard-kill attainment present for the United States, the problem is reciprocal, and in some ways more serious, for the Russians. The seriousness arises from the dual facts that the United States has very active development programs in the targeting accuracy field that will yield true hard-kill capability (in all likelihood, our weapons will be a good deal more accurate than theirs) and the companion fact that, given Soviet force configuration, ICBM vulnerability has more serious strategic consequences for them. A preemption-prone ICBM force would leave the Soviets with a relatively small strategic "monad," making response to American hard-kill capability in the form of concealment and mobility more compelling for them.

Given American stated doctrine basing stable deterrence on mutual maintenance of secure second-strike capability, programs threatening retaliatory forces are inconsistent at the theoretical level, and it is curious that defenses of the U.S. targeting improvements emphasis have spoken only indirectly to this central dilemma. Rather, defenses have taken one of two forms, as Tammen points out: "Two points generally are made in favor of a new hard-target warhead program. First, it would increase efficiency by reducing the number of missiles now targeted against one hard facility. Second, it would allow greater flexibility and hence effectiveness in targeting command posts, communications centers, and other hardened sites that may remain vulnerable in a second-strike environment."[28] Nitze agrees essentially with the first point, in that the program helps redress strategic arsenal imbalances: "All of the options which would be effective in diminishing the one-sided Soviet advantage involve some improvement in the *accuracy* of U.S. missiles" (emphasis in original).[29] In testimony before a Senate Foreign Relations Committee subcommittee on March 4, 1974, then Defense Secretary Schlesinger explained the second justification: "We have to distinguish between disarming first strike, no first use, and counter-force. Counter-force can go against any military target. It can go against IRBM sites as opposed to ICBM sites. It would go against airfields or army camps. It has a range, and one can go counter-force rather than counter-value without necessarily putting himself into a position of having a disarming first-strike capability."[30]

These justifications have a decidedly ex post facto ring about them, and elaborating on them places the defender in an awkward position.

The simple fact, as has been suggested earlier, is that improved targeting accuracy has been an incremental outcome of guidance technology and, akin to MIRV, is a classic case of technology leading doctrine. Doctrinal rationalization and virtue have had to be developed after the fact, and, in all likelihood, no conscious a priori decision was ever made to try to attain hard-target kill capability. Technologically discrete processes like the discovery of terrain contour matching guidance and the application of the MARV evader principles to improving accuracy have simply happened, and justifications have had to be developed for them.

When logically followed through, these justifications have an ideological simplicity that rivals the dogmatism often associated with Soviet stated doctrine. True counterforce capability, as was pointed out in Chapter 2, has meaning primarily when combined with a first-strike firing posture, not for retaliatory purposes. Saving a few warheads because one has greater confidence those used will hit their targets is an advantage mainly if one does not have many excess warheads in the first place, but with a supply in excess of ten thousand, the gain is at best incremental. It is true that destruction of command and control centers would be a major goal in a nuclear war if one wants to cripple the ability of the enemy to continue issuing firing orders, but is it also true that it is *much more useful* to accomplish this goal in a first strike rather than after having absorbed a major attack. At the same time, one does not really need pinpoint accuracy to attack unhardened counterforce targets like airports and military camps.

The primary utility of hard-kill capability is true damage limitation: the ability to destroy enemy offensive weapons before they can be fired and thus diminish retaliatory damage. Other purposes are peripheral or cosmetic, such as attacking empty missile silos so that they cannot be reloaded. The United States recognizes this implication as it attaches to Soviet hard-kill attainment, creating the movement to mobility and concealment. Any Soviet planner who takes the Schlesinger justification and distinction between counterforce and disarming first strike seriously would have to be palpable fool. The certain Soviet response to our targeting accuracy program will undoubtedly be a mirror-image reaction by the Russians because, as Downey correctly observes, *"Our own counterforce will be useful if we plan to start a total nuclear war, but it will do nothing to deter the Soviets from starting one"* (emphasis in original).[31] In fact, it will be preferable from the viewpoint of deterrent stability for the Soviets to counter our hard-target program in the same way we counter theirs, since the alternative is vulnerability that otherwise can be overcome only by a first-strike or launch-on-warning firing doctrine.

That the conversion of a substantial portion of the non-SLBM force to a concealed, mobile mode (characteristics the submarines already have) will create an expensive arms race spiral is obvious. There is, however, dissent from at least one observer that this movement has negative arms control implications: "Silo vulnerability . . . does not become an arms control problem capable of being addressed seriously in SALT, until it is first approached as a serious defense problem. When the Soviet Union observes that the United States is in the process of both solving its silo-vulnerability problem, and . . . is developing a major potential threat to Soviet silo-based ICBMs, then silo vulnerability . . . should become an arms control issue relevant to SALT."[32] The basis of this exceedingly critical commentary on the ability to harness qualitative aspects of the arms competition is that "MX is the system that should persuade very tough-minded Soviet officials that the hard-target counterforce race cannot be won."[33] It is also arguable, of course, that once the United States (and by then, presumably, the Soviets as well) has gone ahead with mobile deployment, there will be little left to control within SALT.

The Prospects for Controlling the Process Improved targeting accuracy is but the latest technological innovation in strategic offensive weaponry to become part of the calculation of nuclear balance and for which doctrinal justification must be fabricated. Soviet testing of a hunter-killer satellite that can destroy orbiting satellites was met in the summer of 1978 with stern American warnings that, if they did not desist, the United States would build its own, and better, version of the weapon. Laser guidance has already been wedded to battlefield artillery firing, and the public literature has begun to speculate on the strategic applications of lasers[34] and charged particles for ballistic missile defense, a breakthrough that would have staggering implications for conventional deterrence thought (both the Americans and the Soviets are reported to be actively pursuing the possibilities). And with the possibility of ICBM vulnerability becoming reality, both sides are searching for means to make the most invulnerable of systems, the ballistic missile submarine, ready prey.

Prospects like these could fundamentally alter the nature of nuclear deterrence as it has developed over the past quarter century without anyone really intending to do so. Given the truly unsettling effects new developments could have, controlling the R&D prisoner's dilemma that is the basis for the qualitative arms race is the central problem facing arms controllers. As stated at the conclusion of Chapter 6, the ultimate success and long-range viability of the arms control process depend upon solving this central dilemma.

Assuming that there is a mutuality of interest between the superpowers in capping the qualitative spiral, the only way this can occur is through a verifiable ban on new systems testing. Rathjens, a leading proponent of this approach, summarizes the problem and the testing solution: "Unless the political climate changes, it is going to be difficult to negotiate agreements based on the kinds of intrusive inspection that would permit one to limit R and D at any earlier point in the research, development, test and evaluation cycle. If intrusive inspection is not in the cards, the point to put a handle on the development problem is at the test stage where there are demonstrable effects that are easily observed."[35] Since intrusive verification is probably not "in the cards," a test ban or limitation would have to be implemented through national technical means of verification, which, as has been pointed out, is not an entirely perfect solution. It is always possible to cheat to some degree, so that, as Newhouse points out, "the issue then turns on the degree of reliability."[36] Since not all testing can be observed, "this means for the present, the only limitations on strategic arms development that are likely to be acceptable are those on certain kinds of tests that can be monitored with some effectiveness with various remote sensors."[37]

Testing limits (or, preferably, total bans) could be effective, despite the ability to cheat somewhat. This would be particularly true of a comprehensive delivery systems test ban, since even the hint of a violation would raise concern. A ban could be developed parallel to the "escape clause" suggested by Brennan for warhead testing whereby the confirmed suspicion of a violation by anybody would abrogate the agreement for everybody.[38] In turn, a testing ban could be effective largely because of the very nature of the new systems under development and the purposes they have. That nature is that targeting accuracy, missile defense, and ASW are all very precise systems that must perform very exactly or they are of little use. Hard-kill capability requires destroying all (or virtually all) of an ICBM force before preemption becomes attractive, ASW that can sink only part of the SLBM force leaves some residue for retaliation, and it is of little comfort to know that an ABM system can intercept some (or even most) of the incoming missiles fired at you.

The high confidence one must have in very precise weapons systems requires extensive testing, evaluation, and redesign and retesting before deployment decisions can be made, and it is for that reason that the inability to test is an effective deterrent. The provisions in the Protocols to the SALT II agreement prohibiting new systems testing for three years are a tentative first step in this direction, and the expansion of the concept in SALT III and through CTBT could institutionalize this aspect of arms control.

In conclusion, it should be noted that testing bans are neither universally welcomed nor accepted. The same kinds of criticisms that Burt anticipated concerning SALT II apply to a weapons system test ban or limitation: "(1) the possible asymmetries in the overall strategic balance that could emerge . . . and (2) the dampening effect that the three-year protocol could have on American technological initiative."[39] Relative force levels, including the technological characteristics defining those capabilities, are an ongoing consideration, and care would have to be taken to "freeze" technology at some equitable level. It is undoubtedly true as well that the inability to conduct systems tests would be dispiriting to weapons scientists on each side (many of whom could be expected to oppose the idea on the ground of self-interest), but such a "dampening" is precisely what a ban would be intended to accomplish.

The Continuing Soviet Challenge The evolution of Soviet nuclear capability has been the subject of an earlier chapter that attempted to describe the basis of public Soviet theorizing and strategic issues, to catalog the characteristics of Soviet forces, and somewhat modestly to assess both of these concerns. The fear that the Russians are intent upon achieving that elusive commodity called strategic superiority and what they might do with it once they attain it is the most basic element in the ongoing strategic debate, and the SALT II agreements and preparations for SALT III will largely be assessed in terms of their contribution to essential equivalence or strategic asymmetry.

For these reasons, it is necessary to reach some judgments on comparative Soviet-American strategic postures and positions, which requires examination of three analytically separate but related concerns: the effects of hard-kill attainment, both in terms of relative capability and the effects of that attainment by one on the other; the overall balance between forces; and the doctrinal asymmetries between the two countries, seeking out the real from the imagined differences and suggesting some means of reconciliation. Making judgments in this area is intellectually treacherous. The dilemma of interpretation arises because assessment must be made on incomplete and thus imperfect information that forces the analyst to substitute indirect evidence such as past behavior and unverified (and usually unverifiable) assumptions for hard factual data. The result is an exercise that is lively and animated, with different observers drawing variant conclusions from the same set of basic materials but organized around diverging assumptions and interpretations.

The problem is most familiar in the dichotomous intentions and capabilities debate. The "hardest" evidence available concerns the physical capabilities each nation possesses (and, as pointed out in Chapter 4, these measurements are the subject of considerable controversy). From these measures, the attempt is made to infer intentions, with capabilities serving as the dependent variable in the causal relationship, for example, "What is the reason (intention) that the Russians have built weapons systems (capabilities) in excess of legitimate second-strike needs, unless?" The exercise is doubtless reciprocal (both sides assess the other in these terms), but the process contains dubious logical foundations.

The Uncertain Future					217

The difficulty arises because neither side really knows, beyond the other's public pronouncements (about which they are naturally suspicious), the intentions either has for their nuclear weapons. The causal element of propositions about why various weapons are developed (for example, if the Soviets have these intentions, then they would develop these capabilities) must be based on inferences and assumptions (often stated implicitly rather than explicitly) about motivations, or capabilities are used as a means from which to derive estimates of intentions.

This latter form of analysis runs the risk of trampling on sound logical inquiry in one or both of two ways. First, the imputation of a specific cause from an effect (unless the relationship is one of necessary and sufficient causation) entails commission of the logical fallacy of affirming the consequent. The frequent assertion, for instance, is that the Soviets can be engaged in a hard-target kill accuracy program only if they have the intention to launch a preemptive strike in a crisis. Seeing that they are moving toward increased accuracy, it is then concluded that their intention (cause) is to adopt a first-strike posture.

The fallacy in this line of reasoning results from the second logical problem and involves the relationship between necessary and sufficient causation. At the risk of oversimplification, a sufficient cause is a precondition to something else occurring (an effect) that can result in the effect, but it is not necessarily the only cause of the effect. Thus, a number of phenomena are sufficient to cause a boulder to roll down a hill (someone pushing it, a tree falling against it, an earthquake), but knowing the boulder is rolling down the hill does not tell one definitively *why* it is doing so. A necessary condition, on the other hand, is a precondition that must exist for an effect to occur, although the cause may not be sufficient by itself to make the effect happen. Thus, atmospheric turbulence is a necessary condition for a tornado to touch the earth, but unless other conditions are also present (such as minimal surface temperatures to produce adequate energy), a tornado cannot be drawn down from the sky.

The logical problems come together. Statements inferring intentions (causes) from capabilities (effects) commit the fallacy of affirming the consequent if they are (as is invariably the case) statements of sufficient condition. The configuration of Soviet (or American) capabilities could have resulted from a number of causes, of which a preemptive intention is but one (if arguably the most likely). Since there are other reasons than preemptive intent for targeting accuracy programs (the official American counterforce stance denying first-strike intentions but advocating the utility of extreme accuracy for retaliatory missions is another sufficient cause), inferring any particular sufficient cause from a given effect is fallacious. The only logically sound manner in which to infer a cause from an effect is if one has a statement of necessary causation (if you have a tornado, you can infer that there was, among other things, atmospheric turbulence). Thus, to be able to conclude that Soviet targeting accuracy is the result of preemptive intentions, one must assert that there is a necessary relationship between the intention and the capability. To do so, of course, would directly contradict the basic American stated purpose for making our missiles more accurate, because, if the relationship is necessary for the Soviets, it is also necessary for the United States; and thus, it would be to argue that the purpose of our accuracy program is preemption (which the Russians, unsurprisingly, suspect anyway).

The purpose of this discursive diversion is both to set the stage for and to provide adequate warnings about this and other assessments of the nature of the Soviet threat. Because the situation is one of imperfect information about the causal element in the intentions-capabilities analogy, any assessment is necessarily conjectural and speculative to some extent, and reasonable people can reach differing conclusions from the evidence available.

The Impact of Targeting Accuracy Soviet attainment of a capability approaching hard-kill accuracy has raised two primary concerns among American strategic analysts. First, the technological sophistication such an accomplishment implies threatens the qualitative superiority of American weaponry that is the primary basis of nuclear parity, given Soviet quantitative advantages accruing from their larger weapons systems. As has been noted, many analysts fear that improved accuracy and full MIRVing of Soviet forces, given their throw-weight and payload advantage, could result in true strategic superiority if technological levels become proximate for the two countries. Second, hard-kill capability raises the ICBM vulnerability problem and hence the difficult, expensive, and potentially destabilizing set of responses discussed in the previous section of this chapter.

Without attempting to denigrate the importance of this issue, two conditioning factors tend to be underemphasized in these discussions. First, Soviet weapons are becoming more sophisticated, but so are American systems, and Soviet technological advances do not necessarily diminish the American qualitative advantage. Second, the discussion has tended to focus almost exclusively on the destabilizing effect Soviet hard-kill attainment has on the nuclear balance, while either downplaying or ignoring the reciprocal impact of parallel American programs.

The first factor is an artifact of the continuing qualitative arms race. As has been emphasized (see especially Chapter 4), the United States has traditionally had a sizable technological advantage in its weapons systems as measured by targeting accuracy, and current and projected American programs will continue that advantage. Cruise missiles are reported to have a CEP in the range of around 100 feet (see discussion in Chapter 4),

The Uncertain Future

and the *Strategic Survey, 1977,* reports a CEP for MX of 200–300 feet. Comparable figures for the newest deployed Soviet ICBMs (the SS-17, SS-18, and SS-19) are "about 1,200 to 1,600 feet and are not expected to come below 1,000 feet until the early 1980s."[40] Since lethality, the most sophisticated measure of a weapon's deadliness, is much more sensitive to accuracy than it is to yield, this kind of factor suggests a continuing U.S. advantage. There is, moreover, no hard indication that the Soviets have solved the guidance problems that have allowed the cruise missile to become a potent strategic system (which is undoubtedly why they have opposed ALCM deployment in SALT), and the fact they have satisfactorily tested multiple warhead missiles launched from nonfixed sites (either SLBMs or mobile ICBMs) indicates that they lag behind in the more sophisticated and difficult technological areas.

The issue of a diminishing U.S. qualitative edge, in other words, will be determined largely by development and deployment decisions both nations make in the upcoming years. If arms control processes embedded in the SALT II test ban moratorium within the protocols are unsuccessful in arresting Soviet deployment of its "fifth generation" ICBMs, then it may be necessary for the United States to implement programs like MX and Trident II to maintain strategic parity. These programs are enormously expensive and thus politically controversial, and the real heart of the concern about the impact of Soviet hard-kill programs is likely to revolve on American political resolve to answer threats that arise.

The second, conditioning factor in the hard-kill debate is its reciprocal effect. It is a curiosity in the current American debate that Soviet hard-kill attainment is universally regarded as undesirable and destabilizing because it would presumably be attached to a preemptive firing doctrine (see discussion above), but there is less concern about the destabilizing impact of American accuracy improvement (because we continue to profess, with limited exceptions, a second-strike doctrine). Official American justification of improved accuracy, of course, is based on the notion of damage limitation after absorbing either a massive or (it is argued more likely) controlled Soviet preemptive attack. Thus, former Secretary Schlesinger argued that improved accuracy could be tied not only to preemptive intent (which, it has been argued previously, is the clearest purpose for such a capability), but also to retaliatory missions.

The difficulty with this justification is that, particularly from the Soviet vantage point, it is unbelievable. Although it is true that hard-kill capable weapons can be used in retaliation, it is equally (and possibly more) true that the ultimate form of damage limitation is the ability to knock out an opponent's retaliatory forces through a first strike. Referring directly to Schlesinger's justification of the U.S. program, Kincade points to this anomaly: "In terms of perceptions, however, his separation of targeting doctrine from hardware and his denial of first-strike intentions were somewhat academic, since the U.S. was pursuing simultaneously a warfighting targeting and employment doctrine and missile accuracy improvements which lend themselves to a first-strike interpretation."[41] Metzger applies this problem specifically to the problem MX deployment would raise for the Soviets: "M-X could appear to the Soviets to constitute a credibly first-strike threat against their land-based missiles, which are the backbone of Soviet strategic forces."[42]

This means that hard-kill attainment is destabilizing if either side has it. The Soviets obviously cannot accept "at face value"[43] American assurances that our emerging systems do not have preemptive intent, because they will have that capability. As a result, the Soviets are faced with the same three strategic choices available to the United States. First, they can relegate themselves to a vulnerable ICBM force and phase it out, which is highly unlikely given their heavy dependence on ICBMs. Second, they can exercise the mobility-concealment option, but both the expense and the additional technological problems that targeting from nonfixed sites pose make this a less then appealing prospect. Their third, and most likely, option is to adopt a first-strike or launched-on-warning doctrine. Obviously, such a decision is maximally destabilizing to the overall deterrence system, but it would guarantee that the Soviets would be able to use their missiles. Additionally, this option supports the movement toward greater accuracy as a damage limitation measure so that, in a curious way, the American accuracy program serves as a major incentive for its Soviet counterpart.

The Relative State of the Balance It is evident that the growth of Soviet arms levels from a state of marked inferiority in the early 1960s to a level of parity and by some measures superiority has been the major variable in U.S. doctrinal reexamination. Despite the capping of delivery systems numbers within SALT, the qualitative arms race has continued, resulting in enormous destructive inventories on both sides. The significance of this change in relative capabilities has and continues to be the source of major contention among analysts and therefore deserves attention.

Overviewing the Soviet weapons program's growth, Polmar states that three trends have emerged: "(1) Soviet strategic arms development has been both independent and responsive to U.S. actions, but has not copied U.S. developments; (2) the Soviet Union has the capability of matching or exceeding the quantity, and on a longer-term basis the quality of U.S. strategic forces; and (3) with the current strategic 'balance,' the possible use of highly selective strategic weapons or tactical nuclear weapons probably is increasing."[44]

The second trend, if it continues, particularly troubles many observers and leads Gray to the assessment that "on every important indicator the Soviet Union is either ahead . . . or is catching up rapidly."[45] Admitting that the Soviet Union possesses comparable nuclear forces and possibly slightly superior capability, the question of what difference such a balance makes becomes crucial.

At the strictly military level of deterring or fighting a nuclear conflict, it is difficult to maintain that the current situation is intolerable. Both sides possess enormous arsenals, enough of which would survive even the most massive preemptive strike to make first-striking incentives minimal. The hostage effect, although subject to considerable criticism, still adheres, and with it the accompanying incentives for second-strike strategies, public pronouncements (especially by the Soviets) notwithstanding. At the theoretical level, the Soviet weapons buildup to a position of secure second-strike capability has in fact stabilized the system to the extent that prior to that buildup the Soviets could reasonably have questioned their ability to absorb an American first strike (see discussion in Chapter 2). In this sense, the balance remains stable and thus acceptable. As Rathjens and Ruina put it:

> For more than a decade the United States and the Soviet Union have been able to destroy each other many times over with neither having the ability to prevent its own destruction. All of the additions to, and modernizations of, the nuclear arsenals in this period have had their primary impact as signals of intent, determination, and priorities. They conveyed political messages to allies, and adversaries, and were responses to internal pressures. However, *they hardly affected the existing nuclear balance.* (Emphasis added.)[46]

It is, of course, easy to be overly comforted by this sort of assessment and to forget that the system has been marked by an extreme dynamism that makes the balance susceptible to dramatic change. This factor is most obviously true in the research and development area. Both sides, for instance, are very actively involved in ABM research that, if it reaches fruition, could dramatically alter conventional calculations.

This is the heart of the R&D prisoner's dilemma problem with which arms control efforts have yet successfully to come to grips and means that, as Secretary Brown implies, continued vigilance and activity are necessary: "Unless one side or the other is careless—and allows a major imbalance to develop—or makes serious miscalculations, a condition of mutual deterrence and essential equivalence is likely to prevail in the future, just as it does today."[47] At the military utility level, one must conclude that it is necessary to be concerned with the growth and continuing evolution of Soviet nuclear strength, but that prudent U.S. action can keep the situation stable.

The primary importance of the growth in Soviet strategic arms and the altered nuclear balance is more indirect, operating at the political and psychological levels. As was discussed in Chapter 5, the Soviets view their nuclear arsenal as supporting other foreign policy goal attainment, at least to the extent that equivalent forces do not place them in a situation where they are forced to back down because of perceived inferiority (the apparent Cuban missile crisis lesson). Thus, as Brennan maintains, "given the character of real decision makers in the real world, it is virtually certain that the question as to who has the 'bigger' force will play an important role in the settlement of a crisis."[48] Weapons systems balance thus may have a psychological impact on the confidence of decision makers in following different courses of actions and be of symbolic value to allies and others concerning strength and resolve quite apart from the direct relevance nuclear balance has on any given issue.

Doctrinal Asymmetry A great deal of attention has been given in the recent literature to differences between American and Soviet deterrence notions. Partially, this emphasis has been a reaction to the realization that the Russians were not as diligent and attentive students of our ideas as we once thought they were, and partly it has been the result of the need to determine Soviet intentions for their growing capabilities. Like the concern over those forces themselves, the debate has sometimes been strident.

The public Soviet nuclear position has been discussed in Chapter 5 and thus need not be repeated here. Two components of that posture, however, bear directly on the U.S. strategic debate and thus warrant reiteration and consideration: the Russian view of the utility of nuclear weapons and Soviet attitudes toward limited nuclear exchange.

The first concern is important in context of the ongoing debate about Soviet strategic intentions. To summarize, the key to the public Soviet view is expressed in the peaceful coexistence notion. As described earlier, that concept maintains that nuclear war is an inappropriate form of intersystem competition and that deterrence is the most important nuclear value. Although there is disagreement about what dynamic maintains deterrence (the Soviet ability to win a nuclear war versus mutual assured destruction), the key element is adherence to deterrence as the primary value.

At the same time, peaceful coexistence contains the basis for continuing competition and conflict at the non-nuclear level and thus creates an indirect role and utility for nuclear weapons possession. Ideologically committed to an evangelical doctrine and hence expansionistic, an equivalent or superior nuclear force removes obstacles present in an inferior position. To this extent, there are linkages between Soviet nuclear and other foreign policy behavior that cannot be ignored. Assessing these factors, Culver reaches a balanced conclusion that represents a sensible assessment of Soviet motivation: "I have always believed that Soviet leaders would do whatever they could to expand their power and influence, though probably not at the risk of war with the United States. I accept the general principle that Soviet leaders will do whatever is in their self-interest, whether that be to build missiles or to abide by an arms control agreement."[49]

The second major concern relates to the issue of limited nuclear war and is important because of the emphasis on this issue in the American debate and in American force planning. As was pointed out in Chapter 3, the "limited nuclear options" position has been praised as a realistic alternative given the catastrophe general exchange would entail and as adding desirable flexibility for proportionate response to limited Soviet initiatives. The concept has been condemned as lowering the nuclear threshold by introducing new and seemingly less dire circumstances in which nuclear weapons might be employed and because the Soviets publicly say limited nuclear war is not feasible.

Since limiting a nuclear war would necessarily have to be a mutual activity, the Soviet position is salient to American planning. The Soviet position has been consistent and is summarized by Snyder: "Soviet commentaries on the U.S. selective options doctrine have left no room for doubt that they take a decidedly jaundiced view of the strategy and continue to believe (or at least continue to wish us to think they believe) that a central nuclear war would brook no possibility of being restrained short of total effort by each side to achieve total victory."[50] If the Soviets truly believe an exchange could not remain at less than a general level (and that assumption is open to question), then much of the current American debate is irrelevant. The idea that nuclear war would be total is certainly compatible with general Soviet doctrine emphasizing winning such a war should it occur. To the extent that such notions have been translated into actual force deployments and weapons missions, the Soviet arsenal and planning mechanisms could well be locked into an escalatory model from which it would be difficult for them to deviate in an actual crisis. Record suggests this latter possibility with regard to Russian tactical weapons: "This 'insensitivity' to what Western military analysts believe are distinctions of crucial significance may well be a function of the character of the Soviet TNW arsenal."[51] Moreover, if the Soviets believe that a nuclear war would quickly escalate to the general level, they have relatively little incentive to launch the controlled preemptive strike that is the keystone of the assumptions built into the Schlesinger doctrine.

Horizontal Nuclear Proliferation The spread of nuclear weapons to states that currently do not possess them may represent the most serious problem facing strategic doctrine in the next decade. Accelerated by the worldwide oil crisis of 1973, nuclear power generation facilities have proliferated rapidly in recent years, bringing with them access to the materials from which nuclear weapons can be fashioned in many countries that previously lacked those materials. The issue, and particularly its solution, is closely linked to arms control generally, as pointed out in Chapter 6.

In order to explicate the problems for the nuclear system presented by horizontal proliferation, the discussion will be organized around three basic concerns: the theoretical challenge to underlying deterrence notions largely developed around the bilateral U.S.-U.S.S.R. nuclear relationship that new and independent powers present and some of the implications these problems create for doctrinal choice; some of the difficulties presented by the kinds of states likely to develop nuclear weapons and the kinds of nuclear forces they will be capable of developing; and steps that have been proposed to halt or slow the prospects of weapons spread.

Theoretical Problems At the level of deterrence theorizing, the basic problem is that the considerable amount of thought that has been devoted to understanding nuclear balance dynamics (as reflected primarily in Chapter 2) has been concerned with stabilizing the Soviet-American strategic relationship. As a result, Rosecrance notes, "Deterrence theory, fundamentally dyadic in character, does not tell us proper strategies for an n-party game. It may be, therefore, that entirely new strategies will have to be found for a multipolar world."[52]

That the problems for which theoretical solutions will need to be found will be different from those with which nuclear analysts have traditionally grappled is obvious. Rosecrance suggests at least three new complications that a proliferated world presents. First, "political disputes in a wider world context may be more serious," an observation based on the likelihood that one or more traditional adversaries in the developing world may acquire and be tempted to use nuclear weapons to settle their differences, as for example, in the Middle East. Second, "a multipolar strategic environment poses new questions for deterrent theory and practice," a number of which will be considered in the remainder of this section. Third, "deterrence will come to depend even more on political factors and alignments than in the bipolar case," a reflection on the fact that many of the potential states are not closely aligned and thus directly controllable by the superpowers.[53]

A nuclear multipolarity represents a much more uncertain strategic environment in which to operate. The environment is complicated by the potential of numerous rather than a single nuclear adversary, where potential enemies may be individual or in coalition and where the behavior of some potential adversaries is difficult to predict in crisis situations. This combination of factors in turn has both strategic and force level implications. The Soviet Union, facing both the United States and the People's Republic of China, has had to deal with the problem of multiple nuclear enemies for some time, and Rosecrance points out that this set of difficulties will confront all nuclear powers in the future:

"Should one plan to retaliate against all members of the international strategic system, against some or only against one? One does not know how large an assured destruction force he needs. The lack of certainty on this point is likely to accentuate the arms race."[54]

Implicit in a world of many nuclear powers, then, is the possibility that nuclear exchange would not necessarily be a systemwide proposition. Rather, some nuclear states, acting individually or in concert, might refrain from engaging in a nuclear conflict between the superpowers, waiting for all to deplete their arsenals, and then use *their* arsenals to blackmail the spent giants. Such a prospect means, on the one hand, that "the criterion of technical deterrent effectiveness in a multipolar system is the ability to retaliate against a coalition of the remaining powers."[55] On the other hand, the force planning uncertainties in dealing with all the possible exigencies in such an environment are not conducive to effective arms control. In a sense, then, the relationship between proliferation and arms control becomes causal in both directions: as pointed out in the last chapter, controls on vertical proliferation appear to be a precondition (necessary, if only arguably sufficient) to avoiding horizontal proliferation; at the same time, horizontal proliferation contains dynamics that run contrary to vertical weapons limitation.

These prospects present a complex and dreary picture, the most salient characteristic of which is the enormous uncertainty and risk enhancement a proliferated world offers. The chilling aspects are made all the more sober if one accepts something like Rosecrance's definition of the basis of a secure multipolar nuclear world: "If strategic deterrence is to be achieved in multipolarity, technical requirements dictate that all nuclear forces be invulnerable, relatively equal in size, and that the number of powers be kept to a minimum."[56] The point here is that, unless restraints are exercised, a proliferated world would consist of a number of states whose forces could meet neither of Rosecrance's basic criteria (invulnerability and equality in size) and where the numbers of nuclear states would be in excess of anyone's definition of a minimal level. To understand the dynamics of the threats presented by such a situation requires examination of some of the specific problems inherent in a horizontally proliferated nuclear world.

Specific Problems The likely patterns in which weapons proliferation would take place raise a number of knotty difficulties to the overall strategic system. For analytical purposes, these can be divided into three categories. The first is the nature of the potential new members, including the political stability of individual members and combinations of members, the dangers of the catalytic nuclear war effect should a regional war become nuclear, and the prospect that either a national government or extremist political group might use nuclear weapons for terrorist purposes. Second and related is the problem of overall system control that exists when a number of independent decision centers have their "fingers on the nuclear button." Third is the problem of the kinds of nuclear forces new system members are likely to be able to develop in terms of force sizes, delivery systems, and warhead technology, all of which have implications for underlying deterrent stability notions.

Nature of New Members: As was pointed out in the last chapter, the spread of nuclear-generating technology since the Arab oil boycott in 1973 has brought nuclear technology to virtually every corner of the globe. Because the safeguarding process to ensure that weapons-grade materials are not diverted from electricity generation to military usage is imperfect, inevitably the potential to join the nuclear weapons club is entailed in the transfer of peaceful atomic power. As evidenced most recently in spring 1978 when a Harvard undergraduate economics student successfully designed a nuclear device from the public literature, the technology for bomb construction is available to any skilled person or group, and students from the developing world have been trained for years in physics and engineering at Western universities to guarantee that expertise base. As a result, access to weapons-grade plutonium and uranium that are by-products of the fuel cycle completes the requisites for weapons acquisition and makes an examination of the nature of new system members vital. That this task has a high degree of urgency is suggested by Falk with regard to purported Israeli activity during the Yom Kippur War:

> Such developments disclose the alarming prospect that easier access to nuclear technology will make it relatively simple and thus more likely for a beleaguered government or a desperate political actor of any sort to acquire and possibly use nuclear weapons. This process is already evidenced by Israel's semi-overt reliance on nuclear weapons . . . at the time of the October 1973 war; these nuclear weapons were reportedly assembled after Israeli leaders had been informed by the military commander on the Suez front that Egyptian forces might soon break through Israeli defense lines.[57]

Political Stability: One of the major concerns about the potential new members of the nuclear club is their political stability and responsibility. As the list of "near-nuclear" states in Chapter 6 indicated, a diverse group of countries spanning the globe and containing representatives from every continent will, before the end of the century, possess the weapons option unless trends are reversed.

Obviously, some of these states are politically stable and responsible, such as the Scandinavian states, which have had but not exercised the option for some years. A greater concern exists over such states in the developing world as Libya or Nigeria, which either have unstable leadership or possess radical elements that, if they came to power, would be questionably responsible with these weapons (Libya's Colonel Qaddafi, for instance, is reported to be interested in procuring nuclear weapons so that he can make them available to the Palestine Liberation Organization).

More troubling than the prospect that individual states that would be questionably responsible with nuclear weapons obtaining them is the prospect that historic enemies might join the nuclear club, a possibility Hoag raises: "The next nuclear powers are likely to come in opposing pairs—Israel versus the United Arab Republic; India versus mainland China, but then Pakistan versus India; Australia versus mainland China; and so on—with resultant strategic programs that are oriented almost exclusively to immediate regional fears."[58] Although the forces these nations are likely to be able to develop (Argentina and Brazil could be added to the list) are unlikely to be strategic in the sense of posing a direct threat to the homelands of the United States and the Soviet Union, they will be strategic in the dyadic relationships between the pairs listed. For instance, the Israelis already possess missiles with a range of around four hundred miles that cannot reach Soviet territory, but can, if tipped with nuclear warheads, level the capital cities of any of her traditional Middle Eastern adversaries, making them strategic in that context. If the Israelis (or Egyptians, or Syrians) were badly losing another Middle Eastern war but had nuclear capability, their ability to refrain from using such devices rather than accepting defeat is at best problematical.

Catalytic Effects: The prospect of regional nuclear conflicts is sufficiently gruesome simply measured in terms of the immediate devastation and residual radiation effect it would have in the affected area. Such an act would have ramifications for the entire system, however, because it would represent crossing the firebreak. This fact in turn raises the dual problems of the nature of the escalatory process and the possibility of what is called "catalytic war."

As has been stated, there is no direct empirical evidence about the nuclear escalatory process in the situation where more than one adversary possesses the weapons. It is possible that, should nuclear conflict break out in, say, the Middle East, the superpowers could agree implicitly or explicitly to remain aloof from it. As Clarence D. Long suggests, however, once the process has begun, it is also possible that the superpowers could be inexorably drawn into the vortex: "More likely than an all-out nuclear war beginning between superpowers is a nuclear exchange between small countries, and a nuclear war anywhere has to be assumed to risk escalation to superpower involvement whether by deliberate intervention, or by miscalculation, bluff or panic."[59]

A variant of this possibility is the deliberate inducement of superpower involvement in regional nuclear conflict by the side that feared it would lose in the absence of such intervention. As Brennan explains,

"One of the traditional 'scenarios' involving the interaction of primary and secondary nuclear states is that of the disguised, or 'catalytic,' attack."[60] Within this possible form, "a state seeks to simulate an attack by some third power, inducing the victim to retaliate upon the innocent party."[61] Thus, were Israel losing a Middle Eastern nuclear war, she might be tempted, either through a one-way flight by one of her American-built aircraft with U.S. markings or by firing an acquired missile at a time when U.S. Poseidon submarines were known to be in or near her waters, to attack a Soviet target in the hopes the Soviets would retaliate against the United States, thereby drawing the Americans into a conflict from which we had otherwise remained aloof. Although improved U.S.–U.S.S.R. communications (the "hot line") and advance satellite reconnaisance are designed to minimize the possibility that such a tactic would work, it is not entirely certain that, in the heat of the moment, decision processes would remain calm, rational, and analytic.

Nuclear Terror: The previous examples are, of course, highly speculative. Another possibility that has more recently attracted public attention, as reflected, among other places, in a growing fictional literature,[62] is the use of nuclear weapons for terrorist purposes. The dual prospects of irresponsible governments making available nuclear weapons to political extremist movements or the terrorist use of weapons by national governments led one observer to conclude, "If strategic nuclear warfare between major powers is avoided, nuclear terrorism may be one of the most important political and social problems of the next fifty years."[63]

The scenarios in which nuclear terror might arise are virtually endless and thoroughly horrifying. Given the absolute ruthlessness of some elements of organizations such as the Black Septembrists of the PLO, the Japanese Red Army, and the Red Brigade within the Italian communist movement, it is not at all difficult to imagine individuals who would make and carry out nuclear threats to achieve their political ends if the requisite paraphernalia were available. Schelling sees nuclear terror as compatible with the activities of some national governments.

> The organizations most likely to engage in nuclear terrorism will be national governments. Passive terrorism on a grand scale we call "deterrence." When it is directed at us, rather than our enemies, we call it "blackmail." At the small end of the scale of magnitude would be an organization possessing only a few weapons, or claiming to possess them or believed to possess them, weapons that might be clandestinely emplaced where they could do, or threaten to do, horrendous civilian damage in the interest of some political cause. But that is probably the image that would fit one of those small nations that might have a small supply of nuclear bombs or a small capability to produce some in a hurry.[64]

The urgency that has attached to this concern has come about partly because the rise in non-nuclear terrorism and the spread of weapons-grade materials have more or less coincided, if independently of one another. The fact that plutonium is available in numerous places presents a problem because of lax safeguarding procedures in some countries (including, according to some allegations, the United States) that makes theft at nuclear power generation sites a possibility. The possibility of stealing enough weapons-grade material to construct a bomb is exacerbated, according to Guhin, because "many hundreds of thousands of kilograms of weapons-usable materials, particularly plutonium, are expected to be in use, transport, or storage in various parts of the globe, with transport of significant quantities of nuclear materials running into several thousands of shipments annually."[65] Thus, controlling access to the materials to wed to available technology will become increasingly a problem in the future.

The Control Problem: Proliferation also inevitably means that more nations will have the ability to initiate nuclear war, thereby decentralizing the control of the nuclear decision process by multiplying what I have elsewhere referred to as the "fingers on the nuclear button."[66] The statistical probability that a larger number of nuclear decision makers are more likely to arrive at the decision to employ these weapons than a smaller number, combined with the arguable level of responsibility of some who might have the opportunity to exercise the option, creates a legitimate concern which Wohlstetter describes vividly: "With the multiplication of national strike forces, the control problem becomes especially acute. If many nations have the power of decision, and if, in addition, each nation decentralizes its control to a multiplicity of subordinates, or—worse—to some electronic automata, it is evident that the situation could get out of hand very easily."[67]

In addition to more people in more countries having the ability to reach the nuclear decision, of course, there are additional conflictual situations in a proliferated worlds where these weapons could be employed. Nuclear possibilities currently do not affect the growing conflict in southern Africa, but would not an increasingly beleaguered white South African government be tempted to build these weapons (which it is perfectly capable of doing with a little effort) and using them if the tide of black African nationalism were to seem otherwise undeniable? Faced with their crushing defeat and loss of half their national territory, would a nuclear Pakistan not have been sorely tempted to use those weapons against Indian forces supporting the Bengali secession in 1970? Even more ominous (and particularly if one accepts the information in the Falk quotation earlier), would the decisive loser in a future Middle Eastern conflict accept defeat if it had the ultimate choice of nuclear weapons usage? The examples, and others one might cite, are chilling and are summarized well by Quester, who asks, "How many n th countries would continue to be 'responsible' in their use of nuclear weapons when they were being overrun?"[68]

The Nature of n th Country Forces: The kinds of nuclear capability that most potential nuclear powers are likely to be able to develop is also a matter of considerable practical and deterrent stability concern. The technology to produce primitive nuclear explosive devices is within reach of numerous countries, but the capacity to develop and build the sophisticated and invulnerable delivery systems characteristic of American and Soviet arsenals is not. Large and technologically advanced weapons delivery systems are enormously expensive to build and maintain and require both technical skill and human resource commitments, in addition to capital outlay, that are beyond the resources of nations other than the nuclear superpowers (although the Chinese are trying diligently to emulate these levels). At the same time, sophistications in warhead design in areas such as fission-fusion and fission-fusion-fission that have allowed both enormous yield and miniaturized warheads have not been made as available as "simple" nuclear designs and are beyond the technical grasp of all but a handful of nations. Limitations on both forms of technology have serious implications for the nuclear balance.

Delivery Systems: The two most salient characteristics of nuclear forces likely to be feasible for additional weapons states are that they will be numerically small and deliverable conventionally (the Israeli force, equipped with American surface-to-surface missiles, as described earlier, are mobile and thus present a limited exception if their conventional warheads are replaced by nuclear devices). The net result of these characteristics is that these arsenals will be vulnerable, as Rosecrance points out: "The difficulty . . . of multipolar strategic systems is that probable technological and political tendencies move inversely to stability. As time goes on, more and more nations will become nuclear strategic powers; thus n will rise. But the countries joining the nuclear system are not likely to have capacities as invulnerable as those possessed by the two charter members of the nuclear club."[69] Hoag suggests that this fact could lead to compromise of control mechanisms in certain situations: "But an n th power that finds a secure retaliatory capability beyond its means may sacrifice positive control. It may use clandestine delivery means, whose merit is cheapness, to put nuclear weapons in peacetime where they are most dangerous—namely within preselected enemy target areas (cities)."[70] His example is a prominently mentioned possibility for terrorist employment.

The theoretical implications, within the context developed in Chapter 2, should be clear. Using the categories discussed there, new nuclear powers will be capable only of developing forces that represent less than a first- or second-strike capability. Their small forces obviously will be in that position vis-à-vis the nuclear superpowers, who will retain a first-strike capability against them. The conflictual pairs are likely to have lesser capabilities against one another as well, since their forces will be

The Uncertain Future

subject to at least some interdiction (antiaircraft defenses) and will have insufficient accuracy for more than limited counterforce missions (with the exception of destroying enemy airfields where adversary nuclear carriers are based before they can become airborne). The result is that new nuclear states are virtually forced to adopt destabilizing first-strike firing doctrines, from the dual motivations of damage limitation (hoping to knock out some substantial portion of the enemy's nuclear retaliatory force) and fear that the failure to strike first could result in serious depletion of their forces if they allowed a preemptive strike against them to occur. In a crisis situation where regional nuclear powers both faced these calculations, the probabilities of misinterpreting signals of the imminence of nuclear exchange would be maximized.

The characteristics these forces will likely have also predispose them to countervalue and thus devastation-maximizing targeting. Clearly this is the case of a small or medium (such as France) nuclear power facing one of the superpowers against whom a disarming first strike is impossible, and Kemp offers a likely scenario: "In a crisis situation the leaders of the medium nuclear powers might be placed in the position of having to initiate a countervalue strike against Soviet cities while the Soviet leadership was pleading for a negotiated settlement to avoid countercity strikes."[71] Faced with a first-strike capability in the hands of the superpower, the motivation would be to strike first to be able to strike at all, but knowing this presents the state with the secure second-strike capability the incentive to strike first to disarm the medium or small power. For small nuclear powers, countervalue targeting, even with a relatively small force, could effectively assure the societal destruction of an adversary (for example, relatively few Israeli weapons aimed at Syrian urban areas could essentially obliterate that nation).

Warheads: The consequences of countervalue targeting are worsened by the fact that the nuclear devices new nuclear states will be able to construct are likely to be crude fission bombs, which are the "dirtiest" kinds of devices. As a result, their employment would maximize residual radiation effects and human suffering within the general area of their detonation and ecosystem disrupting within the region and conceivably worldwide. The fact that designs will likely be fairly primitive can also mean that they could be structurally unstable, leading Schelling to suggest an interesting dilemma for the advanced nuclear powers:

> A tantalizing control problem will arise with respect to countries that do have, or are believed to have, nuclear weapons. . . . Little as we like them to have nuclear weapons, if they are going to have them, we would like them to have the best technical safeguards against accidental or unauthorized detonation and possibly—though this is a tricky one—against military attack. At the same time we shall not want to reward countries that make nuclear explosives by offering them our most advanced technology

to safeguard against misuse especially if some of the most effective safeguards would be in the design of the bomb itself.[72]

In a sense, this same dilemma can be extended to the nuclear physics of the bombs themselves: since some of the ecosystem damage could conceivably affect us either directly (radioactive clouds drifting over our territory) or indirectly (contamination of Middle Eastern oil), it would be in our interest to have weapons as "clean" as possible. At the same time, to accomplish this would reward initial behavior we oppose and add to the capability of any country we assisted.

Efforts to Impede the Process As the foregoing discussion should have made amply clear, the prospect of horizontal nuclear proliferation opens a range of theoretical problems and uncertain possible situations that are best handled by avoiding nuclear weapons spread. The NPT was a first step in that process, but, as pointed out in Chapter 6, the treaty has been less than a total success. In light of the spread of nuclear power facilities and heightened fears of irresponsible governments or terrorist organizations gaining access to nuclear materials, increased emphasis has been placed on inhibiting weapons spread. This movement has had two essential and interrelated thrusts: restrictions on access to weapons-grade materials necessary for bomb construction and creation of disincentives to weapons development.

The effort to inhibit access to weapons-grade material has had two basic emphases: denial of access to technologies that produce weapons-grade material and control by the nuclear suppliers of the enriched fuels from which weapons can be fashioned. These two thrusts have been discussed by the so-called Nuclear Suppliers Group in meetings in London[75] with varying advocacy and opposition for both approaches.

Bans on the export of technologies and facilities for nuclear production have centered on two types of facility: the so-called "breeder" reactor (which uses a physical process that produces more plutonium—of weapon quality—than it expends in the fuel cycle) and reprocessing plants (facilities wherein "spent" fuel rods for power generators have expended materials—including plutonium—removed and replaced by new fuel). Denial of these obviously contracts the availability of plutonium for diversion to weapons use. This approach has met with a lukewarm reception among some nuclear suppliers, because the nuclear power generator industry is a highly lucrative market. France, for instance, has a large investment in breeder technology on which it would like to capitalize, and West Germany expects a handsome profit from its sale of a reprocessing plant to Brazil. Nuclear export thus can partially ameliorate balance-of-payments problems created by the rise in world petroleum prices, and there is a ready market for these controversial technologies, because they lessen consumer dependence for continuing fuel supplies (which has an important symbolic value in the wake of the 1973 oil boycott).

The second and related thrust, advocated most strongly by the United States, involves tight supplier controls of fuel. This approach involves supplier guarantees of continuous supplies of fuel rods that would be reprocessed in the supplying nation if the receiving nation accepts IAEA safeguards. When combined with reaction types like the light water reactor (LWR), the procedure can effectively impede weapons diversion of materials, as Greenwood, Rathjens, and Ruina explain: "A country that imports its enriched fuel and refrains from reprocessing its spent fuel would not have access to weapons materials from an LWR fuel cycle."[74]

Opposition to these limits has taken two forms. First, many question the effectiveness of IAEA safeguards, given both the small staff of investigators the organization has and its lack of intrusive inspection authority (noted in Chapter 6). Second, control of access to nuclear, including weapons-grade, fuel is almost universally opposed in the developing world which offers the large market for the nuclear industry (including nations with no known weapons aspirations) on the grounds of paternalism and unacceptable dependence.

Efforts have also been proposed to reduce nuclear incentives. Through advocacy of alternate forms of power generation and pointing out that nuclear generation is an extremely expensive way to produce energy that is questionably cost-effective, this has included "reducing the nuclear-power-related incentives nations may have to acquire such technology"[75] and thereby restricting the demand for nuclear materials. Success of these efforts has been lessened by the high prestige many nations attach to possessing a nuclear energy industry and the inability to demonstrate the effectiveness of some alternate energy-producing methods on a wide scale (such as solar generation). At the same time, it has become official American policy to try to create weapons disincentives by deemphasizing nuclear weapons themselves, as Nye points out in a semiofficial statement: "As for prestige motivations that might lead states to acquire nuclear weapons, the best answer is to reduce the role of nuclear weapons in world politics. This was a goal that President Carter stated in his Inaugural Address. Efforts to control the vertical proliferation of nuclear arsenals through the SALT and Comprehensive Test Ban negotiations have an important effect on nonproliferation policy."[76]

These sets of concerns have led the Carter administration to propose a six-pronged policy to deal with the proliferation problem. Nye summarizes the elements of the policy:

(1) Making the safeguards systems more effective by insisting upon comprehensive safeguards;
(2) Self-restraint in the transfer of sensitive technologies and materials that can contribute directly to weapons until we have learned to make them more safeguardable;

(3) Creation of nonproliferation incentives through fuel assurances and assistance in the management of spent fuel;
(4) Building consensus about the future structure and management of the nuclear fuel cycle through cooperative studies;
(5) Taking steps at home to ensure that our domestic nuclear policy is consistent with our international objectives;
(6) Taking steps to reduce any security or defense motives that states might have to develop nuclear explosives.[77]

This proposal is the most ambitious and comprehensive statement any U.S. administration has made on this subject. As has been pointed out in the immediately preceding pages, however, there are likely objections to almost every one of its suggestions.

Assessing the Future in a Systematic Manner As has been suggested, if there has been a major shortcoming in the recent strategic literature and in some cases decision making with regard to nuclear matters, it has been a tendency to view specific issues, problems, weapons systems, and the like discretely rather than as part of a system with ramifications that are systemwide, although often in ways not immediately apparent. In a sense this problem is understandable and probably endemic to a subject distinguished by an extreme dynamism, monitoring the pace of which can be a consuming task.

The major driving force in this dynamic environment has been and continues to be an extremely innovative technology that has produced weapons possibilities that would have had a science fiction aura about them only a few years ago, and there is little indication that this pace is slackening. Active research and development processes, fueled by fears from the technological prisoner's dilemma, are working on yet more sophisticated and deadly strategic military accoutrement, as indicated by ongoing work in areas such as laser and charged-particle beams, ABM and ASW. The result of this activity level has been to create a situation where technology dominates the doctrine it is supposed to serve, and much of the current doctrinal debate and confusion can be attributed to the pace with which often piecemeal decisions have been made with regard to individual technological outputs.

This impact is apparent from even a cursory reexamination of the evolution of the strategic environment from the enunciation of massive retaliation to the current muddle. The centerpiece of the Eisenhower administration's new look, massive retaliation, was made incredible by the intercontinental ballistic missile, a weapons innovation that destroyed unconditional viability as a nonzero cell descriptor of defensive possibility and, in the process, made broad nuclear threats too costly to be believed.

The Uncertain Future

The result of the ICBM technological gift was to throw strategic doctrine into a state of flux and debate that spanned the latter Eisenhower and early Kennedy-Johnson years, flirted with various nuclear options (including selective usage within the controlled response doctrine), and ended with the doctrine of assured destruction. As has been argued previously, that doctrine was essentially a description of the destructive capabilities of the American arsenal and the limitations and targeting accuracy and thus targeting doctrine at the time and was expanded to mutual assured destruction when the Russians increased the size of their arsenal to pose a similar threat to the American homeland.

A technological innovation was also responsible for raising basic questions about the doctrine of mutual assured destruction that continues to confound strategic thinking. That innovation was MIRV, a technological spin-off and semiaccident of the space program originally justified as a response to the Soviet ABM program but continued and expanded, particularly in improved warhead and guidance designs resulting in increased accuracy. MIRV has created two strategic doctrinal problems around which much of the current debate has swirled.

First, the American MIRV program and its emerging Soviet counterpart have enormously increased the lethal capacities of both arsenals. This outcome has had dual effects. Most prominently, MIRV has raised the deadly calculus to such a high level as to make the threat of "massive retaliation" in the form of assured destruction counterstrikes seem incredible to many (in essence, the Schlesinger doctrine). Less frequently mentioned but of possibly greater long-term significance, these increases have left the superpowers ever more vulnerable to vertical proliferation charges in the non-nuclear world and thus weakened nonproliferation advocacies.[78] Second, MIRV-induced warhead proliferation has created targeting problems and opportunities. The major problem, of course, is that arsenal sizes are far in excess of the needs of an assured destruction mission, thereby creating the need for additional targeting priorities. The increased accuracy that MIRV extension has produced partially turns this difficulty into an opportunity, however, because the possibility of counterforce doctrine simultaneously allows thinking about damage limitation and selectivity and creates whole new targeting categories at which to aim excess warheads.

The result has been an extended debate that has gone on for nearly a decade about the propriety of doctrinal options and shows little indication of satisfactory resolution in the near future. That the questions have not been followed by answers around which a consensual doctrine can congeal is not the result of a lack of effort or the malevolence of strategic thinkers. Rather, the basic problem has been institutionalized thought patterns that result in discrete and incremental decisions with inadequate attention to systemwide implications of individual decisions.

A Systematic Approach The basic vehicles around which, for exemplary purposes, the discussion of viewing systemic implications of strategic options will be organized is integration of the framework presented in Chapter 1 with the list of primary problems facing the system identified earlier in this chapter. The three basic problems are the impact of ongoing technological innovation, particularly in targeting accuracy but potentially in areas like ABM, the continuing Soviet threat, and horizontal nuclear proliferation. Placing these concerns within the appropriate framework categories facilitates looking at them in terms of their overall system implications.

Figure 7.1 places the various concerns within the framework and suggests a series of causal connections that both show the systematic impact of possible options and outline a potential cycle through which the system could go. The identified factors in the American internal environment are the ongoing debate about the adequacy of American force levels and configurations and the consequences of projected force comparisons between the Soviet Union and the United States arising from the SALT process. External factors include the evolving configuration of Soviet forces in the bilateral relationship, the impact of SALT levels on Soviet capabilities in the context of Russian technological advances, and the prospects of horizontal weapons proliferation. Technological forces are the ongoing American and Soviet accuracy improvement programs and other future "horizon" developments, although

Figure 7.1

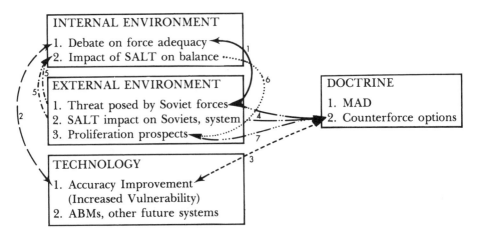

discussion of the latter will be deferred until the final remarks. Doctrine includes both the basic MAD position and the targeting options (with their doctrinal implications) that counterforce capability entails. The series of causal arrows attempts to project the ways in which each concern directly affects others and, in the process, creates a decision cycle.

Although it is somewhat arbitrary, assessment of the growing and changing threat posed by Soviet forces to American strategic security is a convenient first step in the systematic process. This point has some empirical referent in the sense that the ongoing U.S. debate's inception more or less coincided with Soviet achievement of numerical launcher parity and is represented by the solid line number 1 and depicted with causal arrows in both directions. The causal arrow from American debate suggests that the Soviets undoubtedly have noted the American debate in their own force planning as an external environmental factor particularly in light of the coincidence of the American debate about our force adequacy when our MIRV program was greatly increasing our warhead lead. Assigning this as the first step in the current process is arbitrary in this latter sense, because it can be and has been argued that the American MIRV program was the primary stimulus to Soviet force expansions since SALT I began in 1969, in which case a prefatory causal line would have to be drawn between American technology and Soviet force levels.

The second causal step has been the impact of generally negative, or at least cautiously suspicious, force comparisons during the 1970s. As indicated by the dual and reciprocal dashed lines, there have been two major impacts. First, assessment of the adequacy of American security within the U.S.–Soviet comparison affects attitudes toward SALT processes and various SALT outcomes, as discussed extensively in earlier chapters. The line is reciprocal, because SALT as an independent variable, and particularly the projected advantages or disadvantages for force comparisons attaching to various outcomes, have provided major impetus to the need for force adequacy judgments. Second, guarded assessments about American force adequacy have led numerous observers to advocate, for strategic and political reasons, programs emphasizing American technological superiority. Both as a means to correct perceived deficiencies in the current situation and as a hedge against future Soviet force developments, this concern has translated into advocacy of programs such as the B-1 bomber, cruise missiles, Trident I and II, the missile experimental, the Mark 12A warhead, and MARV. A common characteristic of these development efforts has been increasing targeting accuracy in the direction of hard-target kill capability.

The result of these programs, as indicated by the third arrow alternating dots and dashes, has been the emergence of counterforce options as part of strategic doctrine. As argued earlier, discussions of these options were premature and operationally limited in the controlled response era of comparatively high missile CEPs, but both the enormous quantitative increase in warhead numbers and the equally impressive reductions in CEPs have made such discussions meaningful in the current debate. The result has been to resuscitate and elaborate controlled response under the name of limited nuclear options and to give rise to corollary notions like damage limitation. Technological innovation and doctrinal drift mutually reinforce one another at this junction: as increased accuracy permits meaningful discussion of counterforce doctrine, the justification of those options creates incentives and demands for greater accuracy better to carry out counterforce doctrine.

These developments have a direct effect in turn on the Soviets, both in their assessment of the ongoing balance and the attractiveness of various SALT outcomes (as indicated by the dotted lines forming the fourth step). At the level of current and future force configurations, the most obvious implication of increased accuracy for both sides is ICBM vulnerability and what to do about it. For reasons of geography, technological limitations on targeting when firing from nonfixed sites, and probably bureaucratic inertia, Soviet continued disproportionate reliance on fixed-site ICBMs makes this a greater problem for them than for the United States. American disclaimers of a true counterforce intent (first-striking) cannot be accepted by Soviet force planners and strategists, both because they are inherently unbelievable and because, like their American counterparts, Soviet planners must engage in worst-case planning. Among the contingencies for which they must plan is an American preemptive strike made possible by hard-kill-capable weapons.

These realities in turn affect thinking about SALT, both by the Soviets (as indicated in step 4) and by the United States (the alternated long and short dashes identified as the fifth step). As an isolated part of the strategic process, there are appropriate and probably effective countermeasures to the problems posed by hard-kill attainment against fixed systems. As has been suggested, these options basically amount to adoption of one or both of two system variations, mobility and concealment. The two ideas are combined in MX proposals either for silo rotation or missile placement on tracks (thus creating the MAP, or multiple aiming point problem, for the Soviets[79]). The Soviet SS-16 program works on the mobility principle, and American cruise missiles can overcome vulnerability through concealment.

These options create insuperable obstacles to ongoing arms control efforts. Despite attempts to justify the compatibility of emerging systems designed to counter increased accuracy with arms control objectives, it remains a fact that the two ideas are simply incompatible. As we have known it, arms control rests on verification, which has at its heart the ability to know where weapons are in order to count them. That knowledge can, of course, be applied to aiming strategic offensive weapons at targets, the avoidance of which is the very purpose of concealment and mobility programs. One cannot have it both ways in this regard.

There are, of course, various ways to deal with this problem with different arms control implications. An obvious hedge is to keep strategic forces large enough so that, even if some or most weapons are destroyed, some will likely survive, through malfunction of some incoming forces, the fratricide effect, and so forth. If this option is selected, it obviously militates against deep-cut arms reduction and may have been an implicit or explicit motivation for the relatively shallow SALT II cuts (for explanation, see the arms reduction incentives discussion in Chapter 6). At the same time, since allowing vulnerable systems to remain that way is repugnant to force planners, there will undoubtedly be increased pressure to implement mobile and concealed weapons systems. As argued, such decisions would result either in the likely abrogation of arms control progress or the negotiation of unverifiable agreements in which cheating incentives would be maximized, and either possibility is potentially arms race-inducing.

It is at the point where arms control implications and force level and doctrinal decisions intersect that the prospects for horizontal nuclear weapons proliferation are either enhanced or dampened. This relationship, depicted in the line of connected arrows as step 6, derives from the necessity to halt or reverse vertical proliferation as a precondition to potential weapons states denouncing the nuclear option and accepting the safeguards felt necessary in the West to halt access to weapons-grade materials. In many developing countries, Lord Kennet's depiction of SALT holds true: "SALT, be-all and end-all to the superpowers, is to the non-nuclear powers just the son of NPT. SALT is there because the non-nuclear powers would not accept the NPT unless the superpowers undertook to reach nuclear disarmament."[80] Although what constitutes progress toward disarmament is not certain, the superpower record, even within SALT, of capping or diminishing the lethal spiral is not particularly impressive.

Continued SALT success does not, of course, guarantee that additional states, however motivated, will refrain from acquiring nuclear weapons any more than failure of arms control efforts by the superpowers necessarily means that proliferation will occur. At the same time, however, the potential nuclear states argue that their abstinence is unreasonable in the face of the superpowers' inability to limit themselves, leaving nuclear club advocacies of nonproliferation increasingly hard convincingly to sustain.

The possibility that decisions with negative arms control impact will stimulate increased pressure toward proliferation brings the systemic example full circle. As the wavy line (step 7) indicates, a proliferated world presents enormous new difficulties for strategic doctrine and potentially for force levels, which in turn reactivates the force comparison process and begins the cycle anew. Some of the more predictable and widely discussed of those exigencies have been cataloged above, but doubtless there are others that would seriously alter the environment as well.

A cautionary note regarding the precision of this analysis needs to be made. If too much precision is imputed to this more suggestively intended discussion, it would be easy to draw inferences well beyond those the framework allows. The decision to deploy MX, for instance, does not mean that Brazil will build nuclear weapons any more than the failure to build mobile missiles will ensure that the Brazilians will abstain from weapons development. The processes and dynamics underlying the nuclear system are far too complex and our knowledge is far too rudimentary to allow that kind of prediction.

Rather, the purpose of the example was to suggest that environmental factors interact significantly, that the interactions can affect problems and concerns not within the narrow purview of the immediate situation, and that decisions and advocacies based on specific problems alone run the risk of ignoring additional difficulties that could have been foreseen and weighed in the broader systemwide context. Thus, improved targeting accuracy has broader implications than simply countering Soviet throw-weight advantages; counterforce targeting implies more than a means to limit the extent of a nuclear war; and the MX can have effects far broader than relieving the ICBM vulnerability problem. It is to assist conceptually in identifying possible ramifications of situation-specific decisions that the idea of a systemwide approach is put forward.

Facing the 1980s In many ways, the trends and problems of the 1970s strategic environment remain as American strategic doctrine looks toward the last fifth of the twentieth century. The doctrinal controversy that was building as the decade began has continued through two SALT agreements and promises to remain a prominent part of follow-on discussions during the 1980s. At the same time, development and deployment decisions about new and more deadly weapons systems will have to be reached, and weapons laboratories continue to labor on ever more exotic and sophisticated weapons of destruction.

Within this flow of change and uncertainty, one principle remains immutable: the purpose of strategic nuclear endeavors remains the deterrence of nuclear war. Because of the uncertainties involved should the nuclear threshold be crossed, therefore, examinations of strategic options continue to require assessment in terms of their contribution to keeping the nuclear threshold as high as possible. It is not unreasonable, just as the 1970s produced legislation requiring that new weapons systems proposals have an Arms Control Impact Statement attached to them, that strategic proposals for the 1980s be organized around a kind of Nuclear Threshold Impact Statement.

Two major issues, both of which will have to be at least tacitly dealt with mutually by the United States and the Soviet Union, appear to present the major challenges to deterrent stability in the 1980s. The first is gaining control of the technological prisoner's dilemma and in the process more closely aligning doctrinal predisposition with technical productivity. The second is dampening the prospect of nuclear proliferation in the face of increased accessibility to nuclear materials. These two problems, which come face to face in the arms control process, have been discussed in some detail already, but each is important enough to warrant brief concluding attention.

That there is reason to question the extent to which desired doctrinal outcomes and forces derived from doctrinal determinations dictate the products of technology should by now be obvious. On the emerging strategic landscape, the movement toward hard-target capability, if accompanied by success in ASW and depressed trajectory SLBM launch against bomber forces, could make the secure second-strike capability of nuclear forces (regardless of whether one accepts MAD or not) obsolete. At the point where technology overcomes retaliatory invulnerability, the entire conceptual basis on which conventional notions of deterrent stability is premised would be undermined, and assumptions and ideas developed over the past quarter century will have to be rethought, with no readily acceptable alternatives apparent.

If counterforce capability represents the tip of the technological assault on conventional notions, there are other possibilities with even more serious consequences. One hears increasing rumblings, for instance, in the ABM area, both in terms of breakthroughs in the defensive (for example, laser and charged-particle beams either as guidance improvements for ABM or as ABMs) and offensive, ABM-evader (for example MARV) areas. The prospect of truly effective ABMs threatens the second cardinal tenet of retaliatory forces, penetrability, with debilitating effects on the hostage relationship (which, of course, remains the ultimate threat even under limited options doctrine). Threats to either penetrability or survivability allow the calculation of survival of whatever target structure has been identified and thus make the possibility of gain (through successful preemption) or survival (through interdiction of retaliatory forces) conceivable. Obviously, either possibility is conceptually antithetical to deterrence ideas based on the theory that a nuclear attack is always suicidal.

In a sense, the difficulty of dealing with technological developments altering the bipolar American-Russian balance that would eventuate should qualitative arms control efforts fail are conceptually simpler than the theoretical implications of many nuclear weapons possessors. Threats to secure second-strike notions would be upsetting to stability, but at least would take place within the boundaries of a developed body of thought that could help form the basis for doctrinal adjustments that could reformulate a stable deterrent basis.

What forms the basis of deterrent stability in a world where many states and possibly political movements have nuclear weapons is not so clearly understood, partly because we have only limited practical empirical knowledge of the effects when additional states gain nuclear capabilities and partly because, until the last few years, comparatively little thought has been expended on the subject. What is clear, however, is that nuclear proliferation opens a Pandora's Box of grim possibilities that we are not well equipped either conceptually or practically to deal with and that any of those possibilities could lower the nuclear threshold in unanticipated ways.

In the third of a century since man unlocked the power of nuclear reaction, the nuclear genie has remained in its bottle. As the preceding pages have sought to convey, however, a series of issues about possible nuclear futures, weapons, configurations, and doctrinal decisions have come together as the 1970s give way to the 1980s. The ability to keep the nuclear threshold from being crossed will be the result of careful and thoughtful decisions about the complex of concerns raised here and others that will emerge. To ensure that the nuclear future is not a cratered wasteland requires a systematic approach to these problems.

Glossary

The following is a list of terms, abbreviations, and acronyms in the text, and, in cases where the meaning may not be obvious, the definitions of those terms as they are applied in the text.

Accidental war—hypothesized situation whereby nuclear war would occur without conscious decision by political authorities but as a result of some unauthorized or accidental event.
ARP—Action-reaction phenomenon—form of analysis, based on Richardson processes, that hypothesizes that most arms race phenomena are explainable in terms of reactions to initiatives (actions) by the other party.
Active defense—systems, such as ABMs or interceptor aircraft, designed to interdict and destroy incoming strategic weapons before they reach their targets.
Air-breathing forces—designation for the nonballistic missile component of TRIAD; consists of bomber forces and cruise missiles.
ALCM—Air-launched cruise missile—pilotless, subsonic aircraft capable of traveling long distances at low altitudes with high accuracy when launched from a bomber or other aircraft.
ABM—Antiballistic missile—a weapon capable of intercepting and destroying an incoming ballistic missile.
ABM Treaty—one of the major agreements in SALT I, limiting deployment of ABM systems.
ASW—Antisubmarine warfare—systems intended to destroy subsurface combat vessels.
ACIS—Arms Control Impact Statement—statement required to accompany proposals for new weapons systems assessing effect on arms control efforts by deployment.
Arms freeze—agreement to limit armaments at the level of actual deployments at a point in time.
B-1 bomber—proposed supersonic replacement for B-52 bomber force; deployment was deferred in 1977.
B-52 bomber—American strategic bomber deployed during 1950s and early 1960s.
Backfire—new Soviet bomber with strategic capabilities that has been a major issue in SALT.
BOT—Balance of terror—term used to describe the mutual possession of second-strike capabilities by the United States and U.S.S.R.
BMD—Ballistic missile defense—consists of active and passive elements.
Bargaining chip—actual or projected weapons system the purpose of which is to gain some form of concession in arms control negotiations.
Bean counting—euphemistic nickname for process of making strategic force comparisons.
Bear bomber—turboprop Soviet strategic bomber; 100 are in service in LRA.
Bison bomber—jet-propelled, subsonic Soviet bomber; 40 are in service in LRA.

Blast overpressure—the creation of enormous atmospheric pressure by nuclear weapons explosion; measured in pounds per square inch (psi).

Capability—the amount and kind of nuclear power a nation possesses.

Catalytic war—the possibility that a lesser power, either purposely or inadvertently, would draw the nuclear superpowers into a nuclear conflict.

CEP—Circular error probable—the radius around a target in which one has at least 50 percent confidence a missile will land.

CTBT—Comprehensive Test Ban Treaty—proposed agreement to ban all nuclear testing.

Conditional viability—the situation in which a state can be destroyed with nuclear weapons, but the state with that capability refrains from doing so.

Controlled response—early Kennedy administration doctrine aimed at keeping the level of nuclear exchange low enough in a war-fighting environment to allow damage limitation in such a conflict.

Counterforce targeting—aiming one's nuclear forces at an enemy's military capabilities (nuclear, conventional, or support).

Countervalue targeting—aiming one's forces at population centers, productive capacities, and the like.

Damage limitation—defensive planning to minimize the physical damage incurred as a result of nuclear exchange.

Deep-cut arms reduction—significant decreases in the size of strategic arsenals.

Detente—general relaxation of tensions.

Deterrence—the maintaining of vast military power and weaponry in order to discourage war.

Deterrence-only—the strategic school of thought that believes the only utility of nuclear weapons is their deterrent effect.

Deterrence-plus—the strategic school of thought that advocates nuclear war-fighting planning in addition to deterrent roles for nuclear weapons.

Enhanced-radiation warhead—proposed explosive device based on fission-fusion reaction with order of blast effects rearranged to maximize initial neutron radiation.

Escalatory process—the hypothesized sequence by which the initial use of nuclear weapons could eventuate in general homelands exchange between the superpowers.

Essential equivalence—force level policy enunciated by former Secretary Schlesinger, the major thrust of which is that overall American and Soviet forces should have equivalent capabilities.

Ex post–ex ante—dilemma in strategic planning posed because a strategy that might be optimal for deterrence could be undesirable should deterrence fail; has been raised as objection to MAD.

External environment—factors outside the domestic control of the nation-state that influence strategic doctrine.

FB-111H—so-called "stretched" version of the FB-111 fighter-bomber proposed as an alternative to the B-1 bomber.

Firebreak—see nuclear threshold.

First-strike capability—the capacity to launch a preemptive strike against an enemy and in the process destroy the enemy's ability to retaliate.

First-strike strategy—the intention to fire one's nuclear weapons before absorbing a similar attack.

Glossary

Fission—splitting of atoms of unstable elements to produce a nuclear reaction; the simplest form of nuclear reaction.

Fission-fusion—two-step atomic reaction where a fission "trigger" is used to initiate a fusion reaction wherein atoms of deuterium and tritium are fused together; the basis of the "hydrogen" and "enhanced-radiation" bombs.

Fission-fusion-fission—atomic process used to produce multiple MT warheads; involves three-stage reaction wherein fission trigger initiates a fusion reaction, the heat and emitted neutrons from which cause a second fission reaction.

Flexible response—general defense policy of the Kennedy and Johnson administrations; MAD was the nuclear component of the policy.

FBS—Forward-based systems—American nuclear weapons capabilities located outside the United States; specifically those forces stationed in Europe capable of attacking targets in the Soviet Union.

Fratricide effect—the premature detonation of incoming warheads caused by the heat and blast effects of previously exploded warheads.

Galosh—light, area ABM system deployed around Moscow.

GET—Greater than expected threat—method of assessing enemy threat developed by McNamara Defense Department involving estimation beyond projection of enemy capabilities in the National Intelligence Estimate and planning forces based on that estimate.

Hard-target kill capability—the capacity to destroy a missile in a hardened (protected) container, such as an ICBM silo.

IOC—Initial operating capability—the point at which a weapons system is capable of being deployed.

Initial radiation—nuclear effect caused by emission of lethal gamma rays and neutrons while a blast is occurring; is primary effect of enhanced-radiation weapons.

ICBM—Intercontinental ballistic missile.

IOA—Interim Agreement on Offensive Arms—one of the major outcomes of SALT I; placed a five-year freeze on deployment of strategic launchers.

Internal environment—the domestic factors that influence the formulation of strategic doctrine.

IAEA—International Atomic Energy Agency—charged with monitoring safeguards against diversions of nuclear materials to weapons use under NPT.

Intrusive monitoring—the right of inspection to determine violations of agreements, usually by on-site methods and often without prior approval of inspection.

KT—Kiloton—the explosive equivalent of one thousand tons of TNT.

Launcher—any "platform" from which a strategic weapon can be fired (for example, missile silo, submarine missile tube, aircraft).

Lethality—often called K or Kill-factor—force measurement derived from ratio of MTE and accuracy of delivery.

Limited nuclear options—position generally associated with former Defense Secretary Schlesinger proposing development of nuclear force usage options at less than the level of general exchange.

LTBT—Limited Test Ban Treaty—signed in 1963, making it permissible to conduct nuclear tests underground only.

Linkages—the concept that progress in SALT must be viewed in the context of overall Soviet foreign policy behavior.

LRA—Long-range aviation—official designation for Soviet strategic bomber forces; the Russian equivalent of SAC.

Look-down shoot-down capability—the radar capacity to track incoming forces from above and, based on that identification, to destroy the forces.

MARV—Maneuverable reentry vehicle—originally developed as ABM evader; its guidance principles have also been applied to accuracy improvement.

Manhattan Project—the code name of the research effort that created the first atomic reaction in 1942.

Mark 12A warhead—proposed replacement of MK 12 warhead on Minuteman III missiles; is purported to have hard-target kill capability.

Massive retaliation—strategic doctrine associated with the Eisenhower administration.

MT—Megaton—the explosive equivalent of one million tons of TNT.

MTE—Megaton equivalent—the area destructive power of a one MT weapon derived by taking two-thirds of yield in MTs.

Megatonnage—measure of the explosive power of a weapon or, in aggregate, of a total arsenal.

Mininukes—small, low-yield nuclear devices, some with yields lower than those of large conventional (chemical) bombs.

Minuteman (MN)—solid propellant ICBM that is the backbone of U.S. force; current deployment is 450 single warhead MN 2s and 550 MIRVed MN 3s.

Mirror imaging—the assumption that the reasons an adversary does something arise from the same motivations that would cause one to do the same thing.

MX—Missile experimental—a proposed U.S. ICBM that can be deployed either at fixed sites or as a mobile system.

MLBM—Modern large ballistic missile—designation of large payload missiles covered in SALT II agreement.

MAP—Multiple aim points—the number of places from which the proposed MX system could be fired, thereby creating targets at which the Soviets would have to target weapons to ensure destroying the missiles in a first strike.

MIRV—Multiple independently targetable reentry vehicle.

MBFR—Mutual and balanced force reductions—ongoing discussions aimed at reducing military force levels in Europe.

MAD—Mutual assured destruction—U.S. nuclear doctrine formulated during the Kennedy/Johnson administrations.

Mutual hostage effect—the situation in which the populations of the United States and the U.S.S.R. can be destroyed by the thermonuclear arsenals of the other with no ability to protect against such an attack.

$N+1$ problem—additional difficulties created for the nuclear system when the present number of weapons states (N) is expanded ($+1$).

NSC-68—the first comprehensive planning document on nuclear strategy, developed during the Truman administration.

NCA—National Command Authority—those centers involved in reaching decisions about the use of strategic weapons.

National technical means of verification—collective name for nonintrusive monitoring capabilities, notably satellite and electronic surveillance.

NPT—Non-Proliferation Treaty—international agreement signed in 1970 whereby non-nuclear signatories agreed not to produce nuclear weapons and nuclear signatories agreed not to aid efforts to obtain weapons.

Glossary

Nuclear threshold—the point at which nuclear weapons are employed in war (also referred to as the firebreak).

Passive defense—means, such as evacuation, planned dispersal, and sheltering, intended to minimize destruction from a nuclear attack.

Peaceful coexistence—public Soviet position that nuclear war is an inappropriate form of competition between the East and West.

PNET—Peaceful Nuclear Explosions Treaty—companion to TTBT outlawing peaceful detonations above 150 KT except with international observation and inspection.

Polaris—the earliest class of U.S. atomic-powered missile-launching submarines.

Poseidon—the second generation of U.S. SLBM-carrying submarines.

Prisoner's dilemma—a game-theoretical situation the normal outcome of which is for players to engage in mutually injurious behavior; often used to describe the dynamics of the arms race.

RV—Reentry vehicles—that part of a ballistic missile that remains after the last booster stage has been decoupled, including the warheads.

R,D,T,&E—Research, development, testing, and evaluation—the developmental process of a weapons system from ideation to implementation (often referred to as "R&D").

Residual radiation—contamination from fission reaction caused by return of radioactive by-products of nuclear blast into ecosystem.

SS—prefix for designation of Soviet ICBMs (for example, SS-7, SS-13, SS-17).

SAMOS—Satellite and missile observation satellite—satellite system used for monitoring Soviet activities.

SLCM—Sea-launched cruise missile. See ALCM for description.

Second-strike capability—the capacity to absorb any conceivable nuclear attack and retain sufficient retaliatory ability to inflict unacceptable damage on the attacker.

Second-strike strategy—the determination to fire one's nuclear forces only after having absorbed an initial attack.

Shallow-cut arms reductions—modest decreases in the size of nuclear arsenals at levels that do not materially affect destructive capabilities.

SSKP—Single-shot kill probability—the likelihood, expressed as a percentage, that a given strategic weapon will destroy its target once it has arrived at the target.

Spectrum defense—the idea that nuclear weapons can be used to deter both nuclear and non-nuclear (conventional) military threats.

Standoff launch—the firing of ALCMs from outside Soviet air space.

SAC—Strategic Air Command.

SALT—Strategic Arms Limitation Talks.

Strategic parity—a condition of a least rough balance or equality in U.S. and U.S.S.R. strategic nuclear capabilities.

Strategic Rocket Forces—official designation of the ICBM component of Soviet strategic forces.

Strategic sufficiency—nuclear doctrine of the early Nixon administration.

Strategic weapons—those nuclear weapons systems capable of attacking targets in the homeland of an adversary.

Strategy—the plan for use of nuclear weapons in a war-fighting situation.

SLBM—Submarine-launched ballistic missile.

TRIAD—the acronym for describing the three components (or "legs") of American strategic systems.

Technological prisoner's dilemma—the difficulty of placing limits on R&D activity in the absence of effective verification techniques.

TERCOM—Terrain comparison matching—guidance system designed for ALCM.

TNW—Theater (tactical) nuclear weapons—battlefield nuclear weapons planned for use in Europe but not for a homeland attack against the Soviet Union.

TTBT—Threshold Test Ban Treaty—agreement signed in 1974 limiting the yield of any underground weapons test to 150 KT.

Throw-weight—the weight of that part of a missile above the last booster stage.

Titan—largest and oldest ICBM in U.S. arsenal; 54 of these multiple-megaton, liquid-fuel rockets are deployed.

Trident—new U.S. missile-bearing submarines, scheduled for IOC in 1981.

Unconditional viability—the condition in which a state cannot be effectively destroyed with nuclear weapons.

Verification—the process of monitoring compliance with provisions of an agreement; generally applied to arms control treaties and agreements.

Vladivostok Accords—interim agreement on launcher limits to extend IOA, reached by President Ford and Secretary Brezhnev in 1974.

Worst-case analysis—analytic method that involves looking for most dangerous outcome of a strategic situation and planning to counter that worst contingency.

Notes

Introduction

1. "Technology and the Military Balance," *Foreign Affairs*, vol. 56, no. 3, April 1978, p. 562.
2. "Can Nuclear Deterrence Last Out the Century," in Robert J. Pranger and Roger P. Labrie, eds., *Nuclear Strategy and National Security: Points of View*, Washington, D.C.: American Enterprise Institute for Public Policy Research, 1977, p. 72.
3. *Deadly Logic: The Theory of Nuclear Deterrence*, Columbus, Ohio: The Ohio State University Press, 1966, p. 219.

Chapter 1

1. "The Nuclear Test Ban Treaty," in Henry A. Kissinger, ed., *Problems of National Strategy: A Book of Readings*, New York: Frederick A. Praeger, 1965, p. 412.
2. See Ted Greenwood, George W. Rathjens, and Jack Ruina, *Nuclear Power and Weapons Proliferation*, Adelphi Papers No. 130, London: International Institute of Strategic Studies, 1976, p. 4.
3. *Escalation and the Nuclear Option*, Princeton, N.J.: Princeton University Press, 1966, p. 74.
4. *Strategy in the Missile Age*, Princeton, N.J.: Princeton University Press, 1959, p. 225.
5. Philip Green, *Deadly Logic: The Theory of Nuclear Deterrence*, Columbus, Ohio: The Ohio State University Press, 1966, p. 228.
6. See particularly *Arms and Influence*, New Haven: Yale University Press, 1966.
7. *Deadly Logic*, p. 4.
8. Schelling, *Arms and Influence*, pp. 22, 23.
9. *Annual Report, Department of Defense, Fiscal Year 1979*, Washington, D.C.: U.S. Government Printing Office, 1978, p. 33.
10. *Webster's New World Dictionary* (College Edition), New York: World Publishing Company, 1964, p. 399.
11. Michael A. Milstein and Leo S. Semeiko, "Problems of the Inadmissibility of Nuclear Conflict," *International Studies Quarterly*, vol. 20, no. 1, March 1976, p. 102.
12. *On the Uses of Military Power in the Nuclear Age*, Princeton, N.J.: Princeton University Press, 1966, p. 89.
13. "Introduction," in Richard Rosecrance, ed., *The Future of the International Strategic System*, San Francisco: Chandler Publishing Company, 1972, p. 6.
14. *Strategic Deterrence Reconsidered*, Adelphi Papers No. 116, London: International Institute of Strategic Studies, 1975, p. 1.
15. See *On Thermonuclear War*, Princeton, N.J.: Princeton University Press, 1961.
16. See particularly *Strategy in the Missile Age* and *Escalation and the Nuclear Option*.

17. See particularly *The Strategy of Conflict,* Cambridge, Mass.: Harvard University Press, 1960, and *Arms and Influence.*
18. *Conflict and Defense: A General Theory,* New York: Harper and Row, 1963.
19. *Problems of National Strategy.*
20. "Across the Nuclear Divide—Strategic Studies, Past and Present," *International Security,* vol. 2, no. 1, Summer 1977, pp. 24–46.
21. For a discussion of basic game-theoretical concepts and types of games, see Donald M. Snow, *Introduction to Game Theory* (rev. ed.), Columbus, Ohio: Consortium for International Studies Education, 1978.
22. Boulding's *Conflict and Defense* is a particularly good example of an early attempt at this analytic form.
23. An explanation of the arms race as prisoner's dilemma designed for experiential classroom use is Donald M. Snow, *The Shadow of the Mushroom-Shaped Cloud: Basic Ideas and Problems of Nuclear Deterrence,* Learning Packages in National Security Series No. 2, Columbus, Ohio: Consortium for International Studies Education, 1978.
24. *Deadly Logic,* p. 211.
25. Ibid., p. 215.
26. For a cogent introduction, see Boulding, *Conflict and Defense,* especially Chapter 2.
27. New York: Praeger Special Studies, 1973.
28. Ibid., p. 6.
29. *Understanding the Soviet Military Threat: How CIA Estimates Went Astray,* New York: National Security Information Center, 1977, p. 30.
30. "On Estimating and Imputing Intentions," *International Security,* vol. 2, no. 3, Winter 1978, p. 25.
31. *Cold Dawn: The Story of SALT,* New York: Holt, Rinehart and Winston, 1973, pp. 156–57.
32. Brodie, *Escalation and the Nuclear Option,* p. 86.
33. *MIRV and the Arms Race,* p. 60.
34. *Cold Dawn,* p. 33. One of the major strengths of Newhouse's work is its detailed analysis of bureaucratic machinations in the SALT process.
35. For a thoughtful analysis of Soviet cultural influences, see Jack L. Snyder, *The Soviet Strategic Culture: Implications for Limited Nuclear Operations,* A Project AIR FORCE Report Prepared for the United States Air Force, Santa Monica, Calif.: Rand Corporation, September 1977.
36. *The Essence of Security: Reflections in Office,* New York: Harper and Row, 1968, p. 58.
37. *On the Uses of Military Power in the Nuclear Age,* p. 106.
38. "Technical Innovation and Arms Control," in Robert J. Pranger and Roger P. Labrie, eds., *Nuclear Strategy and National Security: Points of View,* Washington, D.C.: American Enterprise Institute for Public Policy Research, 1977, p. 500.
39. Ibid., p. 498.
40. Ibid., p. 481.
41. "General Introduction," in Pranger and Labrie, eds., *Nuclear Strategy and National Security,* p. 4.
42. Tammen, *MIRV and the Arms Race,* p. 93.

43. *The Soviet Strategic Culture,* p. 5.
44. *The Faces of Verification: Strategic Arms Control for the 1980s,* The Rand Paper Series No. P-5986, Santa Monica, Calif.: Rand Corporation, August 1977, p. 23.
45. "Why the Soviet Union Thinks It Could Fight and Win a Nuclear War," *Commentary,* vol. 64, no. 1, July 1977, p. 29.
46. "Technology and the Military Balance," *Foreign Affairs,* vol. 56, no. 3, April 1978, p. 545.
47. In capsule form, this is Tammen's basic thesis (*MIRV and the Arms Race*).

Chapter 2

1. *Webster's New College Dictionary* (1974 edition), New York: World Publishing Co., 1975, p. 399.
2. C. Johnston Conover, *U.S. Strategic Nuclear Weapons and Deterrence,* Rand Paper Series No. P-5967, Santa Monica, Calif.: Rand Corporation, August 1977, p. 1.
3. *Escalation and the Nuclear Option,* Princeton, N.J.: Princeton University Press, 1966, p. 88.
4. "The Need for a New Analytical Framework: Review of *Security in the Nuclear Age*," *International Security,* vol. 1, no. 2, Fall 1976, p. 138.
5. *Conflict and Defense: A General Theory,* New York: Harper and Row, 1963.
6. "The Weapon as an Element in the Social System," in Richard Rosecrance, ed., *The Future of the International Strategic System,* San Francisco: Chandler Publishing Co., 1972, p. 88.
7. Thomas C. Schelling, *Arms and Influence,* New Haven, Conn.: Yale Universtiy Press, 1966, p. 10.
8. "Repeating History: The Civil Defense Debate Renewed," *International Security,* vol. 2, no. 3, Winter 1978, p. 118.
9. "French Defense Planning—The Future in the Past," *International Security,* vol. 1, no. 2, Fall 1976, p. 17.
10. For a thorough and thoughtful discussion, see the extensive remarks of Bernard Brodie in *Strategy in the Missile Age,* Princeton, N.J.: Princeton University Press, 1959.
11. Ibid., p. 201.
12. *From Superiority to Parity: The United States and the Strategic Arms Race, 1961–1971,* Westport, Conn.: Greenwood Press, Inc., 1973, p. 254.
13. Richard Pipes, "Why the Soviet Union Thinks It Could Fight and Win a Nuclear War," *Commentary,* vol. 64, no. 1, July 1977, p. 29.
14. *Escalation and the Nuclear Option,* p. 158.
15. "U.S. Strategy and the Defense of Europe," in Henry A. Kissinger, ed., *Problems of National Strategy: A Book of Readings,* New York: Frederick A. Praeger, 1965, pp. 294, 295.
16. *Nuclear Forces for Medium Powers, Part I: Targets and Weapons Systems,* Adelphi Papers No. 106, London: International Institute of Strategic Studies, 1974, p. 2.
17. *Strategic Deterrence Reconsidered,* Adelphi Papers No. 116, London: International Institute of Strategic Studies, 1975, p. 3.

18. Herbert F. York, "The Origins of MIRV," in David Carlton and Carlo Schaerf, eds., *The Dynamics of the Arms Race*, New York: John Wiley and Sons, 1975, p. 33.

19. *Deadly Logic: The Theory of Nuclear Deterrence*, Columbus, Ohio: The Ohio State University Press, 1966, p. 222.

20. *Cold Dawn: The Story of SALT*, New York: Holt, Rinehart and Winston, 1973, p. 20.

21. "The Delicate Balance of Terror," in Kissinger, ed., *Problems of National Strategy*, p. 40.

22. "SALT II: The Search for a Follow-on Agreement," *Orbis*, vol. 17, 1973, p. 352.

23. *Strategic Weapons: An Introduction*, New York: National Strategy Information Center, 1975, p. 1.

24. "Nuclear Sharing: Nato and the N+1 Country," in Kissinger, ed., *Problems of National Strategy*, p. 199.

25. For a good recent overview of this debate and the rationale for continued Israeli ambiguity on the subject, see Alan Dowty, "Nuclear Proliferation: The Israeli Case," *International Studies Quarterly*, vol. 22, no. 1, March 1978, pp. 79–120.

26. *Strategy in the Missile Age*, p. 303.

27. *Escalation and the Nuclear Option*, p. 195.

28. Ronald L. Tammen, *MIRV and the Arms Race: An Interpretation of Defense Strategy*, New York: Praeger Publishers, Inc., 1973, p. 52.

29. Wolfgang K. H. Panofsky, "The Mutual Hostage Relationship between America and Russia," in Robert J. Pranger and Roger P. Labrie, eds., *Nuclear Strategy and National Security: Points of View*, Washington, D.C.: American Enterprise Institute for Public Policy Research, 1977, p. 79.

30. *Deadly Logic*, p. 92.

31. "The Future of the Strategic Bomber," *AEI Defense Review*, vol. 2, no. 1, 1978, p. 10.

32. Text reprinted as "Foreign Policy and National Security" in *International Security*, vol. 1, no. 1, Summer 1976, p. 187.

33. "Strategic Vulnerability: The Balance between Prudence and Paranoia," *International Security*, vol. 1, no. 1, Summer 1976, p. 141.

34. *Escalation and the Nuclear Option*, p. 38.

35. "The Weapon as an Element," p. 90.

36. "American Deterrent Policy," in Kissinger, ed., *Problems of National Strategy*, p. 124.

37. "Across the Nuclear Divide—Strategic Studies, Past and Present," *International Security*, vol. 2, no. 1, Summer 1977, p. 29.

38. "Escalation as a Strategy," in Kissinger, ed., *Problems of National Security*, p. 25.

39. "The Doctrine of Tactical Nuclear Warfare and Some Alternatives," in Carlton and Schaerf, eds., *Dynamics of the Arms Race*, p. 136.

40. "U.S. Tactical Nuclear Weapons: A Controversial Arsenal," in David T. Johnson and Barry R. Schneider, eds., *Current Issues in U.S. Defense Policy*, New York: Praeger Special Studies, 1976, p. 186.

41. Panofsky, "Mutual Hostage Relationship," p. 76.

42. *On the Uses of Military Power in the Nuclear Age,* Princeton, N.J.: Princeton University Press, 1966, p. 90.
43. *Strategy in the Missile Age,* p. 36.
44. "Problems of the Inadmissibility of Nuclear Conflict," *International Studies Quarterly,* vol. 20, no. 1, March 1976, p. 99.
45. Knorr, *On the Uses of Military Power.*
46. *Arms and Influence,* p. 206.
47. *MIRV and the Arms Race,* p. 42.
48. See Donald M. Snow, "Deterrence Theorizing and the Nuclear Debate: The Methodological Dilemma," *International Studies Notes,* vol. 6, no. 2, Summer 1979, pp. 1-5.

Chapter 3

1. Paul Doty, Albert Carnesale, and Michael Nacht, "The Race to Control Nuclear Arms," in Robert J. Pranger and Roger P. Labrie, eds., *Nuclear Strategy and National Security: Points of View,* Washington, D.C.: American Enterprise Institute for Public Policy Research, 1977, p. 465.
2. Samuel F. Wells, Jr., "America and the 'MAD' World," *The Wilson Quarterly,* vol. 1, no. 5, Autumn 1977, p. 58.
3. Ibid., p. 60.
4. *Strategic Weapons: An Introduction,* New York: National Strategy Information Center, 1975, p. 3.
5. Jerome H. Kahan, *Security in the Nuclear Age: Developing U.S. Strategic Arms Policy,* Washington, D.C.: The Brookings Institution, 1975, p. 11.
6. *From Superiority to Parity: The United States and the Strategic Arms Race, 1961–1971,* Westport, Conn.: Greenwood Press, Inc., 1973, p. 248.
7. *Security in the Nuclear Age,* p. 37.
8. Quoted in "The Evolution of Foreign Policy," *U.S. Department of State Bulletin,* vol. 30, January 25, 1954, p. 108.
9. Polmar, *Strategic Weapons,* p. 3.
10. *From Superiority to Parity,* p. 16.
11. *Security in the Nuclear Age,* p. 15.
12. Mason Willrich, "SALT I: An Appraisal," in Mason Willrich and John B. Rhinelander, eds., *SALT: The Moscow Agreements and Beyond,* New York: Free Press, 1974, p. 257.
13. *Escalation and the Nuclear Option,* Princeton, N.J.: Princeton University Press, 1966, p. 229.
14. *From Superiority to Parity,* p. 15.
15. Brodie, *Escalation and the Nuclear Option,* p. 236.
16. Ibid., p. 259.
17. *Security in the Nuclear Age,* p. 31.
18. Brodie, *Escalation and the Nuclear Option,* p. 28.
19. *From Superiority to Parity,* p. 260.
20. *MIRV and the Arms Race: An Interpretation of Defense Strategy,* New York: Praeger Publishers, Inc., 1973, p. 40.
21. "America and the 'MAD' World," p. 67.

22. *From Superiority to Parity*, p. 251.
23. *Security in the Nuclear Age*, p. 47.
24. Data from *The Military Balance, 1975–1976,* London: International Institute of Strategic Studies, 1976, p. 73.
25. Ibid.
26. Edward J. Ohlert, "Strategic Deterrence and the Cruise Missile," *Naval War College Review*, vol. 30, no. 3, Winter 1978, p. 24. Moulton (*From Superiority to Parity*, p. 93) makes the same point.
27. Fred C. Ikle, "Can Nuclear Deterrence Last Out the Century," in Pranger and Labrie, eds., *Nuclear Strategy and National Security*, p. 68.
28. "Strategic Adaptability," in Pranger and Labrie, eds., *Nuclear Strategy and National Security*, p. 210.
29. "An Assessment of the Bomber–Cruise Missile Controversy," *International Security*, vol. 2, no. 1, Summer 1977, p. 57.
30. Kahan, *Security in the Nuclear Age*, p. 76.
31. Moulton, *From Superiority to Parity*, p. 176.
32. *Security in the Nuclear Age*, p. 91.
33. *MIRV and the Arms Race*, p. 112.
34. This point is developed thoroughly by Kahan, *Security in the Nuclear Age*, pp. 91–94.
35. *From Superiority to Parity*, p. 87.
36. *Security in the Nuclear Age*, p. 91.
37. Ibid., p. 94.
38. Moulton, *From Superiority to Parity*, p. 51.
39. *Annual Report, Department of Defense, Fiscal Year 1979,* Washington, D.C.: U.S. Government Printing Office, 1978, pp. 55, 49.
40. "Across the Nuclear Divide—Strategic Studies, Past and Present," *International Security*, vol. 2, no. 1, Summer 1977, p. 33.
41. "Deterrence in Dyadic and Multipolar Environments," in Richard Rosecrance, ed., *The Future of the International Strategic System*, San Francisco: Chandler Publishing Co., 1972, p. 132.
42. Stefan H. Leader and Barry R. Schneider, "U.S.–Soviet Strategic Forces: SALT and the Search for Parity," in David T. Johnson and Barry R. Schneider, eds., *Current Issues in U.S. Defense Policy*, New York: Praeger Special Studies, 1976, p. 141.
43. Christopher Lehman and Peter C. Hughes, " 'Equivalence' and SALT II," *Orbis*, vol. 20, 1977, p. 1047.
44. *Cold Dawn: The Story of SALT,* New York: Holt, Rinehart and Winston, 1973, p. 72.
45. *The Essence of Security: Reflections in Office,* New York: Harper and Row, 1968, p. 53.
46. "Repeating History: The Civil Defense Debate Renewed," *International Security*, vol. 2, no. 3, Winter 1978, p. 100.
47. *Strategic Deterrence Reconsidered*, Adelphi Papers No. 116, London: International Institute of Strategic Studies, 1975, p. 10.
48. Rosecrance, "Deterrence," p. 126.
49. "Annual Defense Department Report, FY 1975," in Pranger and Labrie, eds., *Nuclear Strategy and National Security*, p. 97. (The Report was issued on March 4, 1974.)

Notes to Chapter 3

50. "The Politics of Twenty Nuclear Powers," in Rosecrance, ed., *The Future of the International Strategic System*, p. 75.
51. *Deadly Logic: The Theory of Nuclear Deterrence*, Columbus, Ohio: The Ohio State University Press, 1966, p. 237.
52. "American Deterrent Policy," in Henry A. Kissinger, ed., *Problems of National Strategy: A Book of Readings*, New York: Frederick A. Praeger, 1965, p. 121.
53. "Annual DOD Report," 1975, p. 98.
54. "The Evolution of American Policy toward the Soviet Union," *International Security*, vol. 1, no. 1, Summer 1976, p. 43.
55. *Strategic Weapons*, p. 110.
56. "The Relationship of Strategic and Theater Nuclear Forces," *International Security*, vol. 2, no. 2, Fall 1977, p. 131.
57. "Beyond SALT II—A Missile Test Quota," *The Bulletin of the Atomic Scientists*, vol. 33, no. 5, May 1977, p. 37.
58. "The Mutual Hostage Relationship between America and Russia," in Pranger and Labrie, eds., *Nuclear Strategy and National Security*, p. 82.
59. Nitze, "Strategic and Theater Nuclear Forces," p. 124.
60. "Assuring Strategic Stability in An Era of Detente," *Foreign Affairs*, vol. 54, no. 2, January 1976, p. 223.
61. "The United States—A Military Power Second to None?" *International Security*, vol. 1, no. 1, p. 51.
62. "Repeating History," p. 117.
63. "Foreign Policy and National Security," reprinted in *International Security*, vol. 1, no. 1, Summer 1976, p. 186.
64. *Understanding the Soviet Military Threat: How CIA Estimates Went Astray*, New York: National Strategy Information Center, 1977, p. 10.
65. " 'Equivalence' and SALT II," p. 1049.
66. *MIRV and the Arms Race*, p. 55.
67. Nitze, "Assuring Strategic Stability", p. 227. Lehman and Hughes (" 'Equivalence' and SALT II," p. 1054) derive much the same conclusion, although phrasing it in "the policy of equivalence."
68. P. 91.
69. "Civil Defense in Limited War—Opposed," in Pranger and Labrie, eds., *Nuclear Strategy and National Securtiy*, p. 379.
70. "Foreign Policy and National Security," p. 186.
71. *Cold Dawn*.
72. "Can Nuclear Deterrence Last Out the Century," p. 58.
73. *From Superiority to Parity*, p. 294.
74. *U.S. Nuclear Weapons in Europe: Issues and Alternatives*, Washington, D.C.: The Brookings Institution, 1974, p. 31.
75. *Annual Report, FY 1975*, p. 92.
76. Testimony reprinted as "U.S.-U.S.S.R. Strategic Policies," in Pranger and Labrie, eds., *Nuclear Strategy and National Security*, p. 105.
77. Reprinted as "Analyses of Effects of Limited Nuclear Warfare," in Pranger and Labrie, eds., *Nuclear Strategy and National Security*, pp. 127–28.
78. *Limited Nuclear Options: Deterrence and the New American Doctrine*, Adelphi Papers No. 121, London: International Institute of Strategic Studies, Winter 1975–76, p. 8.

79. Ibid., p. 5.
80. "Strategic Deterrence," p. 23.
81. *Limited Nuclear Options*, p. 4.
82. *U.S. Strategic Nuclear Weapons and Deterrence*, Rand Paper Series No. P-5967, Santa Monica, Calif.: Rand Corporation, August 1977, p. 28.
83. *Strategic Deterrence Reconsidered*, pp. 16–17.
84. *On the Uses of Military Power in the Nuclear Age*, Princeton, N.J.: Princeton University Press, 1966, p. 98.
85. Lehman and Hughes, " 'Equivalence' and SALT II," p. 1046.
86. Ibid. Their statement of criteria is drawn virtually verbatim from Schles— inger, "Annual DOD Report, 1975," pp. 102–3.
87. "Strategic Adaptability," p. 219.
88. "The Future of the Strategic Bomber," *AEI Defense Review*, vol. 2, no. 1, 1978, p. 8.
89. "Annual Defense Department Report, FY 1978," in Pranger and Labrie, eds., *Nuclear Strategy and National Security*, p. 196.
90. *MIRV and the Arms Race*, p. 131.
91. "Analyses of Effects of Limited Nuclear Warfare," p. 137.
92. "Modernizing the Strategic Bomber Force without Really Trying—A Case against the B-1 Bomber," *International Security*, vol. 1, no. 2, Fall 1976, p. 113.
93. "Future Limitations of Strategic Arms," in Willrich and Rhinelander, eds., *SALT*, p. 226.
94. "U.S.–Soviet Strategic Forces," p. 141.
95. "Second Lecture," in James L. Buckley and Paul C. Warnke, *Strategic Sufficiency: Fact or Fiction?* Washington, D.C.: American Enterprise Institute for Public Policy Research, 1972. p. 23.
96. "Strategic Adaptability," p. 207.
97. "SALT under the Carter Administration," Paper presented to the 19th Annual Meeting of the International Studies Association, Washington, D.C., February 22–25, 1978, p. 20.
98. *Annual Report, FY 1979*, p. 58.
99. Ibid., p. 54.
100. Ibid., p. 32.
101. "The Strategic Nuclear Balance," *Commanders Digest*, vol. 21, no. 4, March 9, 1978, p. 2.
102. *Annual Report, FY 1979*, p. 105.
103. Ibid., p. 53.
104. Ibid., p. 56.
105. Ibid., pp. 56 and 103.
106. Ibid., p. 34.
107. "Strategic Deterrence," p. 25.

Chapter 4

1. Frank Barnaby, "Crossing the Nuclear Threshold," *New Scientist*, January 19, 1978, p. 151.
2. The basic physics of these reaction forms, written so as to be comprehensible to the nonphysicist, can be found in Albert Legault and George Lindsey, *The*

Dynamics of the Nuclear Balance, Ithaca, N.Y.: Cornell University Press, 1974.

3. Ibid., p. 30.

4. *Strategic Weapons: An Introduction,* New York: National Strategy Information Center, 1975, p. 24.

5. "The Verification of Arms Control Agreements," *Arms Control Today,* vol. 7, no. 7–8, July–August 1977, p. 1.

6. "The Case for a Modern Strategic Bomber," *AEI Defense Review,* vol. 2, no. 1, 1978, p. 19.

7. "Technological Innovation and Arms Control," in Robert J. Pranger and Roger P. Labrie, eds., *Nuclear Strategy and National Security: Points of View,* Washington, D.C.: American Enterprise Institute for Public Policy Research, 1977, pp. 479–80.

8. This process is described in detail by Ronald L. Tammen in *MIRV and the Arms Race: An Interpretation of Defense Strategy,* New York: Praeger Publishers, Inc., 1973.

9. "Technological Innovation and Arms Control," p. 480.

10. Ibid., p. 498.

11. Tammen, *MIRV and the Arms Race,* p. 32.

12. "Technology and the Military Balance," *Foreign Affairs,* vol. 56, no. 3, April 1978, p. 550.

13. *Cold Dawn: The Story of SALT,* New York: Holt, Rinehart and Winston, 1973, p. 19.

14. "The Case for the B-1 Bomber," *International Security,* vol. 1, no. 2, Fall 1976, p. 80.

15. Harold Brown, *Annual Report, Department of Defense, Fiscal Year 1979,* Washington, D.C.: U.S. Government Printing Office, 1978, p. 58.

16. "Effective Military Technology in the 1980's," *International Security,* vol. 1, no. 2, Fall 1976, p. 60.

17. *Strategic Weapons,* p. 108.

18. *Escalation and the Nuclear Option,* Princeton, N.J.: Princeton University Press, 1966, p. 285.

19. "The Case for a Modern Strategic Bomber," p. 18.

20. "Modernizing the Strategic Bomber Force without Really Trying—A Case against the B-1 Bomber," *International Security,* vol. 1, no. 2, Fall 1976, p. 99.

21. McCarthy, "The Case for the B-1 Bomber," p. 82.

22. *Strategic Weapons,* p. 100.

23. "SALT II: The Search for a Follow-on Agreement," *Orbis,* vol. 17, 1973, p. 350.

24. *Strategic Weapons,* p. 65.

25. "Strategic Vulnerability: The Balance between Prudence and Paranoia," *International Security,* vol. 1, no. 1, Summer 1976, p. 169.

26. Ibid., p. 149.

27. *Annual Report, FY 1979,* p. 63.

28. *Cold Dawn,* p. 25.

29. "SALT II," p. 343.

30. "Soviet-American Strategic Competition: Instruments, Doctrine, and Purposes," in Pranger and Labrie, eds., *Nuclear Strategy and National Security,* p. 291.

31. "Effective Military Technology," p. 54.

32. "Soviet-American Strategic Competition," p. 287.
33. "Can Nuclear Deterrence Last Out the Century?" in Pranger and Labrie, eds., *Nuclear Strategy and National Security*, p. 65.
34. Polmar, *Strategic Weapons*, p. 62.
35. George S. Brown, "The Strategic Nuclear Balance," *Commanders Digest*, vol. 21, no. 4, March 9, 1978, p. 9.
36. "U.S.–Soviet Strategic Forces: SALT and the Search for Parity," in David T. Johnson and Barry R. Schneider, eds., *Current Issues in U.S. Defense Policy*, New York: Praeger Special Studies, 1976, p. 134.
37. "U.S. Strategy and the Defense of Europe," in Henry A. Kissinger, ed., *Problems of National Security: A Book of Readings*, New York: Frederick A. Praeger, 1965, p. 295.
38. "The Direct Payoff in SALT II," *Arms Control Today*, vol. 8, no. 3, March 1978, p. 4.
39. *Cold Dawn*, p. 22.
40. "Nuclear Missile Submarines and Nuclear Strategy," in Johnson and Schneider, eds., *Current Issues*, p. 158.
41. Brown, "The Strategic Nuclear Balance," p. 8.
42. P. 113.
43. Polmar, *Strategic Weapons*, p. 104.
44. "Anti-Submarine Warfare and Missile Submarines," in David Carlton and Carlo Schaerf, eds., *The Dynamics of the Arms Race*, New York: John Wiley and Sons, 1975, p. 40.
45. Ibid., p. 45.
46. "Future Limitations of Strategic Arms," in Mason Willrich and John B. Rhinelander, eds., *SALT: The Moscow Agreements and Beyond*, New York: Free Press, 1973, p. 247.
47. "SALT II," p. 353.
48. Harold Brown, *Department of Defense Annual Report, Fiscal Year 1980*, Washington, D.C.: U.S. Government Printing Office, 1979, p. 66.
49. John C. Culver, "The Future of the Strategic Bomber," *AEI Defense Review*, vol. 2, no. 1, 1978, p. 7.
50. "The Strategic Nuclear Balance," p. 10.
51. "The Case for the B-1 Bomber," p. 87.
52. "The B-1 Bomber," in Johnson and Schneider, eds., *Current Issues*, p. 168.
53. "The Case for the B-1 Bomber," p. 82.
54. Brown, "The Strategic Nuclear Balance," p. 11.
55. *Strategic Weapons*, p. 95.
56. *Annual Report, FY 1979*, p. 51.
57. Ibid., p. 115.
58. Culver, "The Future of the Strategic Bomber," p. 6.
59. "Cruise Missiles: Different Missions, Different Arms Control Impact," *Arms Control Today*, vol. 8, no. 1, January 1978, p. 1.
60. "The Strategic Nuclear Balance," p. 11.
61. "The Future of the Strategic Bomber," p. 8.
62. *Annual Report, FY 1979*, p. 119.
63. Robert Perry makes this point in *The Faces of Verification: Strategic Arms Control for the 1980's*, The Rand Paper Series No. P-5986, Santa Monica, Calif.: Rand Corporation, August 1977, p. 22.

64. "The Future of the Strategic Bomber," p. 12.
65. Richard Burt, *New Weapons Technologies: Debate and Directions*, Adelphi Papers No. 126, London: International Institute of Strategic Studies, 1976, p. 24.
66. "Strategic Deterrence and the Cruise Missile," *Navel War College Review*, vol. 30, no. 3, Winter 1978, p. 26.
67. "Effective Military Technology," p. 59.
68. Culver, "The Future of the Strategic Bomber." McLucas ("The Case for a Modern Strategic Bomber," p. 19) disagrees with this assessment.
69. "Strategic Deterrence," p. 22.
70. Ibid., p. 31. Garwin ("Effective Military Technology," p. 64) makes the same basic point.
71. *New Weapons Technologies*, p. 24.
72. "U.S.–Soviet Strategic Forces," p. 140.
73. *Strategic Weapons*, p. 91.
74. "Detente and the Military Balance," *Bulletin of the Atomic Scientists*, vol. 33, no. 4, April 1977, p. 14.
75. *The Military Balance, 1975–1976*, London: International Institute of Strategic Studies, 1976, p. 73.
76. "The Direct Payoff in SALT II," p. 3.
77. *Strategic Nuclear Weapons: The Deadly Calculus*, Learning Materials in National Security Education, Columbus, Ohio: Consortium for International Studies Education (forthcoming).
78. "U.S.–Soviet Strategic Forces," p. 130.
79. *Strategic Nuclear Weapons*.
80. Leader and Schneider, "U.S.–Soviet Strategic Forces," p. 134.
81. "Foreword," in Donald G. Brennan, *Arms Treaties with Moscow: Unequal Terms Unevenly Applied?* New York: National Strategy Information Center, 1975, p. x.
82. Leader and Schneider, "U.S.–Soviet Strategic Forces," p. 136.
83. Downey, "The Direct Payoff in SALT II," pp. 174–75. Leader and Schneider ("U.S.–Soviet Strategic Forces") sophisticate this for large weapons: for weapons over one megaton, one takes the square root of the yield rather than the two-thirds power.
84. Downey, "The Direct Payoff in SALT II," p. 139.
85. *Strategic Nuclear Weapons*.
86. This concept is discussed in Steinbruner and Garwin, "Strategic Vulnerability," p. 143.
87. Using the Leader and Schneider definition for multiple megaton weapons, the calculation would be even further to dramatize the affect of accuracy.
88. Geoffrey Kemp, *Nuclear Forces for Medium Powers: Part I, Targets and Weapons Systems*, Adelphi Papers No. 106, London: International Institute of Strategic Studies, 1974, p. 15.
89. For authoritative and sophisticated analyses of these issues and problems, see Samuel Glasstone, ed., *The Effects of Nuclear Weapons* (rev. ed.), Washington, D.C.: U.S. Government Printing Office, 1964; and National Academy of Sciences, *Long-Term Worldwide Effects of Multiple Nuclear-Weapons Detonations*, Washington, D.C.: U.S. Government Printing Office, 1975.
90. *MIRV and the Arms Race*, p. 81.

91. "The Origins of MIRV," in Carlton and Schaerf, eds., *The Dynamics of the Arms Race,* p. 35.
92. Ibid., p. 24.
93. *MIRV and the Arms Race,* p. 105.
94. "Superpower Strategic Postures for a Multipolar World," in Richard Rosecrance, ed., *The Future of the International Strategic System,* San Francisco: Chandler Publishing Co., 1972, p. 37.
95. *MIRV and the Arms Race,* p. 139.
96. "Future Limitations of Strategic Arms," p. 228. Newhouse (*Cold Dawn,* p. 101) makes the same point in relation to both MIRV and ABM, saying that "neither bore much relationship to the really significant buildup of enemy strength."
97. See York, "The Origins of MIRV," p. 25.
98. *MIRV and the Arms Race,* p. 91.
99. Ibid., p. 91.
100. *Strategic Weapons,* p. 66.
101. *MIRV and the Arms Race,* p. 133.
102. *From Superiority to Parity: The United States and the Strategic Arms Race, 1961–1971,* Westport, Conn.: Greenwood Press, Inc., 1973, p. 293.
103. Polmar, *Strategic Weapons,* p. 104.
104. "The Dangers of a New Cold War," *Bulletin of the Atomic Scientists,* vol. 33, no. 3, March 1977, p. 38.
105. Tammen, *MIRV and the Arms Race,* p. 105. The reference to reduced megatonnage arises because the more space and weight is devoted to the casing for multiple warheads, the less is available for explosives.
106. Polmar, *Strategic Weapons,* pp. 85, 66.
107. Tammen, *MIRV and the Arms Race,* p. 54. Rathjens, "Future Limitations of Strategic Arms," p. 227, makes the identical point.
108. *Escalation and the Nuclear Option,* p. 180.
109. Both missiles are described in Newhouse, *Cold Dawn,* p. 154.
110. "Some Remarks on Multipolar Nuclear Strategy," in Rosecrance, ed., *The Future of the International Strategic System,* p. 28.
111. "Superpower Strategic Postures," p. 40.
112. *On the Uses of Military Power in the Nuclear Age,* Princeton, N.J.: Princeton University Press, 1966, p. 103.
113. *Escalation and the Nuclear Option,* p. 221.
114. Newhouse, *Cold Dawn,* p. 230.
115. "The Arms Race and SALT," in Carlton and Schaerf, eds., *The Dynamics of the Arms Race,* p. 52.
116. *Escalation and the Nuclear Option,* p. 203.
117. Ibid., p. 213.
118. *U.S. Nuclear Weapons in Europe: Issues and Alternatives,* Washington, D.C.: The Brookings Institution, 1974, p. 20.
119. *Annual Report, FY 1979,* p. 130.
120. Ibid., p. 71.
121. *U.S. Nuclear Weapons in Europe,* p. 7.
122. Ibid., p. 47.

123. *Strategic Deterrence Reconsidered,* Adelphi Papers No. 116, London: International Institute of Strategic Studies, 1975, p. 26.
124. "Disarmament: 30 Years of Failure," *International Security,* vol. 2, no. 3, Winter 1978, p. 42.
125. *Escalation and the Nuclear Option,* pp. 81–82.
126. "U.S. Tactical Nuclear Weapons: A Controversial Arsenal," in Johnson and Schneider, eds., *Current Issues,* p. 176.
127. Record, *U.S. Nuclear Weapons in Europe,* p. 61.
128. "French Defense Planning—The Future in the Past," *International Security,* vol. 1, no. 2, Fall 1976, p. 23.
129. *U.S. Nuclear Weapons in Europe,* p. 64.
130. "U.S. Tactical Nuclear Weapons," p. 188.
131. *New Weapons Technologies,* p. 23.
132. *U.S. Nuclear Weapons in Europe,* p. 15.
133. Leader and Schneider, "U.S.–Soviet Strategic Forces," p. 140. Record (*U.S. Nuclear Weapons in Europe,* p. 7) refers to these as "semistrategic forces."
134. "Some Remarks," p. 28.
135. Rathjens, "Future Limitations of Strategic Arms," p. 241.
136. "SALT II," p. 345.
137. "Strategic Vulnerability," p. 159.
138. "Negotiating with the Russians: Some Lessons from SALT," *International Security,* vol. 1, no. 4, Spring 1977, p. 20.
139. "Second Lecture," in James L. Buckley and Paul C. Warnke, *Strategic Sufficiency: Fact or Fiction?* Washington, D.C.: American Enterprise Institute for Public Policy Research, 1972, p. 36.
140. Brown, *Annual Report, FY 1979,* p. 37.
141. "SALT under the Carter Administration," Paper presented to the 19th Annual Meeting of the International Studies Association, Washington, D.C., February 22–25, 1978, p. 35.
142. "How to Avoid Monad—and Disaster," *Foreign Policy,* no. 24, Fall 1976, pp. 172–201.
143. "The Case for a Modern Strategic Bomber," p. 22.
144. "The Development of Nuclear Strategy," *International Security,* vol. 2, no. 4, Spring 1978, p. 71.
145. Brown, *Annual Report, FY 1979,* p. 65.
146. "Cruise Missiles," p. 1. Steinbruner and Garwin ("Strategic Vulnerability," p. 170) raise cost as a major factor.
147. "How to Avoid Monad—and Disaster," pp. 176–77.
148. Leader and Schneider, "U.S.–Soviet Strategic Forces," p. 131.
149. "Nuclear Strategy and Nuclear Weapons," in Pranger and Labrie, eds., *Nuclear Strategy and National Security,* p. 234.
150. Ibid., p. 233.
151. "How to Avoid Monad—and Disaster," p. 177.
152. "Neutron Bomb is Almost 20 Years Old: Was Focus of Heated Controversy in 1961," *Los Angeles Times,* July 13, 1977, p. 10.
153. "A Primer on Enhanced Radiation Weapons," *Bulletin of the Atomic Scientists,* vol. 33, no. 10, December 1977, p. 7.

154. Jorma K. Miettinen, "Enhanced Radiation Warfare," *Bulletin of the Atomic Scientists,* vol. 33, no. 7, September 1977, p. 33.
155. Agnew, "A Primer on Enhanced Radiation Weapons," pp. 7-8.
156. Legault and Lindsey, *The Dynamics of the Nuclear Balance,* p. 31.
157. "The Neutron Bomb Arms Control Impact Statement," *Congressional Record,* August 3, 1977, p. H-8500.
158. Barnaby, "Crossing the Nuclear Threshold."
159. Bernard T. Feld, "The Neutron Bomb," *Bulletin of the Atomic Scientists,* vol. 33, no. 7, September 1977, p. 11.
160. Barnaby, "Crossing the Nuclear Threshold."
161. "Neutron Bomb Poses Dilemma for Congress," *Congressional Quarterly Weekly Report,* July 9, 1977, p. 1407.

Chapter 5

1. *Escalation and the Nuclear Option,* Princeton, N.J.: Princeton University Press, 1966, p. 171.
2. "On Estimating and Imputing Intentions," *International Security,* vol. 2, no. 3, Winter 1978, p. 25.
3. *The Soviet Strategic Culture: Implications for Limited Nuclear Operations,* A Project AIR FORCE Report Prepared for the United States Air Force, Santa Monica, Calif.: Rand Corporation, September 1977, p. 5.
4. "Soviet-American Strategic Competition: Instruments, Documents and Purposes," in Robert J. Pranger and Roger P. Labrie, eds., *Nuclear Strategy and National Security: Points of View,* Washington, D.C.: American Enterprise Institute for Public Policy Research, 1977, p. 280.
5. Leon Goure, Foy D. Kohler, and Mose L. Harvey, *The Role of Nuclear Forces in Current Soviet Strategy,* Miami, Fla.: Center for Advanced International Studies, 1974, p. 42.
6. *The Soviet Strategic Culture,* p. 22.
7. Ibid., p. 33.
8. "Why the Soviet Union Thinks It Could Fight and Win a Nuclear War," *Commentary,* vol. 64, no. 1, July 1977, p. 29.
9. Ibid., p. 28.
10. *The Soviet Strategic Culture,* p. 11.
11. *From Superiority to Parity: The United States and the Strategic Arms Race, 1961-1971,* Westport, Conn.: Greenwood Press, Inc., 1973, p. 93.
12. "How to Avoid Monad—and Disaster," *Foreign Policy,* no. 24, Fall 1976, p. 187.
13. *Security in the Nuclear Age: Developing U.S. Strategic Arms Policy,* Washington, D.C.: The Brookings Institution, 1975, p. 112.
14. *The Role of Nuclear Forces,* p. ix.
15. Ibid., p. 10.
16. Michael A. Milstein and Leo S. Semeiko, "Problems of the Inadmissibility of Nuclear Conflict." *International Studies Quarterly,* vol. 20, no. 1, March 1976, p. 98.
17. Snyder, *The Soviet Strategic Culture,* p. 22.

Notes to Chapter 5

18. *Understanding the Soviet Military Threat: How CIA Estimates Went Astray,* New York: National Strategy Information Center, 1977, p. 31.
19. Pipes, "Why the Soviet Union Thinks," p. 30.
20. C. G. Jacobsen, "Soviet Strategic Capabilities: The Superpower Balance," *Current History,* vol. 73, no. 430, October 1977, p. 99.
21. *The Role of Nuclear Forces,* p. 14.
22. "The Enigma of Soviet Strategic Policy," *The Wilson Quarterly,* vol. 1, no. 5, Autumn 1977, p. 87.
23. *The Role of Nuclear Forces,* pp. 48, ix.
24. "Political Realism and the 'Realistic Deterrence' Strategy," in Pranger and Labrie, eds., *Nuclear Strategy and National Security,* p. 46.
25. "The Dangers of a New Cold War," *Bulletin of the Atomic Scientists,* vol. 33, no. 3, March 1977, p. 35.
26. Goure, Kohler, and Harvey, *The Role of Nuclear Forces,* p. 5.
27. "On Estimating and Imputing Intentions," p. 28.
28. "The Relationship of Strategic and Theater Nuclear Forces," *International Security,* vol. 2, no. 2, Fall 1977, p. 125.
29. "First Lecture," in James L. Buckley and Paul C. Warnke, *Strategic Sufficiency: Fact or Fiction?* Washington, D.C.: American Enterprise Institute for Public Policy Research, 1972, p. 3.
30. Goure, Kohler, and Harvey, *The Role of Nuclear Forces,* p. 67.
31. "Assuring Strategic Stability in an Era of Detente," *Foreign Affairs,* vol. 54, no. 2, January 1976, p. 212.
32. Richard G. Head, "Technology and the Military Balance," *Foreign Affairs,* vol. 56, no. 3, April 1978, p. 549.
33. Pipes, "Why the Soviet Union Thinks," p. 25.
34. Head, "Technology and the Military Balance," p. 547.
35. "Why the Soviet Union Thinks," p. 34.
36. Lee, *Understanding the Soviet Military Threat,* p. 44.
37. Pipes, "Why the Soviet Union Thinks," p. 21.
38. Ibid., p. 31.
39. *The Role of Nuclear Forces,* pp. 77, 4.
40. Ibid., p. 16.
41. "The Arms Race as Posturing," in David Carlton and Carlo Schaerf, eds., *The Dynamics of the Arms Race,* New York: John Wiley and Sons, 1975.
42. *The Soviet Strategic Culture,* p. 38.
43. Pipes, "Why the Soviet Union Thinks," p. 33.
44. Lee, *Understanding the Soviet Military Threat,* p. 10.
45. *The Soviet Strategic Culture,* p. 20.
46. Buckley, "First Lecture," p. 12.
47. "Why the Soviet Union Thinks," p. 30.
48. Goure, Kohler, and Harvey, *The Role of Nuclear Forces,* p. 18.
49. *U.S. Nuclear Weapons in Europe: Issues and Alternatives,* Washington, D.C.: The Brookings Institution, 1974, p. 48.
50. "The Enhanced Radiation Warhead: A Military Perspective," *Arms Control Today,* vol. 8, no. 6, June 1978, p. 5.
51. *U.S. Nuclear Weapons in Europe,* p. 42.

52. *Arms and Influence,* New Haven: Yale University Press, 1966, p. 158.
53. *Annual Report, Department of Defense, Fiscal Year 1979,* Washington, D.C.: U.S. Government Printing Office, 1978, p. 53.
54. *Arms Treaties with Moscow: Unequal Terms Unevenly Applied?* New York: National Strategy Information Center, 1975, p. 2.
55. *Understanding the Soviet Military Threat,* p. 11.
56. *The Role of Nuclear Forces,* p. 3.
57. "Technology and the Military Balance," p. 559.
58. "Annual Defense Department Annual Report, FY 1975," in Pranger and Labrie, eds., *Nuclear Strategy and National Security,* p. 89.
59. "The Development of Nuclear Strategy," *International Security,* vol. 2, no. 4, Spring 1978, p. 75.
60. Ibid., p. 74.
61. "Soviet-American Strategic Competition," pp. 300, 279.
62. "Technology and the Military Balance," p. 547.
63. George S. Brown, "The Strategic Nuclear Balance," *Commanders Digest,* vol. 21, no. 4, March 9, 1978, p. 3.
64. Head, "Technology and the Military Balance," p. 553.
65. *Strategic Weapons: An Introduction,* New York: National Strategy Information Center, 1975, p. 85.
66. *Annual Report, FY 1979,* p. 63.
67. "Technology and the Military Balance," p. 557.
68. *Annual Report, FY 1979,* p. 50.
69. "Beyond SALT II—A Missile Test Quota," *Bulletin of the Atomic Scientists,* vol. 33, no. 5, May 1977, p. 39.
70. *Annual Report, FY 1979,* p. 63.
71. "The Strategic Nuclear Balance," pp. 4–5.
72. Kahan, *Security in the Nuclear Age,* p. 166.
73. Detailed in John B. Rhinelander, "The SALT I Agreements," in Mason Willrich and John B. Rhinelander, eds., *SALT: The Moscow Agreements and Beyond,* New York: Free Press, 1974, p. 148.
74. *Annual Report, FY 1979,* pp. 50–51.
75. Ibid.
76. Ibid., p. 5l.
77. "The Strategic Nuclear Balance," p. 10.
78. *Strategic Weapons,* p. 52.
79. *Security in the Nuclear Age,* p. 50.
80. All actual force figures are January 1, 1978, estimates from the *Annual Report, FY 1979,* p. 47.
81. Polmar, *Strategic Weapons,* p. 60.
82. Arbatov, "The Dangers of a New Cold War," p. 37.
83. *Annual Report, FY 1979,* p. 125.
84. These programs are detailed by Leon Goure in *War Survival in Soviet Strategy: USSR Civil Defense,* Miami, Fla.: Center for Advanced International Studies, 1976. Educational and training program elements and extent are summarized in his rejoinder to Fred Kaplan, "Another Interpretation," *Bulletin of the Atomic Scientists,* vol. 34, no. 4, April 1978, pp. 48–51.
85. "Why the Soviet Union Thinks," p. 33.

86. *Annual Report, FY 1979*, p. 64.
87. " 'Equivalence' and SALT II," *Orbis*, vol. 20, 1977, p. 1053.
88. "Assuring Strategic Stability in an Era of Detente," *Foreign Affairs*, vol. 54, no. 2, January 1979, p. 223.
89. "Soviet-American Strategic Competition," p. 285.
90. "Civil Defense in Limited War—In Favor," in Pranger and Labrie, eds., *Nuclear Strategy and National Security*, p. 371. Sidney D. Drell makes the same point in "Civil Defense in Limited War—Opposed," ibid., p. 383.
91. "Another Interpretation," p. 48.
92. "An Assessment of the Bomber–Cruise Missile Controversy," *International Security*, vol. 2, no. 1, Summer 1977, p. 48.
93. "The Dangers of a New Cold War," p. 39.
94. "The Soviet Civil Defense Myth, Part II," *Bulletin of the Atomic Scientists*, vol. 34, no. 4, April 1978, p. 48.
95. "The Future of the Strategic Bomber," *AEI Defense Review*, vol. 2, no. 1, 1978, p. 11.
96. "Repeating History: The Civil Defense Debate Renewed," *International Security*, vol. 2, no. 3, Winter 1978, pp. 114–15.
97. "The Soviet Civil Defense Myth, Part I," *Bulletin of the Atomic Scientists*, vol. 34, no. 3, March 1978, p. 15.
98. "Soviet Civil Defense: Myth and Reality," *Arms Control Today*, vol. 6, no. 9, September 1976, p. 2.
99. Kincade, "Repeating History," p. 102.
100. Aspin, "Soviet Civil Defense," p. 1.
101. Kaplan, "Part I," p. 16.
102. "Repeating History," p. 112.
103. Henry S. Rowen, "The Need for a New Analytical Framework: Review of *Security in the Nuclear Age*," *International Security*, vol. 1, no. 2, Fall 1976, p. 143.
104. Culver, "The Future of the Strategic Bomber." Citing "the most significant conclusions of a series of studies of nuclear war effects," Kincade ("Repeating History," p. 109) makes the same point.
105. *From Superiority to Parity*, p. 314.
106. *Understanding the Soviet Military Threat*, p. 12.
107. *The Faces of Verification: Strategic Arms Control for the 1980's*, The Rand Paper Series No. P-5986, Santa Monica, Calif.: Rand Corporation, August 1977, p. 25.
108. Head, "Technology and the Military Balance," p. 555.
109. "Linking SALT to Ethiopia or Unlinking It from Detente," *Bulletin of the Atomic Scientists*, vol. 34, no. 6, June 1978, p. 39.
110. Excerpts quoted in "Moscow's Failures," *Newsweek*, vol. 91, no. 25, June 19, 1978, p. 43.

Chapter 6

1. "Arms Control: A Global Imperative," *Bulletin of the Atomic Scientists*, vol. 34, no. 6, June 1978, p. 32.
2. "SALT II: The Search for a Follow-on Agreement," *Orbis*, vol. 17, 1973, p. 361.

3. "Arms Control," p. 33.

4. This basic discussion, with the addition of the self-fulfilling prophecy notion, is also found in Donald M. Snow, *The Shadow of the Mushroom-Shaped Cloud: Basic Ideas and Problems of Nuclear Deterrence,* Learning Materials in National Security Education, No. 2, Columbus, Ohio: Consortium for International Studies Education, 1978, pp. 35–47.

5. "Verification and Control," in David Carlton and Carlo Schaerf, eds., *The Dynamics of the Arms Race,* New York: John Wiley and Sons, 1975, p. 120.

6. "A Leap Forward in Verification," in Mason Willrich and John B. Rhinelander, eds., *SALT: The Moscow Agreements and Beyond,* New York: Free Press, 1974, p. 163.

7. Ibid., p. 162.

8. *The Faces of Verification: Strategic Arms Control for the 1980's,* The Rand Paper Series No. P-5986, Santa Monica, Calif.: Rand Corporation, August 1977, p. 9.

9. George S. Brown, "The Strategic Nuclear Balance," *Commanders Digest,* vol. 21, no. 4, March 9, 1978, p. 5.

10. *On the Uses of Military Power in the Nuclear Age,* Princeton, N.J.: Princeton University Press, 1966, p. 107.

11. *The Faces of Verification,* p. 6.

12. Harland B. Moulton, *From Superiority to Parity: The United States and the Strategic Arms Race, 1961–1971,* Westport, Conn.: Greenwood Press, Inc., 1973, p. 266.

13. *Arms Treaties with Moscow: Unequal Terms Unevenly Applied?* New York: National Strategy Information Center, 1975, p. 15.

14. "Can Deterrence Last Out the Century," in Robert J. Pranger and Roger P. Labrie, eds., *Nuclear Strategy and National Security: Points of View,* Washington, D.C.: American Enterprise Institute for Public Policy Research, 1977, p. 58.

15. "Arms Control 'the American Way,' " *The Wilson Quarterly,* vol. 1, no. 5, Autumn 1977, pp. 94–99.

16. "Future Limitation of Strategic Arms," in Willrich and Rhinelander, eds., *SALT,* p. 239.

17. "SALT II," p. 343.

18. "A Leap Forward in Verification," p. 166.

19. "America and the 'MAD' World," *The Wilson Quarterly,* vol. 1, no. 5, Autumn 1977, p. 67.

20. *From Superiority to Parity,* p. 108,

21. *Security in the Nuclear Age: Developing U.S. Strategic Arms Policy,* Washington, D.C.: The Brookings Institution, 1975, p. 139.

22. "Disarmament: 30 Years of Failure," *International Security,* vol. 2, no. 3, Winter 1978, p. 40.

23. Scoville, "A Leap Forward in Verification," p. 178.

24. "The U.S.–India Safeguards Dispute," *Bulletin of the Atomic Scientists,* vol. 34, no. 6, June 1978, p. 50.

25. Warnke, "Arms Control," p. 34.

26. John Maddox, *Prospects for Nuclear Proliferation,* Adelphi Papers No. 113, London: International Institute of Strategic Studies, 1975, p. 3.

27. "The Race to Control Nuclear Arms," in Pranger and Labrie, eds., *Nuclear Strategy and National Security,* p. 476.

28. "Nuclear 'Gray Marketeering,' " *International Security,* vol. 1, no. 3, Winter 1976, p. 117.

29. *Nuclear Paradox: Security Risks of the Peaceful Atom,* Washington, D.C.: American Enterprise Institute for Public Policy Research, 1976, p. 3.

30. "Nuclear Proliferation: The Israeli Case," *International Studies Quarterly,* vol. 22, no. 1, March 1978, pp. 79–120.

31. "Nonproliferation: A Long-Term Strategy," *Foreign Affairs,* vol. 56, no. 3, April 1978, p. 605.

32. *Nuclear Power and Weapons Proliferation,* Adelphi Papers No. 130, London: International Institute of Strategic Studies, 1976, p. 6.

33. Ibid. Guhin (*Nuclear Paradox,* p. 55) reaches the same conclusion.

34. "Nuclear Sharing: Nato and the N + 1 Country," in Henry A. Kissinger, ed., *Problems of National Strategy: A Book of Readings,* New York: Frederick A. Praeger, 1965, p. 189.

35. *Nuclear Paradox,* pp. 12–13.

36. "A Historical Survey of Nonproliferation Politics," *International Security,* vol. 2, no. 1, Summer 1977, p. 76.

37. "The Politics of Twenty Nuclear Nations," in Richard Rosecrance, ed., *The Future of the International Strategic System,* San Francisco: Chandler Publishing Company, 1972, p. 65.

38. "Disarmament," p. 40.

39. *Nuclear Paradox,* p. 60. Greenwood, Rathjens, and Ruina (*Nuclear Power and Weapons Proliferation,* p. 20) raise this same concern in the context of monitoring reprocessing facilities.

40. "The U.S.–India Safeguards Dispute," p. 51.

41. *Prospects for Nuclear Proliferation,* p. 8.

42. "The U.S.–India Safeguards Dispute," p. 51.

43. Guhin, *Nuclear Paradox,* p. 4.

44. "The New Test Ban Treaties: What Do They Mean? Where Do They Lead?" *International Security,* vol. 1, no. 3, Winter 1976, p. 169.

45. Ibid., p. 170.

46. "Commentary on the New Test Ban Treaties," *International Security,* vol. 1, no. 3, Winter 1976, p. 180.

47. Doty, Carnesale, and Nacht, "The Race to Control Nuclear Arms," p. 468.

48. "The New Test Ban Treaties," p. 174.

49. "A Comprehensive Test Ban Treaty: Everybody or Nobody," *International Security,* vol. 1, no. 1, Summer 1976, p. 116.

50. "The Verification of Arms Control Agreements," *Arms Control Today,* vol. 7, no. 7–8, July–August 1977, p. 1. Scoville ("A Leap Forward in Verification") makes the same point.

51. Suggested in Sweet, "The U.S.–India Safeguards Dispute."

52. "A New World Order Problem" *International Security,* vol. 1, no. 3, Winter 1976, p. 86.

53. "Arms Control," p. 34.

54. "A Comprehensive Test Ban Treaty," p. 113.

55. Ibid., p. 101.

56. "Strategic Vulnerability: The Balance between Prudence and Paranoia," *International Security,* vol. 1, no. 1, Summer 1976, p. 161.

57. "A Comprehensive Test Ban Treaty," p. 94.

58. *The Essence of Security: Reflection in Office,* New York: Harper and Row, 1968, p. 62.

59. *MIRV and the Arms Race: An Interpretation of Defense Strategy,* New York: Praeger Publishers, Inc., 1973, p. 134.

60. Brennan, *Arms Treaties with Moscow,* p. 5.

61. John Newhouse, *Cold Dawn: The Story of SALT,* New York: Holt, Rinehart and Winston, 1973.

62. Complete texts of the various documents can be found in the appendixes of Willrich and Rhinelander, eds., *SALT.*

63. "SALT II," p. 335.

64. John B. Rhinelander, "The SALT I Agreements," in Willrich and Rhinelander, eds., *SALT,* p. 143.

65. Ibid., pp. 144, 145.

66. "SALT II," p. 343.

67. Henry S. Rowen, "The Need for a New Analytical Framework: Review of *Security in the Nuclear Age,*" *International Security,* vol. 1, no. 2, Fall 1976, p. 144.

68. "The SALT I Agreements," p. 147.

69. *Strategic Weapons: An Introduction,* New York: National Strategy Information Center, 1975, p. 89.

70. Discussed in Newhouse, *Cold Dawn,* p. 177.

71. "SALT II," p. 349.

72. Rhinelander, "The SALT I Agreements," p. 151.

73. Scoville, "A Leap Forward in Verification," p. 171.

74. *Cold Dawn,* p. 194.

75. Kruzel, "SALT II," pp. 3-4.

76. "The SALT I Agreement," p. 126.

77. Scoville, "A Leap Forward in Verification," p. 169.

78. *Arms Treaties with Moscow,* p. 12.

79. *The Role of Nuclear Forces in Current Soviet Strategy,* Miami, Fla.: Center for Advanced International Studies, 1974, p. 17. Brennan (*Arms Treaties with Moscow,* p. 14) makes the same point, adding that the Russians may not have wanted to make the technological difference any "more manifest than it was."

80. "The Arms Race and SALT," in Carlton and Schaerf, eds., *The Dynamics of the Arms Race,* p. 51.

81. "Annual Defense Department Report, FY 1975," in Pranger and Labrie, eds., *Nuclear Strategy and National Security,* p. 96. (Report originally issued March 4, 1974.)

82. "SALT I: An Appraisal," in Willrich and Rhinelander, eds., *SALT,* pp. 263, 264. Rhinelander ("The SALT I Agreements") reaches the same conclusion regarding the effects on strengthening retaliatory ability.

83. "Negotiating with the Russians: Some Lessons from SALT," *International Security,* vol. 1, no. 4, Spring 1977, p. 21.

84. "Problems of the Inadmissibility of Nuclear Conflict," *International Studies Quarterly,* vol. 20, no. 1, March 1976, p. 88.

85. *Cold Dawn,* p. 18.

86. Ibid., p. 168. Moulton (*From Superiority to Parity,* p. xi) makes the same point.

87. Stefan H. Leader and Barry R. Schneider, "U.S.–Soviet Strategic Forces: SALT and the Search for Parity," in David T. Johnson and Barry R. Schneider,

Notes to Chapter 6

eds., *Current Issues in U.S. Defense Policy*, New York: Praeger Special Studies, 1976, p. 146.

88. Ibid., p. 147.

89. Paul Nitze, "Foreword," in Brennan, *Arms Treaties with Moscow*, p. ix.

90. Text of speech delivered in Dallas, Texas, on March 22, 1976, and reprinted as "Foreign Policy and National Security," *International Security*, vol. 1, no. 1, Summer 1976, p. 188.

91. " 'Equivalence' and SALT II," *Orbis*, vol. 20, 1977, p. 1048.

92. Steinbruner and Garwin, "Strategic Vulnerability," p. 138.

93. Polmar, *Strategic Weapons*, p. 91.

94. U.S.–Soviet Strategic Forces," pp. 147–49.

95. "A Leap Forward in Verification," p. 175.

96. Brown, "The Strategic Nuclear Balance," p. 12.

97. Rathjens, "Future Limitations on Strategic Arms," p. 254.

98. Harold Brown, *Annual Report, Department of Defense, Fiscal Year 1979*, Washington, D.C.: U.S. Government Printing Office, 1978, p. 46.

99. "SALT or No SALT," *Bulletin of the Atomic Scientists*, vol. 34, no. 6, June 1978, p. 38.

100. "Soviet-American Strategic Competition: Instruments, Doctrines, and Purposes," in Pranger and Labrie, eds., *Nuclear Strategy and National Security*, p. 294.

101. "The Scope and Limits of SALT," *Foreign Affairs*, vol. 56, no. 4, July 1978, p. 756.

102. Ibid., p. 758.

103. "SALT or No SALT," p. 36. Burt ("The Scope and Limits of SALT," pp. 757–58) confirms these figures. The complete text is available in *SALT II Agreement, Vienna, June 18, 1979*, Washington, D.C.: U.S. Department of State, 1979.

104. Burt, "The Scope and Limits of SALT," p. 759. J. I. Coffey, in "SALT and the Carter Administration," Paper presented to the 19th Annual Meeting of the International Studies Association, Washington, D.C., February 22–25, 1978, p. 8, lists most of these provisions.

105. *The Faces of Verification*, p. 13. Coffey ("SALT and the Carter Administration," p. 5) presents a similar list.

106. *Annual Report, FY 1979*, p. 61.

107. "SALT or No SALT," p. 37.

108. Discussed thoughtfully by Thomas Schelling in "The Importance of Agreements," in Carlton and Schaerf, eds., *The Dynamics of the Arms Race*, pp. 69–70.

109. Lehman and Hughes (" 'Equivalence' and SALT II," p. 1050) state this as a particular problem for "grey area" systems like SS-20 and Backfire, which are potentially either theater or strategic weapons.

110. "Cruise Missiles: Different Missions, Different Arms Control Impact," *Arms Control Today*, vol. 8, no. 1, January 1978, p. 3.

111. As an example, Kosta Tsipis proposes designating "areas of the ocean accessible only to submarines of one nation" as a hedge against ASW. See "Anti-Submarine Warfare and Missile Submarines," in Carlton and Schaerf, eds., *The Dynamics of the Arms Race*, p. 45.

112. "The Race to Control Nuclear Arms," p. 469.

113. "SALT I: An Appraisal," in Willrich and Rhinelander, eds., *SALT*, p. 256.
114. "Negotiating with the Russians," p. 22.
115. "Arms Control 'the American Way,' " p. 97.
116. "The United States—A Military Power Second to None?" *International Security*, vol. 1, no. 1, Summer 1976, p. 49.
117. "Superpower Strategic Postures for a Multipolar World," in Rosecrance, ed., *The Future of the International Strategic System*, p. 34.
118. *The Faces of Verification*, p. 28.
119. "Technical Innovation and Arms Control," in Pranger and Labrie, eds., *Nuclear Strategy and National Security*, p. 479.
120. "Assuring Strategic Stability in an Era of Detente," *Foreign Affairs*, vol. 54, no. 2, January 1976, p. 220. Rathjens makes the same point in "The Verification of Arms Control Agreements," p. 3.
121. Rathjens, "The Verification of Arms Control Agreements," p. 3.
122. "The Nuclear Test Ban Treaty," in Kissinger, ed., *Problems of National Strategy*, p. 419.
123. Rathjens, "Future Limitations on Strategic Arms," p. 252.
124. "Slowing Down the Arms Race," in Carlton and Schaerf, eds., *The Dynamics of the Arms Race*, p. 87.
125. "Technical Innovation and Arms Control," p. 503.
126. "Nuclear Strategy and Nuclear Weapons," in Pranger and Labrie, eds., *Nuclear Strategy and National Security*, pp. 240–41.
127. "The Direct Payoff in SALT II," *Arms Control Today*, vol. 8, no. 3, March 1978, p. 4.
128. "Nuclear Strategy and Nuclear Weapons," p. 240.
129. *The Faces of Verification*, p. 18.
130. "Effective Military Technology in the 1980's," *International Security*, vol. 1, no. 2, Fall 1976, p. 63.
131. "Arms Control," p. 33.
132. "The Direct Payoff in SALT II," p. 3.
133. "The SALT I Agreements," p. 153.
134. "The Arms Race and SALT," p. 50.

Chapter 7

1. Paul C. Warnke, "Arms Control: A Global Imperative," *Bulletin of the Atomic Scientists*, vol. 34, no. 6, June 1978, p. 33.
2. This point is made by Thomas J. Downey in "How to Avoid Monad—and Disaster," *Foreign Policy*, no. 24, Fall 1976, p. 185.
3. "Anti-Submarine Warfare and Missile Submarines," in David Carlton and Carlo Schaerf, eds., *The Dynamics of the Arms Race*, New York: John Wiley and Sons, 1975, p. 42.
4. "The Strategic Nuclear Balance," *Commanders Digest*, vol. 21, no. 4, March 9, 1978, p. 4.
5. *MIRV and the Arms Race: An Interpretation of Defense Strategy*, New York: Praeger Publishing Co., 1973, p. 119.
6. "U.S.–Soviet Strategic Forces: SALT and the Search for Parity," in David T. Johnson and Barry R. Schneider, eds., *Current Issues in U.S. Defense Policy*, New York: Praeger Special Studies, 1976, p. 132.

7. London: International Institute of Strategic Studies, May 1978. Excerpted as "The Technological Trap Beyond SALT: New Technology and Deterrence," in *The New York Review of Books*, vol. 25, no. 12, June 20, 1978, p. 27.

8. Ibid., p. 28.

9. Ibid., p. 27.

10. Ibid.

11. Colin S. Gray, "Soviet-American Strategic Competition: Instruments, Doctrines, and Purposes," in Robert J. Pranger and Roger P. Labrie, eds., *Nuclear Strategy and National Security: Points of View*, Washington, D.C.: American Enterprise Institute for Public Policy Research, 1977, p. 285.

12. "Second Lecture," in James L. Buckley and Paul C. Warnke, *Strategic Sufficiency: Fact or Fiction?* Washington, D.C.: American Enterprise Institute for Public Policy Research, 1972, p. 30.

13. Tammen, *MIRV and the Arms Race*, p. 55.

14. *U.S. Strategic Nuclear Weapons and Deterrence*, Rand Paper Series No. P-5967, Santa Monica, Calif.: Rand Corporation, August 1977, p. 28.

15. "How to Avoid Monad—and Disaster," pp. 172–73, 182.

16. "Slowing Down the Arms Race," in Carlton and Schaerf, eds., *The Dynamics of the Arms Race*, p. 85.

17. *Security in the Nuclear Age: Developing U.S. Strategic Arms Policy*, Washington, D.C.: The Brookings Institution, 1975, p. 330.

18. "The Strategic Forces TRIAD: End of the Road?" *Foreign Affairs*, vol. 56, no. 4, July 1978, p. 785.

19. Joseph Kruzel, "SALT II: The Search for a Follow-on Agreement," *Orbis*, vol. 17, 1973, p. 344.

20. Barry Carter, "Nuclear Strategy and Nuclear Weapons," in Pranger and Labrie, eds., *Nuclear Strategy and National Security*, p. 240.

21. *Cold Dawn: The Story of SALT*, New York: Holt, Rinehart and Winston, 1973, p. 26.

22. "Beyond SALT II—A Missile Test Quota," *Bulletin of the Atomic Scientists*, vol. 33, no. 5, May 1977, p. 39.

23. *U.S. Strategic Nuclear Weapons and Deterrence*, p. 39.

24. "The Case for a Modern Strategic Bomber," *AEI Defense Review*, vol. 2, no. 1, 1978, p. 20.

25. " 'Equivalence' and SALT II," *Orbis*, vol. 20, 1977, p. 1050.

26. "The Dangers of a New Cold War," *Bulletin of the Atomic Scientists*, vol. 33, no. 3, March 1977, p. 38.

27. *U.S. Strategic Nuclear Weapons and Deterrence*, p. 48.

28. *MIRV and the Arms Race*, p. 126.

29. "Assuring Strategic Stability in an Era of Detente," *Foreign Affairs*, vol. 54, no. 2, January 1976, p. 227.

30. "U.S.–U.S.S.R. Strategic Policies," in Pranger and Labrie, eds., *Nuclear Strategy and National Security*, p. 115.

31. "How to Avoid Monad," p. 188.

32. Gray, "The Strategic Forces TRIAD," p. 779.

33. Ibid., p. 788.

34. See, for instance, Barry J. Smernoff, "Strategic and Arms Control Implications of Laser Weapons," *Air University Review*, vol. 29, no. 2, January–February 1978, pp. 38–50.

35. "Slowing Down the Arms Race," p. 91.
36. *Cold Dawn,* p. 14.
37. George W. Rathjens, "Future Limitations of Strategic Arms," in Mason Willrich and John B. Rhinelander, eds., *SALT: The Moscow Agreements and Beyond,* New York: Free Press, 1974, p. 249.
38. See discussion in Donald G. Brennan, "A Comprehensive Test Ban Treaty: Everybody or Nobody," *International Security,* vol. 1, no. 1, Summer 1976, pp. 92–117.
39. "The Scope and Limits of SALT," *Foreign Affairs,* vol. 56, no. 4, July 1978, p. 760.
40. "The Technological Trap Beyond SALT," p. 27.
41. "Repeating History: The Civil Defense Debate Renewed," *International Security,* vol. 2, no. 3, Winter 1978, p. 106.
42. "Cruise Missiles: Different Missions, Different Arms Control Impacts," *Arms Control Today,* vol. 8, no. 1, January 1978, p. 2.
43. This argument is made by Harland B. Moulton in *From Superiority to Parity: The United States and the Strategic Arms Race, 1961–1971,* Westport, Conn.: Greenwood Press, Inc., 1973, p. 300.
44. *Strategic Weapons: An Introduction,* New York: National Strategy Information Center, 1975, p. 7.
45. "The Strategic Forces TRIAD," p. 778.
46. "Commentary on the New Test Ban Treaties," *International Security,* vol. 1, no. 3, Winter 1976, p. 179.
47. *Annual Report, Department of Defense, Fiscal Year 1979,* Washington, D.C.: U.S. Government Printing Office, 1978, p. 45.
48. *Arms Treaties with Moscow: Unequal Terms Unevenly Applied?* New York: National Strategy Information Center, 1975, pp. 16–17.
49. "The Future of the Strategic Bomber," *AEI Defense Review,* vol. 2, no. 1, 1978, p. 9.
50. *The Soviet Strategic Culture: Implications for Limited Nuclear Operations,* A Project AIR FORCE Report Prepared for the United States Air Force, Santa Monica, Calif.: Rand Corporation, September 1977, p. 20.
51. *U.S. Nuclear Weapons in Europe: Issues and Alternatives,* Washington, D.C.: The Brookings Institution, 1974, p. 41.
52. "Introduction," in Richard Rosecrance, ed., *The Future of the International Strategic System,* San Francisco: Chandler Publishing Co., 1972, p. 8.
53. *Strategic Deterrence Reconsidered,* Adelphi Papers No. 116, London: International Institute of Strategic Studies, 1976, p. 27.
54. "Deterrence in Dyadic and Multipolar Environments," in Rosecrance, ed., *The Future of the International Strategic System,* p. 135.
55. Rosecrance, *Strategic Deterrence Reconsidered,* p. 35.
56. Ibid., p. 33.
57. "A New World Order Problem," *International Security,* vol. 1, no. 3, Winter 1976, p. 80.
58. "Superpower Strategic Postures for a Multipolar World," in Rosecrance, ed., *The Future of the International Strategic System,* p. 31.
59. "Nuclear Proliferation: Can Congress Act in Time?" *International Security,* vol. 1, no. 4, Spring 1977, p. 52.

60. "Some Remarks on Multipolar Nuclear Strategy," in Rosecrance, ed., *The Future of the International Strategic System,* p. 20.

61. Rosecrance, "Deterrence in Dyadic and Multipolar Environments," p. 135.

62. As examples of this genre, see Robin Moore and Lewis Perdue, *The Trinity Implosion,* New York: Manor Books Inc., 1976, and Leonard Sanders, *The Hamlet Warning,* New York: Charles Scribner and Sons, 1976.

63. David M. Rosenbaum, "Nuclear Terror," *International Security,* vol. 1, no. 3, Winter 1976, p. 141.

64. "Who Will Have the Bomb?" *International Security,* vol. 1, no. 1, Summer 1976, p. 84.

65. *Nuclear Paradox: Security Risks of the Peaceful Atom,* Washington, D.C.: American Enterprise Institue for Public Policy Research, 1976, p. 2.

66. *The Shadow of the Mushroom-Shaped Cloud: Basic Ideas and Problems of Nuclear Deterrence,* Learning Materials in National Security Education No. 2, Columbus, Ohio: Consortium for International Studies Education, 1978, p. 33.

67. "Nuclear Sharing: Nato and the N+1 Country," in Henry A. Kissinger, ed., *Problems of National Strategy: A Book of Readings,* New York: Frederick A. Praeger, 1965, p. 194.

68. "The Politics of Twenty Nuclear Powers," in Rosecrance, ed., *The Future of the International Strategic System,* p. 66.

69. "Deterrence in Dyadic and Multipolar Environments," p. 137.

70. "Superpower Strategic Postures," p. 46.

71. *Nuclear Forces for Medium Powers: Part I: Targets and Weapons Systems,* Adelphi Papers No. 106, London: International Institute of Strategic Studies, 1974, p. 29.

72. "Who Will Have the Bomb?" p. 89.

73. For a description, see Bertrand Goldschmidt, "A Historical Survey of Nonproliferation Politics," *International Security,* vol. 2, no. 1, Summer 1977, especially pp. 79–80.

74. *Nuclear Power and Weapons Proliferation,* Adelphi Papers No. 130, London: International Institute of Strategic Sutides, 1976, p. 12.

75. Ibid., p. 28.

76. "Nonproliferation: A Long-Term Strategy," *Foreign Affairs,* vol. 56, no. 3, April 1978, p. 620.

77. Ibid., p. 611.

78. This is a major argument in Tammen, *MIRV and the Arms Race.* Newhouse (*Cold Dawn*) also maintains that had the United States not insisted on completing MIRV testing before SALT I began, a ban might have been negotiable, thereby avoiding much of the vertical warhead proliferation that has since occurred.

79. Discussed in some detail in Gray, "The Strategic Forces TRIAD."

80. Wayland Young (Lord Kennet), "Disarmament: 30 Years of Failure," *International Security,* vol. 2, no. 3, Winter 1978, p. 49.

Cited Bibliography

The following is a list of books, journal articles, articles in edited volumes, newspaper stories, official documents, and reports that are cited in the notes to this book. Full citations of the books from which chapters are cited are listed under the editors' names.

Agnew, Harold M. "A Primer on Enhanced Radiation Weapons." *Bulletin of the Atomic Scientists,* vol. 33, no. 10, December 1977, pp. 6–8.
Arbatov, Georgi. "The Dangers of a New Cold War." *Bulletin of the Atomic Scientists,* vol. 33, no. 3, March 1977, pp. 33–40.
Aspin, Les. "SALT or No SALT." *Bulletin of the Atomic Scientists,* vol. 34, no. 6, June 1978, pp. 34–38.
———. "Soviet Civil Defense: Myth and Reality." *Arms Control Today,* vol. 6, no. 9, September 1976, pp. 1–2, 4–5.
Barnaby, Frank. "Crossing the Nuclear Threshold." *New Scientist,* January 19, 1978, p. 151.
Berman, Robert. "The B-1 Bomber." In Johnson and Schneider, eds., pp. 166–75.
Boulding, Kenneth E. *Conflict and Defense: A General Theory.* New York: Harper and Row, 1963.
———. "The Weapon as an Element in the Social System." In Rosecrance, ed., pp. 81–92.
Brennan, Donald G. *Arms Treaties with Moscow: Unequal Terms Unevenly Applied?* New York: National Strategy Information Center, 1975.
———. "A Comprehensive Test Ban Treaty: Everybody or Nobody." *International Security,* vol. 1, no. 1, Summer 1976, pp. 92–117.
———. "Some Remarks on Multipolar Nuclear Strategy." In Rosecrance, ed., pp. 13–28.
Brodie, Bernard. "The Development of Nuclear Strategy." *International Security,* vol. 2, no. 4, Spring 1978, pp. 65–83.
———. *Escalation and the Nuclear Option.* Princeton, N.J.: Princeton University Press, 1966.
———. *Strategy in the Missile Age.* Princeton, N.J.: Princeton University Press, 1959.
Brown, George S. "The Strategic Nuclear Balance." *Commanders Digest,* vol. 21, no. 4, March 1978, pp. 2-12.
Brown, Harold. *Annual Report, Department of Defense, Fiscal Year 1979,* Washington, D.C.: U.S. Government Printing Office, 1978.
———. *Department of Defense Annual Report, Fiscal Year 1980.* Washington, D.C.: U.S. Government Printing Office, 1979.
Broyles, Arthur A., and Eugene P. Wigner. "Civil Defense in Limited War—In Favor." In Pranger and Labrie, eds., pp. 370–76.
Buckley, James L. "First Lecture." In Buckley and Warnke, pp. 1–20.
———, and Paul C. Warnke. *Strategic Sufficiency: Fact or Fiction?* Washington, D.C.: American Enterprise Institute for Public Policy Research, 1972.

Cited Bibliography

Burt, Richard. *New Weapons Technologies: Debate and Directions.* Adelphi Papers No. 126, London: International Institute of Strategic Studies, 1976.

———. "The Scope and Limits of SALT." *Foreign Affairs,* vol. 56, no. 4, July 1978, pp. 751–70.

Carlton, David. "The Doctrine of Tactical Nuclear Warfare and Some Alternatives." In Carlton and Schaerf, eds., pp. 135–42.

———, and Carlo Schaerf, eds. *The Dynamics of the Arms Race.* New York: John Wiley and Sons, 1975.

Carter, Barry. "Nuclear Strategy and Nuclear Weapons." In Pranger and Labrie, eds., pp. 233–45. (Originally published in *Scientific American,* May 1974, pp. 20–31.)

Carter, Jimmy. "Moscow's Failures." *Newsweek,* vol. 91, no. 25, June 19, 1978, p. 43.

Coffey, J. I. "SALT under the Carter Administration." Paper presented to the 19th Annual Meeting of the International Studies Association, Washington, D.C., February 22–25, 1978.

Collins, Arthur S., Jr. "The Enhanced Radiation Warhead: A Military Perspective." *Arms Control Today,* vol. 8, no. 6, June 1978, pp. 1, 5.

Conover, C. Johnston. *U.S. Strategic Nuclear Weapons and Deterrence.* Rand Paper Series No. P-5967, Santa Monica, Calif.: Rand Corporation, August 1977.

Culver, John C. "The Future of the Strategic Bomber." *AEI Defense Review,* vol. 2, no. 1, 1978, pp. 2–12.

Davis, Lynn Ethridge. *Limited Nuclear Options: Deterrence and the New American Doctrine.* Adelphi Papers No. 121, London: International Institute of Strategic Studies, Winter 1975–76.

Doty, Paul, Albert Carnesale, and Michael Nacht. "The Race to Control Nuclear Arms." In Pranger and Labrie, eds., pp. 465–78. (Originally published in *Foreign Affairs,* vol. 55, no. 1, October 1976, pp. 119–32.)

Downey, Thomas J. "The Direct Payoff in SALT II." *Arms Control Today,* vol. 8, no. 3, March 1978, pp. 3–5.

———. "How to Avoid Monad—and Disaster." *Foreign Policy,* vol. 24, Fall 1976, pp. 172–201.

Dowty, Alan. "Nuclear Proliferation: The Israeli Case." *International Studies Quarterly,* vol. 22, no. 1, March 1978, pp. 79–120.

Drell, Sidney D. "Beyond SALT II—A Missile Test Quota." *Bulletin of the Atomic Scientists,* vol. 33, no. 5, May 1977, pp. 34–42.

———. "Civil Defense in Limited War—Opposed." In Pranger and Labrie, eds., pp. 377–84.

Dunn, Lewis A. "Nuclear 'Gray Marketeering.' " *International Security,* vol. 1, no. 3, Winter 1976, pp. 107–18.

Enthoven, Alain C. "American Deterrent Policy." In Kissinger, ed., pp. 120–34.

Falk, Richard. "A New World Order Problem." *International Security,* vol. 1, no. 3, Winter 1976, pp. 79–93.

Feld, Bernard T. "The Neutron Bomb." *Bulletin of the Atomic Scientists,* vol. 33, no. 7, September 1977, p. 11.

Gallois, Pierre M. "French Defense Planning—The Future in the Past." *International Security,* vol. 1, no. 2, Fall 1976, pp. 15–31.

———. "U.S. Strategy and the Defense of Europe." In Kissinger, ed., pp.

288–312. (Originally published in *Orbis*, vol. 7, no. 2, Summer 1963.)

Garrett, Stephen A. "Detente and the Military Balance." *Bulletin of the Atomic Scientists*, vol. 33, no. 4, April 1977, pp. 11–20.

Garthoff, Raymond L. "Negotiating with the Russians: Some Lessons from SALT." *International Security*, vol. 1, no. 4, Spring 1977, pp. 3–24.

———. "On Estimating and Imputing Intentions." *International Security*, vol. 2, no. 3, Winter 1978, pp. 22–32.

Garwin, James L. "Effective Military Technology in the 1980's." *International Security*, vol. 1, no. 2, Fall 1976, pp. 50–77.

Gelber, Harry G. "Technological Innovation and Arms Control." In Pranger and Labrie, eds., pp. 478–510. (Originally published in *World Politics*, vol. 26, no. 4, July 1974, pp. 509–41.)

Gillette, Robert. "Neutron Bomb is Almost 20 Years Old; Was Focus of Heated Controversy in 1961." *Los Angeles Times*, July 13, 1977, p. 10.

Glasstone, Samuel, ed. *The Effects of Nuclear Weapons*, rev. ed., Washington, D.C.: U.S. Government Printing Office, 1964.

Goldschmidt, Bertrand. "A Historical Survey of Nonproliferation Politics." *International Security*, vol. 2, no. 1, Summer 1977, pp. 69–87.

Goure, Leon. "Another Interpretation." *Bulletin of the Atomic Scientists*, vol. 34, no. 4, April 1978, pp. 48–51.

———, Foy D. Kohler, and Mose L. Harvey. *The Role of Nuclear Forces in Current Soviet Strategy*. Miami, Fla.: Center for Advanced International Studies, 1974.

———. *War Survival in Soviet Strategy: USSR Civil Defense*. Miami, Fla.: Center for Advanced International Studies, 1976.

Gray, Colin S. "Across the Nuclear Divide—Strategic Studies, Past and Present." *International Security*, vol. 2, no. 1, Summer 1977, pp. 24–46.

———. "Arms Control 'the American Way.' " *The Wilson Quarterly*, vol. 1, no. 5, Autumn 1977, pp. 94–99.

———. "Soviet-American Strategic Competition: Instruments, Doctrines, and Purposes." In Pranger and Labrie, eds., pp. 278–301.

———. "The Strategic Forces TRIAD: End of the Road?" *Foreign Affairs*, vol. 56, no. 4, July 1978, pp. 771–89.

Green, Philip. *Deadly Logic: The Theory of Nuclear Deterrence*. Columbus, Ohio: The Ohio State University Press, 1966.

Greenwood, Ted, George W. Rathjens, and Jack Ruina. *Nuclear Power and Weapons Proliferation*. Adelphi Papers No. 130, London: International Institute of Strategic Studies, 1976.

Guhin, Michael A. *Nuclear Paradox: Security Risks of the Peaceful Atom*. Washington, D.C.: American Enterprise Institute for Public Policy Research, 1976.

Head, Richard G. "Technology and the Military Balance." *Foreign Affairs*, vol. 56, no. 3, April 1978, pp. 544–63.

Helm, Robert, and Donald Westervelt. "The New Test Ban Treaties: What Do They Mean? Where Do They Lead?" *International Security*, vol. 1, no. 3, Winter 1976, pp. 162–78.

Hoag, Malcolm W. "Superpower Strategic Postures for a Multipolar World." In Rosecrance, ed., pp. 29–48.

Ikle, Fred Charles. "Can Nuclear Deterrence Last Out the Century." In Pranger and Labrie, eds., pp. 57–74. (Originally published in *Foreign Affairs*, vol. 51, no. 2, January 1973, pp. 267–85.)

Jacobsen, C. G. "Soviet Strategic Capabilities: The Superpower Balance." *Current History*, vol. 73, no. 430, October 1977, pp. 97–99, 134–36.
Johnson, David T., and Barry R. Schneider, eds. *Current Issues in U.S. Defense Policy*. New York: Praeger Special Studies, 1976. (Study sponsored by the Center for Defense Information.)
Kahan, Jerome H. *Security in the Nuclear Age: Developing U.S. Strategic Arms Policy*. Washington, D.C.: The Brookings Institution, 1975.
Kahn, Herman. "Escalation as a Strategy." In Kissinger, ed., pp. 17–33.
———. *On Escalation: Metaphors and Scenarios*. New York: Frederick A. Praeger, 1965.
———. *On Thermonuclear War*. Princeton, N.J.: Princeton University Press, 1961.
Kaplan, Fred M. "The Soviet Civil Defense Myth, Part I." *Bulletin of the Atomic Scientists*, vol. 34, no. 3, March 1978, pp. 14–20.
———. "The Soviet Civil Defense Myth, Part II." *Bulletin of the Atomic Scientists*, vol. 34, no. 4, April 1978, pp. 41–48.
Kemp, Geoffrey. *Nuclear Forces for Medium Powers, Part I: Targets and Weapons Systems*. Adelphi Papers No. 106, London: International Institute of Strategic Studies, 1974.
Kincade, William H. "Repeating History: The Civil Defense Debate Renewed." *International Security*, vol. 2, no. 3, Winter 1978, pp. 99–120.
———. "The View from the Pentagon." *Arms Control Today*, vol. 7, no. 10, October 1977, pp. 1–4.
Kissinger, Henry A. "Foreign Policy and National Security." *International Security*, vol. 1, no. 1, Summer 1976, pp. 182–91. (Article is text of speech given on March 22, 1976, in Dallas, Texas.)
———, ed. *Problems of National Strategy: A Book of Readings*. New York: Frederick A. Praeger, 1965.
Knorr, Klaus. *On the Uses of Military Power in the Nuclear Age*. Princeton, N.J.: Princeton University Press, 1966.
Kruzel, Joseph. "SALT II: The Search for a Follow-on Agreement." *Orbis*, vol. 17, 1973, pp. 334–63.
Leader, Stefan H., and Barry R. Schneider. "U.S.–Soviet Strategic Forces: SALT and the Search for Parity." In Johnson and Schneider, eds., pp. 129–52.
Lee, William T. *Understanding the Soviet Military Threat: How CIA Estimates Went Astray*. New York: National Strategy Information Center, 1977.
Legault, Albert, and George Lindsey. *The Dynamics of the Nuclear Balance*. Ithaca, N.Y.: Cornell University Press, 1974.
Lehman, Christopher, and Peter C. Hughes. " 'Equivalence' and SALT II." *Orbis*, vol. 20, 1977, pp. 1045–54.
Long, Clarence D. "Nuclear Proliferation: Can Congress Act in Time?" *International Security*, vol. 1, no. 4, Spring 1977, pp. 52–76.
Maddox, John. *Prospects for Nuclear Proliferation*. Adelphi Papers No. 113, London: International Institute of Strategic Studies, 1975.
McCarthy, John F., Jr. "The Case for the B-1 Bomber." *International Security*, vol. 1, no. 2, Fall 1976, pp. 78–97.
McLucas, John L. "The Case for a Modern Strategic Bomber." *AEI Defense Review*, vol. 2, no. 1, 1978, pp. 13–24.
McNamara, Robert S. *The Essence of Security: Reflections in Office*. New York: Harper and Row, 1968.

Metzger, Robert S. "Cruise Missiles: Different Missions, Different Arms Control Impacts." *Arms Control Today,* vol. 8, no. 1, January 1978, pp. 1–4.

Miettinen, Jorma K. "Enhanced Radiation Warfare." *Bulletin of the Atomic Scientists,* vol. 33, no. 7, September 1977, pp. 32–37.

The Military Balance. 1975–1976. London: International Institute of Strategic-Studies, 1976.

Milstein, Michael A., and Leo S. Semeiko. "Problems of the Inadmissibility of Nuclear Conflict." *International Studies Quarterly,* vol. 20, no. 1, March 1976, pp. 87–103.

Moch, Jules. "Verification and Control." In Carlton and Schaerf, eds., pp. 116–22.

Moulton, Harland B. *From Superiority to Parity: The United States and the Strategic Arms Race, 1961–1971.* Westport, Conn.: Greenwood Press, Inc., 1973.

National Academy of Sciences, *Long-Term Worldwide Effects of Multiple Nuclear Weapons Detonations.* Washington, D.C.: U.S. Government Printing Office, 1975.

"The Neutron Bomb Arms Control Impact Statement." *Congressional Record,* August 3, 1977, pp. H8498–H8502.

Newhouse, John. *Cold Dawn: The Story of SALT.* New York: Holt, Rinehart and Winston, 1973.

Nitze, Paul H. "Assuring Strategic Stability in an Era of Detente." *Foreign Affairs,* vol. 54, no. 2, January 1976, pp. 207–32.

———. "Foreword." In Brennan, *Arms Treaties with Moscow.*

———. "The Relationship of Strategic and Theater Nuclear Forces." *International Security,* vol. 2, no. 2, Fall 1977, pp. 122–32.

Nye, Joseph S. "Nonproliferation: A Long-Term Strategy." *Foreign Affairs,* vol. 56, no. 3, April 1978, pp. 601–23.

Ohlert, Edward J. "Strategic Deterrence and the Cruise Missile." *Naval War College Review,* vol. 30, no. 3, Winter 1978, pp. 21–32.

Panofsky, Wolfgang K. H. "The Mutual Hostage Relationship between America and Russia." In Pranger and Labrie, eds., pp. 74–84. (Originally published in *Foreign Affairs,* vol. 52, no. 1, October 1973, pp. 109–18.)

Perry, Robert. *The Faces of Verification: Strategic Arms Control for the 1980s.* The Rand Paper Series No. P-5986, Santa Monica, Calif.: Rand Corporation, August 1977.

Pipes, Richard. "Why the Soviet Union Thinks It Could Fight and Win a Nuclear War." *Commentary,* vol. 64, no. 1, July 1977, pp. 21–34.

Polmar, Norman, *Strategic Weapons: An Introduction.* New York: National Strategy Information Center, 1975.

Pranger, Robert J., and Roger P. Labrie. "General Introduction." In Pranger and Labrie, eds., pp. 1–4.

———, eds. *Nuclear Strategy and National Security: Points of View.* Washington, D.C.: American Enterprise Institute for Public Policy Research, 1977.

Quester, George H. "The Politics of Twenty Nuclear Powers." In Rosecrance, ed., pp. 56–77.

Rathjens, George W. "Future Limitations of Strategic Arms." In Willrich and Rhinelander, eds., pp. 225–55.

———. "Slowing Down the Arms Race." In Carlton and Schaerf, eds., pp. 82–91.

———. "The Verification of Arms Control Agreements." *Arms Control Today,* vol. 7, no. 7–8, July–August 1977, pp. 1–4.

———, and Jack Ruina. "Commentary on the New Test Ban Treaties." *International Security*, vol. 1, no. 3, Winter 1976, pp. 179–81.

Record, Jeffrey. *U.S. Nuclear Weapons in Europe: Issues and Alternatives.* Washington, D.C.: The Brookings Institution, 1974.

Rhinelander, John B. "The SALT I Agreements." In Willrich and Rhinelander, eds., pp. 125–59.

Rosecrance, Richard. "Deterrence in Dyadic and Multipolar Environments." In Rosecrance, ed., pp. 125–40.

———, ed. *The Future of the International Strategic System.* San Francisco: Chandler Publishing Company, 1972.

———. "Introduction." In Rosecrance, ed., pp. 1–4.

———. *Strategic Deterrence Reconsidered.* Adelphi Papers No. 116, London: International Institute of Strategic Studies, 1975.

Rosenbaum, David M. "Nuclear Terror." *International Security*, vol. 1, no. 3, Winter 1976, pp. 140–61.

Rowen, Henry S. "The Need for a New Analytical Framework: Review of *Security in the Nuclear Age.*" *International Security*, vol. 1, no. 2, Fall 1976, pp. 130–46.

Ruina, Jack. "The Arms Race and SALT." In Carlton and Schaerf, eds., pp. 47–56.

Rumsfeld, Donald H. "Annual Defense Department Report, FY 1978." In Pranger and Labrie, eds., pp. 188–202. (The report was originally released January 17, 1977; excerpted section is pp. 66–84.)

SALT II Agreement, Vienna, June 18, 1979. Washington, D.C.: U.S. Department of State, 1979.

Schelling, Thomas C. *Arms and Influence.* New Haven, Conn.: Yale University Press, 1966.

———. "The Importance of Agreements." In Carlton and Schaerf, eds., pp. 65–77.

———. *The Strategy of Conflict.* Cambridge, Mass.: Harvard University Press, 1960.

———. "Who Will Have the Bomb?" *International Security*, vol. 1, no. 1, Summer 1976, pp. 77–91.

Schlesinger, James R. "Analyses of Effects of Limited Nuclear Warfare." In Pranger and Labrie, eds., pp. 126–44. (Article is drawn from testimony before subcommittee of Senate Foreign Relations Committee, August 1975.)

———. "Annual Defense Department Report, FY 1975." In Pranger and Labrie, eds., pp. 85–104. (Report originally issued March 4, 1974; excerpted section is pp. 25–45.)

———. "The Evolution of American Policy toward the Soviet Union." *International Security*, vol. 1, no. 1, Summer 1976, pp. 37–48.

———. "U.S.–U.S.S.R. Strategic Policies." In Pranger and Labrie, eds., pp. 104–26. (Article is drawn from testimony before subcommittee of Senate Foreign Relations Committee on March 4, 1974.)

Schneider, Barry R. "U.S. Tactical Nuclear Weapons: A Controversial Arsenal." In Johnson and Schneider, eds., pp. 176–94.

Scoville, Herbert, Jr. "A Leap Forward in Verification." In Willrich and Rhinelander, eds., pp. 160–84.

Smernoff, Barry J. "Strategic and Arms Control Implications of Laser Weapons." *Air University Review*, vol. 29, no. 2, January–February 1978, pp. 38–50.

Snow, Donald M. "Deterrence Theorizing and the Nuclear Debate: The Methodological Dilemma." *International Studies Notes*, vol. 6, no. 2, Summer 1979, pp. 1–5.

———. *Introduction to Game Theory* (rev. ed.). Columbus, Ohio: Consortium for International Studies Education, 1978.

———. *The Shadow of the Mushroom-Shaped Cloud; Basic Ideas and Problems of Nuclear Deterrence*. Learning Materials in National Security Education No. 2, Columbus, Ohio: Consortium for International Studies Education, 1978.

Snyder, Jack. "The Enigma of Soviet Strategic Policy." *The Wilson Quarterly*, vol. 1, no. 5, Autumn 1977, pp. 86–93.

———. *The Soviet Strategic Culture: Implications for Limited Nuclear Operations*. A Project AIR FORCE Report Prepared for the United States Air Force, Santa Monica, Calif.: Rand Corporation, September 1977.

Stanford, Phil. "Nuclear Missile Submarines and Nuclear Strategy." In Johnson and Schneider, eds., pp. 153–65.

Steinbruner, John D., and Thomas M. Garwin. "Strategic Vulnerability: The Balance between Prudence and Paranoia." *International Security*, vol. 1, no. 1, Summer 1976, pp. 138–81.

Stone, Jeremy. "Linking SALT to Ethiopia or Unlinking It from Detente." *Bulletin of the Atomic Scientists*, vol. 34, no. 6, June 1978, pp. 38–39.

Sweet, William. "The U.S.–India Safeguards Dispute." *Bulletin of the Atomic Scientists*, vol. 34, no. 6, June 1978, pp. 50–52.

Tammen, Ronald L. *MIRV and the Arms Race: An Interpretation of Defense Strategy*. New York: Praeger Special Studies, 1973.

Taylor, Maxwell D. "The United States—A Military Power Second to None?" *International Security*, vol. 1, no. 1, Summer 1976, pp. 49–55.

"The Technological Trap Beyond SALT: New Technology and Deterrence." *The New York Review of Books*, vol. 25, no. 12, June 20, 1978, pp. 27–30. (Material is excerpted from *Strategic Survey, 1977*, London: International Institute of Strategic Studies, May 1978.)

Teller, Edward. "The Nuclear Test Ban Treaty." In Kissinger, ed., pp. 411–23. (Article is derived from testimony before the Senate Foreign Relations Committee, August 20, 1963.)

Towell, Pat. "Neutron Bomb Poses Dilemma for Congress." *Congressional Quarterly Weekly Report*, July 9, 1977, pp. 1403–7.

Trofimenko, A. "Political Realism and the 'Realistic Deterrence' Strategy." In Pranger and Labrie, eds., pp. 38–53. (Source is Foreign Broadcast Information Service, Soviet Union, December 10, 1971.)

Tsipis, Kosta. "Anti-Submarine Warfare and Missile Submarines." In Carlton and Schaerf, eds., pp. 36–46.

———. "The Arms Race as Posturing." In Carlton and Schaerf, eds., pp. 78–81.

Van Cleave, William R., and Roger W. Barnett. "Strategic Adaptability." In Pranger and Labrie, eds., pp. 203–22. (Originally published in *Orbis*, vol. 18, no. 3, Fall 1974, pp. 655–76.)

Walker, Paul F. *Strategic Nuclear Weapons: The Deadly Calculus*. Learning Materials in National Security Education, Columbus, Ohio: Consortium for International Studies Education, forthcoming.

Warnke, Paul C. "Arms Control: A Global Imperative." *Bulletin of the Atomic Scientists*, vol. 34, no. 6, June 1978, pp. 32–34.

———. "Second Lecture." In Buckley and Warnke, pp. 21–40.
Wells, Samuel F., Jr. "America and the 'MAD' World." *The Wilson Quarterly*, vol. 1, no. 5, Autumn 1977, pp. 57–75.
Willrich, Mason. "SALT I: An Appraisal." In Willrich and Rhinelander, eds., pp. 256–76.
———, and John B. Rhinelander, eds. *SALT: The Moscow Agreements and Beyond.* New York: Free Press, 1974.
Wohlstetter, Albert. "The Delicate Balance of Terror." In Kissinger, ed., pp. 34–58. (Originally published in *Foreign Affairs*, vol. 37, no. 2, January 1959 pp. 211–34).
———. "Nuclear Sharing: Nato and the N+1 Country." In Kissinger, ed., pp. 186–212. (Originally published in *Foreign Affairs*, vol. 39, no. 3, April 1961.)
Wood, Archie L. "Modernizing the Strategic Bomber Force without Really Trying—A Case against the B-1 Bomber." *International Security*, vol. 1, no. 2, Fall 1976, pp. 98–116.
York, Herbert F. "The Origins of MIRV." In Carlton and Schaerf, eds., pp. 23–35.
Young, Wayland (Lord Kennet). "Disarmament: 30 Years of Failure." *International Security*, vol. 2, no. 3, Winter 1978, pp. 33–50.
Zumwalt, Elmo R. "An Assessment of the Bomber–Cruise Missile Controversy." *International Security*, vol. 2, no. 1, Summer 1977, pp. 47–58.

Index

Accidental war, 45–46, 185
Action-reaction phenomenon, 7, 9–10, 59, 144
Active defenses, 63, 86, 98, 117–18; Soviet programs, 151–55
Affirming the consequent, logical fallacy of, 217
Air breathing forces, U.S., 102–07
Anti-ballistic missiles (ABM), 61–62, 63, 77, 90, 98, 117–18, 188, 200, 221, 234, 235, 241
Anti-Ballistic Missile Treaty, 28, 188–89, 190, 199, 200, 202
Anti-submarine warfare (ASW), 102, 197, 206, 215, 234
Arms control, 162–74; dynamics of, 162–68; forms of, 168–74
Arms Control Impact Statement, 162, 202, 241
Arms freeze, 169–70
Arms race, 8, 163
Arms reduction, 170–73; deep cut, 171–72; shallow cut, 171–72

B-1 bomber, 82, 103, 237; compared to cruise missiles, 104–05
B-52 bomber, 102–05
Backfire bomber, 149, 150, 191, 194
Balance of terror, 40, 46, 131
Bargaining chips, 82, 124, 160, 190, 196
Baruch Plan, 49, 161, 175
Bean counting, 31, 86, 107–12
Breeder reactor, 232

Capability, definition of, 31
Capitalist encirclement, 135, 136, 156
Catalytic war, 225, 227–28
Circular error probable (CEP), 61, 91, 207
Cold launch, 89
Comprehensive Test Ban Treaty, 162, 177, 183–84, 198, 199, 215
Conditional viability, 27, 28; insecure, 29; secure, 29, 30, 65
Controlled response, 62, 63–64, 235
Counterforce targeting, 32, 33, 39, 55, 61, 65, 82
Countervailing strategy, 49, 82, 83

Countervalue targeting, ix, 32, 33, 35, 37, 55, 65, 231
Cruise missile, 76, 105–07, 126, 174, 191, 194, 196, 211, 218, 237
Cuban missile crisis, 43, 60–61, 175, 176, 222

Damage Limitation, 55, 63, 64, 213
Depressed trajectory launch, 95, 103, 206
Detente, 134–35, 159
Deterrence, 4, 23–24
"Deterrence-only," 5, 44, 71, 72, 73, 79
"Deterrence-plus," 5–6, 44, 69, 71, 72
Disarmament, 162, 173–74

Enhanced-radiation warhead, 21, 43, 52, 87, 88, 122, 124, 127–29; as ABM, 128; and Soviet civil defense, 128
Escalation, 29, 45
Escalatory process, 43–45, 71, 227
Essential equivalence, 49, 68, 79–81, 82, 84–85, 199, 216
Ex Post-Ex Ante dilemma, 69–71
Extended deterrence, 38

Fingers on the nuclear button, 229
Firebreak (nuclear threshold), 43, 46, 204, 227
First-strike capability, 31–32, 33, 230, 231; and strategy, 55, 231
Fission-fusion-fission weapons, 52, 88, 230
Fission-fusion weapons, 52, 88, 230
Fission weapons, 87–88, 231
Flexible nuclear response, 68
Flexible response, 48, 57, 62, 78–79
Force frappe, 37
Forward-based systems, 15, 86, 123, 185, 187, 188, 195
Fratricide effect, 97, 206, 239

Galosh ABM, 113, 151, 188
Game theory, 7–9
Greater-Than-Expected Threat, 12, 66

Hard-kill capability, 126–27, 196, 197, 201, 203, 205, 206–14, 215, 216, 237; basis for, 206–08; effects on deter-

Index 283

rence, 208–09; justifications for, 212–14; solutions to, 209–12
Horizontal proliferation, 162, 177, 205, 223–24, 230, 239
Hostage effect, 4, 65, 72, 118, 152
Hot launch, 89
Hot Line Agreement, 176, 185
Hunter-killer satellite, 151

Initial Operating Capability (IOC), 17
Intercontinental ballistic missile (ICBM), 51, 52, 56, 94, 96–99, 167, 172, 176, 185, 186, 205; alternatives to, 97–99; vulnerability of, 96–97, 124, 140, 146, 147, 201, 238, 240
Interim Agreement on Offensive Arms (IOA), 185–88, 190, 199
International Atomic Energy Agency (IAEA), 180, 181, 233
Israeli nuclear force, 37

Jackson amendment, 80

Korean War, 50, 51

Laser and particle-beam weapons, 214, 234, 241
Lethality (K-factor), 111, 207
Light water reactor, 233
Limited nuclear options, 45, 68, 78–79, 81, 83, 141, 223
Limited Test Ban Treaty (LTBT), 175, 176, 177, 197, 198, 200
Linkages, 20–21, 159–60
Look-down–shoot-down capability, 104

Maneuverable reentry vehicle (MARV), 62, 76, 90, 91, 124, 126, 127, 213, 237, 241
Manhattan Project, 1
Massive retaliation, 48–57, 58, 69, 76, 142, 234
Megatonnage, 109–11; "raw," 109–10; equivalent, 110–11
Military force, purposes of, 4–5
Minimax, 9, 164
Minimum deterrence, 37, 64, 210
Mininukes, 122, 129
Minuteman, 96
Mirror-imaging, 9–11, 131
Missile gap, 59, 60
MK12 and 12A, 96, 126, 127, 237

Multiple independently targetable reentry vehicle (MIRV), 17, 21, 62, 90, 91, 112–16, 145, 158, 167, 190, 191, 202, 235, 237; "bus," 90, 92, 113; effects of, 114–16; "footprint," 90, 114; mechanics of, 113–14; origins of, 112–13; SALT sublimits on, 194, 196
Mutual assured destruction (MAD), ix, 5, 19, 22, 36, 49, 55, 57–68, 69, 204, 235; Soviet views on, 135–36
MX missile, 99, 124–26, 186, 187, 202, 210, 219, 220, 237, 238, 240; Multiple Protective Shelters for, 125

NATO, 15–16, 23, 54
National Command Authority (NCA), 91, 188
"Near nuclear" states, 178, 226
New Look, 48, 234
Non-Proliferation Treaty (NPT), 175, 177–82, 198, 232
N+1 Problem, 177–79
NSC-68, 50
Nuclear multipolarity, 224
Nuclear parity, 80, 218
Nuclear terror, 225, 228–29
Nuclear Threshold Impact Statement, 241
Nuclear war-fighting, 69, 71; Soviet capability for, 73

"One-way freedom to mix," 187
Open skies proposal, 175

Passive defense, 63, 86, 118–19; Soviet programs, 151–55
Peaceful coexistence, 134–35, 156, 159, 190, 222
Peaceful Nuclear Explosions Treaty, 177, 182, 183, 199
Penaids, 62
Penetrability, 31, 40, 208
Pershing II missile, 16, 122
Polaris/Poseidon, 99, 187
Prevention of Nuclear War Agreement, 184, 185, 189, 199
Prisoners Dilemma, 8, 163–66; and R,D,T,&E, 168, 169, 174, 196, 205, 214, 220, 221, 234

Reentry vehicle (RV), 89, 91, 96
Research, Development, Testing and En-

gineering (R,D,T,&E), 16–27, 86, 92–93, 183, 192, 200
Richardson processes, 8

Satellite and Missile Observation System (SAMOS), 166
Second-strike capability, 31–32, 46, 65, 82, 172, 208, 212; and strategy, 19, 82
Single-shot kill probability (SSKP), 206
Soviet Union: and basis of deterrence, 135–38; and Cuban missile crisis, 133–34; and defense, 150–55, 221; and limited war, 141, 223; and MAD, 135–36; and peaceful coexistence, 134–35, 143, 222; and nuclear war-winning, 138–41, 142, 156; nuclear forces of, 143–50; threat posed by, x, 88, 155–60, 205, 216–23
Spectrum defense, 53, 58
Standing Consultative Commission, 202
Standoff launch, of cruise missiles, 106
Strategic Arms Limitation Talks (SALT), x, 16, 68, 123, 150, 159, 162, 174, 175, 177, 181, 184, 196, 198, 202, 236, 237, 238, 239
SALT I, 28, 74, 146, 167, 169, 175, 184, 185–91, 194, 198, 199, 237
SALT II, 14, 82, 160, 162, 171, 172, 175, 184, 188, 190, 192–97, 198, 199, 200, 215, 216, 219, 239
SALT III, 162, 188, 215, 216
Strategic bombing, 27, 61
Strategic Sufficiency, 49, 56, 68, 76, 78, 81
Submarine-launched ballistic missiles (SLBM), 36, 57, 94, 95, 99–102, 167, 185, 186–87

Technological drift, 92, 112
TERCOM, 91, 105, 213
Theater nuclear weapons (TNW), 2, 78, 119–23; effects on Europe, 121; and escalation, 122–23; nature and purpose, 119–21
Threshold Test Ban Treaty (TTBT), 177, 182, 183, 198, 199
Throw-weight, 110, 191, 218, 240
Titan missile, 96, 187
TRIAD, xi, 20, 86, 93–96, 171, 172, 209
Trident, 22, 100, 109, 148, 187, 202, 219, 237

Unconditional viability, 26–27, 28, 234

Verification, 166–68, 169, 170, 173, 174, 180, 183, 185, 187, 192, 197, 199–200
Vertical proliferation, 177, 225, 235
Vladivostok accords, 169, 175, 184, 191–92, 194

Warhead numbers, comparisons of, 108–09
"World balance of forces," 132
Worst-case analysis, 11–12, 67